Nanocomposite Sorbents for Multiple Applications

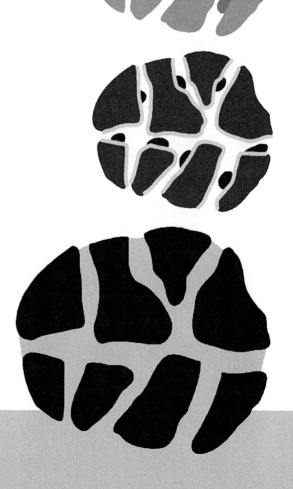

NANOCOMPOSITE SORBENTS FOR MULTIPLE APPLICATIONS

Yu. I. Aristov

JENNY STANFORD PUBLISHING

Published by

Jenny Stanford Publishing Pte. Ltd.
Level 34, Centennial Tower
3 Temasek Avenue
Singapore 039190

Email: editorial@jennystanford.com
Web: www.jennystanford.com

British Library Cataloguing-in-Publication Data
A catalogue record for this book is available from the British Library.

Nanocomposite Sorbents for Multiple Applications

Copyright © 2020 by Jenny Stanford Publishing Pte. Ltd.
All rights reserved. This book, or parts thereof, may not be reproduced in any form or by any means, electronic or mechanical, including photocopying, recording or any information storage and retrieval system now known or to be invented, without written permission from the publisher.

For photocopying of material in this volume, please pay a copying fee through the Copyright Clearance Center, Inc., 222 Rosewood Drive, Danvers, MA 01923, USA. In this case permission to photocopy is not required from the publisher.

ISBN 978-981-4267-50-2 (Hardcover)
ISBN 978-1-315-15650-7 (eBook)

I dedicate this book to the memory of my parents,
***Ivan** and **Anastasia**,*
my first and the most important teachers

Contents

Preface		xv
1.	**Introductory Remarks**	**1**
	1.1 "Old" and "New" Adsorbents	2
	1.2 Adsorption Technologies	3
	1.3 Harmonization of Adsorbent and Process	4
	1.4 Screening versus Nanotailoring	5
	1.5 Target-Oriented Synthesis of Sorbents	6
	1.6 Nanocomposite Sobents "Salt in Porous Matrix"	7
2.	**Description of Adsorption Equilibrium and Dynamics**	**15**
	2.1 Theoretical Adsorption Equations	16
	2.1.1 Langmuir-Type Isotherms	16
	2.1.2 Gibbs Approach	17
	2.2 Empirical Approaches to Adsorption Equilibrium	18
	2.2.1 Polanyi Potential Theory	19
	2.2.2 Trouton's Rule	20
	2.2.3 BET Equation	22
	2.3 Classification of Adsorbents	23
	2.4 Isosteric Heat of Sorption	24
	2.5 Mono-variant versus Bi-variant Systems	24
	2.6 Adsorption Kinetics	26
	2.6.1 Gas Transport Mechanisms	26
	2.6.2 Isothermal Adsorption Kinetics in a Single Grain	27
	2.6.2.1 Surface resistance control	28
	2.6.2.2 Pore diffusion control	28
	2.6.2.3 Linear driving force model	30
	2.6.3 Kinetics of Gas–Solid Reactions	30
3.	**Optimal Adsorbent: Basic Requirements**	**35**
	3.1 How to Formulate Thermodynamic Requirements?	36

3.2		Particular Adsorption Technologies	36
	3.2.1	Gas Drying	37
	3.2.2	Heat Transformation and Storage	39
	3.2.3	Maintaining Relative Humidity	43
	3.2.4	Shifting Chemical Equilibrium	44
	3.2.5	Extraction of Water from the Atmospheric Air	45
3.3		Dynamic Requirements	48
3.4		Other Requirements	53

4. Basic Synthesis Methods   **57**

4.1		Impregnation of the Porous Matrix with an Active Salt	58
	4.1.1	Dry (Incipient Wetness, Pore Volume) Impregnation	58
	4.1.2	Wet (Adsorption, Equilibrium Deposition) Impregnation	61
	4.1.3	Combined Impregnation	62
4.2		Co-precipitation	62
4.3		Mechanical Mixing and Thermal Dispersion	64
4.4		Dry Impregnation: Tools for CSPM Nanotailoring	66
	4.4.1	Active Salt	67
		4.4.1.1 Chemical nature of the salt	67
		4.4.1.2 Crystalline versus amorphous salt phases	68
	4.4.2	Host Matrix	72
		4.4.2.1 Pore size effect	73
		4.4.2.2 Guest–host effect	74
	4.4.3	Synthesis Conditions	74
		4.4.3.1 Solution concentration	74
		4.4.3.2 Solution pH	76
		4.4.3.3 Calcination temperature	77

5. Composite Sorbents of Water Vapor: Effect of a Confined Salt   **83**

5.1		Salts and Hydrates in Bulk	83
	5.1.1	Water-Sorbing Ability of Bulk Salts	84
	5.1.2	Mono-variant Equilibrium for Bulk Salts/Hydrates	85

5.2	Various Salts Confined to the Mesoporous Silica Gel KSK	86
	5.2.1 Effect of Salt's Nature on Water Sorption Equilibrium	87
	5.2.2 Effect of the Salt Content on Water Sorption Equilibrium	88
	5.2.3 Water Sorption Properties of the Salt/Matrix Composites: Non-additivity	89
	5.2.4 Synthesis/Decomposition Hysteresis	91
	5.2.5 Dynamics of Hydrate Synthesis/ Decomposition	92
5.3	Mechanisms of Water Sorption on CSPMs	93
	5.3.1 Isobars of Water Sorption on SWS-1L	94
	5.3.2 Isosteric Chart of Water Sorption on SWS-1L	96
	5.3.3 Universal Description of Water Sorption on SWS-1L	97
	5.3.4 Mechanisms of Water Sorption on SWS-1L	99

6. Composite Sorbents of Water Vapor: Effect of a Host Matrix — **107**

6.1	"Calcium Chloride–Water" System in Bulk	109
6.2	$CaCl_2$ in Matrices with Large Pores	112
	6.2.1 $CaCl_2$/Vermiculite (SWS-1V)	112
	6.2.2 $CaCl_2$/MWCNT	116
	6.2.3 $CaCl_2$/Attapulgite	118
6.3	$CaCl_2$ in Matrices with Middle Size Pores	118
	6.3.1 $CaCl_2$/SBA-15	118
	6.3.2 $CaCl_2$/Alumina	121
	6.3.3 $CaCl_2$/Carbon Sibunit	123
6.4	$CaCl_2$ in Matrices with Small Pores	124
	6.4.1 $CaCl_2$/RD Silica KSM	125
	6.4.2 $CaCl_2$/Activated Carbon	127
	6.4.3 $CaCl_2$/Zeolite 13X	128
6.5	Calcium and Lithium Nitrates in the Silica Pores of Various Size	129
6.6	"Water–Salt" Sorption Equilibrium inside the Matrix Pores	131
	6.6.1 Salt Hydrates in Middle Pores	132
	6.6.2 Salt Solutions in Small Pores	135

x | Contents

7. Composite Sorbents of Methanol   **143**
7.1   Common Methanol Adsorbents   144
7.2   Intent Design of CSPMs for Methanol Sorption   145
    7.2.1   Effect of Active Salt   145
    7.2.2   Effect of Host Matrix   148
    7.2.3   Effect of Synthesis Condition   150
    7.2.4   Effect of Supplementary Salt   151
7.3   Methanol Sorption Properties of CSPMs   152
    7.3.1   Composites "$CaCl_2$–Silica Gel"   153
    7.3.2   Composites "Lithium Halides–Silica Gel"   158

8. Composite Sorbents of Ammonia   **169**
8.1   CSPMs for Ammonia Sorption   170
    8.1.1   Effect of Active Salt   170
    8.1.2   Effect of Host Matrix   172
        8.1.2.1 $CaCl_2$-based composites   172
        8.1.2.2 $BaCl_2$-based composites   175
8.2   Ammonia Sorption Properties of New CSPMs   177
    8.2.1   Composites "$CaCl_2$ inside Porous Matrices"   177
    8.2.2   Composite Sorbents "$SrCl_2$ inside Porous Matrices"   182
    8.2.3   Composite Sorbents "$BaCl_2$ inside Porous Matrices"   183
    8.2.4   Summary of New Ammonia Sorbents "Salt–Matrix"   185
8.3   Effect of Supplementary Salt   187
    8.3.1   Composite Sorbents ($BaCl_2$ + $BaBr_2$)/Silica   188
    8.3.2   Composite Sorbents ($BaCl_2$ + $BaBr_2$)/Vermiculite   189
    8.3.3   Discussion   191

9. Composite Sorbents of Carbon Dioxide   **197**
9.1   Effect of the Nature of Matrix   198
    9.1.1   Breakthrough Curves of Carbon Dioxide under Moist Conditions   199
    9.1.2   Composite Sorbent K_2CO_3/Alumina   201
9.2   Brief Comparison with Common CO_2 Adsorbents   204
9.3   Conclusion   205

10. Thermal Properties of CSPMs — **209**

10.1 Heat Capacity — 209

10.1.1 Low-Temperature Capacity of CaCl$_2$/Silica Gel (SWS-1L) — 210

10.1.2 Heat Capacity of SWS-1L as Function of Water Uptake and Temperature — 214

10.2 Thermal Conductivity — 216

10.2.1 Thermal Conductivity of CSPMs at Atmospheric Pressure — 217

10.2.2 Thermal Conductivity of CaCl$_2$/Alumina Composite — 220

10.2.3 Thermal Conductivity under Real Conditions of AHT Cycle — 221

11. Melting–Solidification of Salt Solution/Hydrates in Pores — **227**

11.1 "CaCl$_2$–Water" System inside Silica Gel Pores — 228

11.1.1 Solidification–Melting Phase Diagram — 229

11.1.2 Effect of Pore Filling — 233

11.1.3 Supercooling and Vitrification of Salt Solutions in Silica Micropores — 235

11.2 Other Salts and Matrices — 238

11.2.1 LiBr Solution in Pores of Silica Gel KSK — 238

11.2.2 Various Hydrates of CaCl$_2$ in Pores of SBA-15 — 239

11.3 Overview — 242

12. Molecular Dynamics of Sorbed Water — **245**

12.1 ^{1}H NMR Spectroscopy — 246

12.1.1 Water in CaCl$_2$ Hydrates Confined to Silica Pores of Middle Size — 246

12.1.2 Water in CaCl$_2$ Hydrates Confined to Silica Pores of Small Size — 248

12.2 ^{2}H NMR Spectroscopy — 249

12.2.1 ^{2}H NMR spectra of CaCl$_2 \cdot n$D$_2$O Hydrates in Bulk and Confined States — 249

12.2.2 ^{2}H NMR T_1 and T_2 Relaxation Times Analysis — 253

12.3 Neutron Scattering — 256

12.3.1 QENS Domain — 256

xii | *Contents*

13. Sorption Dynamics: An Individual Composite Grain   **263**

13.1   Brief Comparison of Water Sorption Dynamics by Bulk and Confined Calcium Chloride   264

13.2   Isothermal Dynamics of Water Sorption in CSPMs   265

     13.2.1   Composite $CaCl_2$/Silica Gel (SWS-1L)   266

     13.2.2   Enhancement of Vapor Transport in Partially Saturated CSPM Pores   270

     13.2.3   Other Composites   273

13.3   NMR Imaging Study of Water Sorption   274

     13.3.1   Sorption Profiles in Individual CSPM Pellet   274

14. Sorption Dynamics: A Composite Bed   **281**

14.1   Water Sorption Profiles Measured by NMR Method   281

     14.1.1   Experimental Methodology: Synthesis of Consolidated Layers and NMR Measurements   282

     14.1.2   Effect of Binder Content   283

     14.1.3   Effect of Primary Grain Size   285

     14.1.4   Effect of Salt Content   287

14.2   Water Sorption Profiles Measured by Gamma Ray Microscopy   287

14.3   Summary   289

15. Isobaric Sorption Dynamics: A Temperature Initiation   **291**

15.1   The Large Temperature Jump Method   292

15.2   Sorption Dynamics: Monolayer of Loose CSPM Grains   294

     15.2.1   Exponential Kinetics   294

     15.2.2   Effect of Isobar Shape   296

     15.2.3   Effect of Grain Size   299

     15.2.4   Effect of Salt Content   301

     15.2.5   Effect of Cycle Boundary Temperatures   303

     15.2.6   Effect of Layers Number N 305

15.3   Summary   306

16. Adsorptive Transformation of Heat: Temperature-Driven Cycles   **313**

16.1   Thermodynamic Harmonization of Adsorbent and Working Conditions of TD AHT Cycle   314

	16.1.1 Various TD AHT Cycles	314
	16.1.2 Adsorbent Optimal for TD AHT	317
	16.1.2.1 The first law efficiency	318
	16.1.2.2 The second law efficiency	318
	16.1.2.3 The dynamic efficiency	320
16.2	Adsorptive Cooling	322
	16.2.1 Composite Sorbents of Water	322
	16.2.2 Composite Sorbents of Methanol	327
	16.2.3 Composite Sorbents of Ammonia	329
16.3	Desiccant Cooling	331
16.4	Adsorption Heat Storage	332

17. Adsorptive Transformation of Heat: Pressure-Driven Cycles — **345**

17.1	Pressure-Driven Adsorptive Cycles	345
	17.1.1 Thermodynamic Charts of a PD Cycle	346
	17.1.1.1 Isothermal PD cycle	346
	17.1.1.2 Non-isothermal PD cycle	347
	17.1.2 First Law Analysis	347
	17.1.3 Second Law Analysis	348
	17.1.4 Adsorbent Optimal from Dynamic Point of View	349
17.2	New Cycle "Heat from Cold"	352
	17.2.1 Evaluation of Useful Heat	353
	17.2.2 Evaluation of Threshold Ambient Temperature	354
	17.2.3 Evaluation of Maximal Heating Temperature	355
	17.2.4 First Testing of HeCol Prototype	356
	17.2.4.1 Design of the first HeCol prototype	356
	17.2.4.2 Testing HeCol unit with the selected "salt/matrix" composites	358

18. Regeneration of Heat and Moisture in Ventilation Systems — **363**

18.1	Climatic Features of Cold Countries	364
18.2	The VENTIREG Approach: Description and Testing	365

	18.2.1	Description of the Approach	365
	18.2.2	Adsorbent Selection: Intuition	366
	18.2.3	Adsorbent Selection: Experiment	367
18.3		Experimental Testing of VENTIREG Prototypes	371
18.4		Other Aspects	373

19. Maintaining Relative Humidity — **379**

19.1		Selection of Salt and Matrix	380
	19.1.1	Requirements of Optimal Adsorbent	381
	19.1.2	Effect of the Matrix	383
19.2		ARTIC-1 Testing	385
	19.2.1	Laboratory Tests	385
	19.2.2	Tests in the Museum and the Library	388
		19.2.2.1 Tests in the museum	388
		19.2.2.2 Tests in the library	390
19.3		New CSPM Buffers for RH < 45%	392

20. Shifting Chemical Equilibrium — **395**

20.1		Introduction to the Problem	395
20.2		Theoretical Considerations	397
	20.2.1	Formulating Requirements to Optimal Adsorbent	397
	20.2.2	Selection of Proper Salts	398
20.3		Experimental Study of Sorbent-Assisted Methanol Synthesis	399
	20.3.1	Composites Preparation	399
	20.3.2	Testing Facilities	399
	20.3.3	Methanol Sorption under Reaction Conditions	400
	20.3.4	Stability of Composites	402

21. Active Heat Insulation — **407**

21.1	Experimental Details	408
21.2	Heat Front Propagation	409
21.3	Further Considerations	413

Postface	417
Index	421

Preface

Looking back over the past decades, it is safe to say that the interest in developing nanomaterials is booming, and this has contributed invaluably to the technological progress of humankind. Many new books exclusively devoted to nanoscale materials have since emerged to deliver rapidly increasing information about newly developed nanomaterials. Because of this, it is becoming extremely difficult to write a book assembling information that has never appeared in other books. This book endeavors to give a first comprehensive survey of a new family of nanoComposite sorbents "**S**alt in **P**orous **M**atrix" (CSPMs). These composites have recently been developed for selective sorption of water, alcohols, ammonia, and carbon dioxide. They owe their origin to a very catchy idea of target-oriented tailoring of materials with predetermined adsorption properties harmonized with a particular adsorption process. This idea runs throughout this book, and it is particularized in Chapter 1.

Chapter 2 addresses the basic definitions and quantities concerning adsorption equilibrium and dynamics that are important for further considerations in this book. It includes the theoretical adsorption equations, IUPAC classification of adsorption isotherms, adsorption system variance, isosteric heat of sorption, and main approaches to analyzing the adsorption kinetics. This chapter also includes two empirical approaches, namely, the Polanyi principle of temperature invariance and Trouton's rule, which have been widely used in the book.

The harmonization of an adsorbent and an adsorption process implies that for each specific process, there is an adsorbent which is optimal. Chapter 3 provides rules for the selection/tailoring of the adsorbent optimal for gas drying, heat transformation/storage, maintaining relative humidity, shifting chemical equilibrium, and harvesting water from the atmosphere.

The formulated requirements serve as a guiding star for the synthesis of real adsorbents with properties as close as possible to those predicted for the optimal adsorbent. The basic approaches to designing CSPMs with pre-required properties are comprehensively considered in Chapter 4.

Chapter 5 addresses CSPMs for adsorption of water vapor. It considers the ability of various inorganic salts to form solid complexes with water molecules in bulk. It then analyzes what happens when salt is introduced into the pores of a host matrix and whether sorption properties of the "salt–matrix" composite are a linear addition of the salt and matrix properties.

Chapter 6 analyzes how changes in the salt sorption properties caused by the salt's confinement to a host matrix depend on the matrix nature. In this chapter, the salt calcium chloride ($CaCl_2$) is investigated in a variety of host matrices with different chemical nature and pore structure. Similar data are presented for composite sorbents of methanol, ammonia, and carbon dioxide in Chapters 7, 8, and 9, respectively.

The heat capacity and heat conductivity considered in Chapter 10 are fundamental characteristics of CSPMs because they create a base for calculating their thermodynamic and dynamic performances in various applications. In Chapter 11, melting and solidification of the salt inside the CSPM pores are surveyed. It includes solubility diagrams, depression of melting point, and subcooling of the confined solution.

Chapter 12 studies the changes in the mobility of water linked with the confined salt. These changes can be a reason behind the significant difference in the sorption properties of salt hydrates and solution confined to matrix pores from their bulk properties.

In the next three chapters, the dynamics of vapor (water, methanol, ammonia) sorption on the CSPMs is comprehensively considered both for an individual composite grain (Chapter 13) as well as a bed of loose or consolidated grains (Chapter 14). Two cases, sorption initiated by *small* and *large* pressure deviations from the equilibrium, have been analyzed for a single grain. Chapter 15 addresses a special case when adsorption is initiated by a drop in the temperature of the metal support on which the composite is mounted. This non-traditional case is realized in temperature-driven units for adsorptive heat transformation and storage. This case is also discussed in the final chapters of the book along with other practical applications of CSPMs.

Chapters 16 and 17 analyze adsorptive heat transformation and storage; Chapter 18 discusses regeneration of heat and moisture in ventilation system of dwellings; Chapter 19 talks about maintaining relative humidity in museums, libraries, and archives; Chapter 20 examines shifting of the equilibrium of an endothermic chemical process toward the formation of desired products; and Chapter 21 explores active heat insulation for heat protection and fire extinguishing systems.

These chapters exhibit the tremendous potential of CSPMs for a number of important applications in various fields. The CSPMs that have been developed largely in the author's laboratory are now being intensively studied further and used in many laboratories all over the world. Thus, this book comprises various topics, including requirements of an optimal sorbent; tools for its synthesis; sorption equilibrium and dynamics of the new composites with water, methanol, ammonia, and carbon dioxide; the effect of changing the salt properties at the nanoscale; and a survey of the possible applications.

I hope the book will be helpful for all those students and non-specialists who are interested in the current progress in the development of nanocomposite materials for multiple applications. In addition, I believe that certain sections of the book will be of interest to those scientists, engineers, and other technologists who are already concerned, either directly or indirectly, with the study or use of the CSPMs.

Many of the ideas and results expressed in this book were developed in collaborative research done over the past 20 years. It is hardly possible to name all my coworkers here but they can be identified in the references listed at the end of each chapter. I wish to express my special thanks to my present and former colleagues from the Laboratory of Energy Accumulating Materials and Processes of the Boreskov Institute of Catalysis, Novosibirsk, whose experimental contributions to CSPM synthesis, study, and applications made this book possible, especially L. G. Gordeeva, M. M. Tokarev, A. D. Grekova, I. S. Glaznev, Zh. V. Veselovskaya, I. A. Simonova, A. Z. Sibgatulin, V. E. Sharonov, and A. G. Okunev. I also thank many colleagues from other laboratories from all over the world, especially A. Freni, G. Giovanni, L. I. Heifets, T. A. Krieger, I. V. Mezentsev, B. N. Okunev, B. Dawoud,

A. G. Stepanov, B. Saha, S. D. Kirik, S. Vasta, A. Jarzebski, I. V. Koptyug, A. Frazzica, N. M. Ostrovskii, G. Maggio, V. N. Alekseev, G. Cacciola, and A. Chakraborty. During my research on new nanocomposite sorbents and while organizing my thoughts for this book, I benefited greatly from discussions with a number of leading researchers in the field, particularly, R. E. Critoph, F. Meunier, R. Wang, L. L. Vasiliev, F. Ziegler, and S. Garimella.

Special thanks are due to A. S. Shorin whose intelligent and funny drawings smoothened the serious style of the book.

It is my great pleasure to thank the team at Jenny Stanford Publishing and especially Jenny Rompas and Shivani Sharma for their patience and highly professional help in editing and publishing the book.

Finally, I would like to thank my wife Oxana, who took on many extra responsibilities so that I could attend to this monograph. She remains a constant source of support and strength.

Yu. I. Aristov
Winter 2020

Chapter 1

Introductory Remarks

"Nobody can embrace the unembraceable" [1]. Therefore, a comprehensive description of the huge world of adsorbents and adsorption technologies is beyond the scope of this book. In this chapter, we intend just to provide an overview in order to explain the motivation behind writing this book and outline its place.

Many adsorbents have been found since the earliest times when the Egyptians and Sumerians used wood chars in medicine [2], water purification, and manufacture of bronze [3]. Such a vast number of novel porous materials with diverse structure and properties are synthesized every year that their counting and classification have become difficult. By now, the main technological application of adsorbents, namely, separation of gases, has been brought to near perfection [4]. Many "secondary" but still important applications, such as ion exchange, gas storage, catalysis, drug delivery, heat transformation/storage, life support systems on a spacecraft, air conditioning, etc., are quite advanced as well. Nevertheless, there is still a room for further improvement in adsorbents and adsorption technologies. This chapter gives a brief overview of the current state of the art in the adsorbent materials and processes as well as future trends and possible strategies for the development of novel adsorbents.

Nanocomposite Sorbents for Multiple Applications
Yu. I. Aristov
Copyright © 2020 Jenny Stanford Publishing Pte. Ltd.
ISBN 978-981-4267-50-2 (Hardcover), 978-981-4303-15-6 (eBook)
www.jennystanford.com

1.1 "Old" and "New" Adsorbents

Although various porous materials were known and used as early as in ancient times, scientific research on their sorption properties started only at the end of the 18th century [5–8]. However, until the late half of the 19th century, mostly natural adsorbents (charcoals, clays, lime, volcanic ash, zeolitic tuffs, etc.) were the subjects of study. At first, synthetic porous solids (metal oxides) were prepared by the thermal decomposition of metal salts, calcination of hydroxides, or precipitation [3]. The next step was the development of the sol–gel process and the synthesis of silica aerogels [9]. Then the VICOR® process was discovered to produce high-silica porous glasses [10]. After the pioneering work of Barrer et al. [11], a wealth of zeolites have been synthesized by the hydrothermal procedure, although natural zeolites have been employed as raw materials for millennia. The first industrially produced activated carbons were developed at the very beginning of the 20th century [12]. Useful chronological tables of the early theoretical and experimental ages of the adsorption science can be found in the literature [13, 14].

Only four types of sorbents have occupied a dominant position in the market: activated carbon, zeolites, silica gel, and activated alumina. An estimated worldwide sale of activated carbons was reported to be $1 billion [15]. In 2001, the industry demands for zeolites, silica gel, and activated alumina were $1,070 million, $71 million, and $63 million, respectively [16]. These synthetic adsorbents along with the aforementioned natural porous materials have been known for many decades and widely used for various applications (see Sec. 1.2), and hence, they can be conventionally called "old" adsorbents.

Classification of adsorbents as "old" and "new" is very relative because in the short run, many new classes of adsorbents progressed from the initial idea to complete development and passed the way from "new" to "old" materials. For instance, the development of crystalline porous phosphates was started in 1982 with the synthesis of alumino-phosphate adsorbents [17]. Due to the incorporation of different transition metals next to phosphorus and aluminum, by 1994 the variety of phosphate-based porous materials, metalloalumino-phosphates, had become comparable with that of the common silicate-based ("old") zeolites [18]. Both structure and adsorption properties of the "new" alumino-phosphate materials depend on the type and amount of incorporated heterometal. Extremely fast progress was made with mesostructured mesoporous materials (MCM, SBA, FSM, etc.) [19, 20], pillared clays [21], carbon nanotubes [22], nanohorns [23], carbon aerogels [24], activated carbon fibers (ACF) [25], fullerenes [26], and many others. Nowadays, similar progress is being achieved with a new family of supra-molecular materials (the so-called metal–organic frameworks, MOFs for short) [27, 28]. Many techniques for the synthesis of versatile structures starting from molecular building blocks have been developed so far.

A comprehensive multi-authored review of adsorbents, which covers in considerable detail the classification, synthesis, characterization, and application of a wide range of porous materials, including both the "old" and "new" adsorbents, can be found in Ref. [13].

1.2 Adsorption Technologies

The demand for new adsorbents from the industry is continuously increasing. The most widespread adsorption technology is gas and liquid separations, which includes trace gas removal, wastewater treatment, respiratory protection, groundwater remediation, bulk gas separations, chromatography, etc. [4]. The separation is needed in chemical, petrochemical, and pharmaceutical industries, for environmental protection, stable water supply, etc. Separation is the process opposite to mixing and, therefore, unfavorable in terms of the second law of thermodynamics. Because of this, adsorptive separation is difficult to be accomplished, and its efficiency crucially depends on the properties of the sorbent.

Many less diffused but important applications of adsorbents are present, such as ion exchange [13], adsorptive transformation and storage of heat [29], maintenance of relative humidity [30], active heat insulation [31], shifting chemical equilibrium [32], pharmaceutics, medicine (drug delivery, blood cleaning, etc.) [13], freeze drying, conditioning of indoor air [33], life support system on a spacecraft [34], soil amendment, animal husbandry, and many others. In this book, we highlight only a few selected applications that are minutely considered below.

These "secondary" applications differ significantly from separation processes and, generally speaking, different adsorbents are needed for their best implementation. For instance, for deep gas drying, a solid desiccant has to bind strongly and retain water molecules at very low relative vapor pressure ($P/P_o < 0.005–0.0005$) that corresponds to the air dew point of $-40 \div -60°C$. On the contrary, to maintain the needed conditions in museums, archives, and libraries (having relative humidity 50–60%, or $P/P_o = 0.5–0.6$), an optimal adsorbent should bind water moderately so that it can give the water back when the P/P_o value gets lower than 0.5. In Chapter 3, we will give more striking examples of diverse demands imposed on the adsorbent desired for specific applications.

1.3 Harmonization of Adsorbent and Process

Despite advanced properties of modern adsorbents and impressive progress in many adsorption technologies, further improvement can be achieved. In our opinion, this improvement should be directed toward better matching of the adsorbent with the process.

As stated above, activated carbon, zeolites, silica gel, and activated alumina occupy a dominant part of the market. Since they are relatively cheap and readily available, it is easy to succumb to the temptation of using these materials for an adsorptive application. One example concerns adsorptive heat pumps (AHPs), which are considered an alternative to common vapor compression systems [29]. The development of AHPs was urged by the Montreal Protocol (1988), and the first progress was summarized in Ref. [35]. Experts from seven European laboratories selected the following "adsorbate–solid sorbent" pairs as the most suitable for AHP applications: "H_2O–zeolite 4A," "H_2O–zeolite NaX," "CH_3OH–active carbon," and "NH_3–active carbon." At the same time, a commercial silica gel Fuji type RD was used in the first commercial adsorption chillers, which utilized low-temperature heat (50–80°C) [36]. All the five selected adsorbents were large-scale commercial products easily available in the market. During the next 15 years of research activity, many novel materials, the sorption properties of which better fit to particular AHP cycles, have been revealed and, moreover, intently prepared [37].

Step by step, for many applications, a similar kind of gap was arising between the demands that were imposed by the application of a desirable adsorbent and the properties of a common adsorbent traditionally used in this application. More often, the mentioned demands are somewhat vague and more intuitive than rational, so that no clear idea exists about the optimal adsorbent for the given application. Thus, it is proper time for considerable efforts to be directed toward closing or reducing this gap. The idea of harmonization of the adsorbent and the process runs all through this book.

1.4 Screening versus Nanotailoring

A common way of harmonization is screening the already available sorbents and selecting the one with thermodynamic properties that fit better (even if not perfectly) to the particular application. This choice is usually made among solid sorbents that are already well known and commercially produced. This concept of screening can be very fruitful because the majority of "new" sorbents have never been tested for a body of possible applications. An encouraging example comes from the family of metal-aluminophosphates (MAPOs). Their main field of application was considered to be catalysis, until in the 21st century when they were suddenly proposed for low-grade heat utilization [38, 39] and a new commercial product soon appeared in the market.

On the other hand, testing a huge number of available adsorbents under conditions of a particular application is an extremely time-consuming process as it requires careful measuring of sorption equilibrium curves in a wide range of temperature and pressure. Moreover, this approach is inherently palliative and gives an intermediate solution rather than an ultimate one, because the selection is made among a restricted set of available materials.

Fortunately, the current level of materials science allows a more efficient approach based on the concept of nanotailoring, which means a target-oriented synthesis of the novel adsorbent with predetermined properties adapted to a given application. The main idea of this approach is that for each adsorptive technology (process, cycle) and its particular conditions, there is an optimal adsorbent, the properties of which would allow perfect performance of this process or cycle.

Thus, it is first necessary to fully realize exactly what an optimal adsorbent is. The next, practical, step is the synthesis of a real material with sorption properties close or even identical to those perfectly fitting the application. In contrast to the concept of screening, the latter approach, in principle, can give an ultimate solution to the harmonization problem. The main content of this book focuses on the nanotailoring approach and its application in selected adsorption processes.

1.5 Target-Oriented Synthesis of Sorbents

Traditionally, sorbents were developed empirically [4, 13, 40] by systematic manipulation of synthesis parameters rather than through rational understanding of what is a desirable adsorbent and how to prepare it. Such empiricism can be as time-consuming as the ingenuous screening mentioned above. It is somewhat similar to the combinatorial approach for preparing catalysts that needs expensive apparatuses and an enormous timetable.

Truly, the more rational basis for the sorbent design should rely on fundamentals of surface chemistry and adsorption theory. This is especially important for the first part of the nanotailoring concept, namely, formulating clear and, advisably, quantitative demands to the optimal adsorbent. In Chapter 3, we enunciate the requirements for several selected applications. On the contrary, synthesis of adsorbents relies partially on a rational base and, to a considerable degree, is still a kind of art similar to cookery where intuition and experience play a significant role (Chapter 4).

For a single-component adsorbent, a list of tools to manipulate its sorption properties is restricted to tuning the pore size and shape and, only slightly, to managing the surface activity. A detailed description of modified single-component adsorbents, including sol–gel oxides, carbons, zeolites, pillared clays, ordered oxides, carbons, metal–organic frameworks, etc., can be found elsewhere (see, e.g., Ref. [13] and references cited therein).

Tuning the properties of a two-component adsorbent is much easier because one can extra-vary at least two parameters, namely, the nature of the second component and its content in the composite. This book addresses a new family of solid sorbents—composites "salt in porous matrix" (CSPMs), which are very convenient objects for realizing the nanotailoring concept in practice. A typical CSPM is a composite material consisting of a porous host matrix and a guest salt confined inside the matrix pores (Fig. 1.1). Even a trivial combination of various hosts and guests can give an enormous number of composites. Synergetic (non-linear) effects (Fig. 1.2) might give more tools for designing a desirable sorbent. For instance, dependence of salt properties on pore size creates challenging opportunities, which are considered further on in this book.

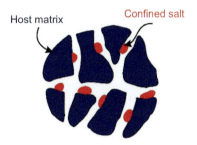

Figure 1.1 Composites "salt in porous matrix."

1.6 Nanocomposite Sobents "Salt in Porous Matrix"

Many porous solids that are used as host matrices are good adsorbents themselves, such as silica gel and alumina for water and activated carbon for methanol vapor. Therefore, an introduced salt has to be rather active to make a valuable contribution to the overall sorption capacity of the composite.

Many salts (S) interact with a sorptive A giving crystalline complexes called solvates

$$S(\text{solid}) + nA(\text{gas}) = S \cdot nA(\text{solid}) \qquad (1.1)$$

in which the sorptive is present in a definite ratio (n). For instance, one molecule of calcium chloride is able to absorb six water molecules according to the reaction

$$CaCl_2(s) + 6H_2O(g) = CaCl_2 \cdot 6H_2O(s) \qquad (1.2)$$

Being related to 1 g of the salt, it corresponds to 0.973 g of H_2O. One molecule of lithium chloride can bind three methanol molecules

$$LiCl(s) + 3CH_3OH(g) = LiCl \cdot 3CH_3OH(s) \qquad (1.3)$$

resulting in a specific sorption of 2.26 g/g. Five molecules of ammonia enter into the solvate with sodium chloride

$$NaCl(s) + 5NH_3(g) = NaCl \cdot 5NH_3(s) \qquad (1.4)$$

which corresponds to 1.47 g of NH_3 per 1 g of the salt. So large values of specific sorption are far superior to those of common adsorbents (typically, 0.2–0.5 g/g), probably because the vapor molecules are *absorbed* in the salt volume, not just *adsorbed* on its surface. Hence, it is reasonable to expect that inorganic salts combined with traditional adsorbents can enhance the total sorption capacity toward water, methanol, ammonia, and other sorptives that are able to form solvates according to reaction (1.1).

Figure 1.2 Schematics of synergetic "guest–host" interaction.

However, the direct use of bulk salts is limited by the following reasons:

1. Reaction (1.1) between the salt crystallites and vapor results in the formation of solid solvate on the surface of the salt. Further reaction requires the diffusion of vapor molecules through the layer of the solvate, which could be a very slow process [41].
2. There can be a pronounced "pressure–temperature" region near equilibrium where reaction (1.1) in bulk is inhibited

[42]. This inhibition region results in a noticeable hysteresis between synthesis and decomposition reactions.

3. Swelling/shrinkage of the solid during the synthesis/decomposition of solvate can lead to a mechanical destruction of the solid and dust formation.

4. According to phase diagrams of "salt–water" systems [43], at increasing relative humidity, hygroscopic salts possess a tendency to deliquescence, thus formatting corrosion solution.

These factors strongly restrict the application of bulk salts for enhancing sorption by simple mixing with traditional adsorbents. That is why the salt was suggested [44–47] to be inserted inside the pores of common adsorbent, as shown in Fig. 1.1.

As calcium chloride is highly hygroscopic, composite "$CaCl_2$ in porous alumina" was first patented as a dehydrating material in 1929 [44]. Then composite "$CaCl_2$ in porous carbon" was used in gas masks for respiratory protection; "LiCl in alumina" [48] and "LiCl in Torvex" [45, 49] were tested for air dehumidification, and "$CaCl_2$ in Celite" was tested for heat storage [46]. However, no systematic investigation was carried out to learn sorption properties of the composites because the goal of those studies was technologically focused and restricted to particular applications. Anyway, it became clear that modification of common porous solids by hygroscopic salts can give an efficient tool for improving sorption properties. This idea was then almost forgotten.

It was reanimated at the end of the 20th century when the systematic study of composites "salt in porous matrix" began, though they were called chemical heat accumulators (CHAs). The CHAs were assumed to be "based on granulated open-porous matrix filled with a hygroscopic substance" and suggested for storage of low-temperature heat [47]. The authors considered the phase diagram of the binary system "$CaCl_2$–water" and concluded that calcium chloride can be one of the most promising salts for storage of low-temperature heat. A new composite "$CaCl_2$ in mesoporous silica gel" (CHA-1) was estimated to be able to sorb up to 0.8 g of H_2O per 1 g of the dry composite and to store 1850 kJ/kg. Isobars of water sorption by this material were then reported in Ref. [50]. It was renamed as SWS-1L (SWS stood for selective water sorbent) because it became clear that the contribution of chemical bonds is only a part of the total enthalpy change during water sorption. The SWS-1L is still the most famous material of the SWS family. After that, many other studies of various composites "salt in porous matrix" were performed in research laboratories all over the world. Since these materials were studied not only as sorbents of water vapor, but also of methanol, ethanol, and ammonia vapors as well as of carbon dioxide, in this book we shall hold a more general abbreviation of CSPM as introduced above.

References

1. Prutkov, K. (1884) *Thoughts and Aphorisms* (Moscow).

2. Joachim, H. (1890) *Papyrus Ebers* (Berlin).

3. Schueth, F., Sing, K. S. W., and Weitkamp, J. (eds.) (2002) *Handbook of Porous Solids* (Wiley-VCH, Weinheim, Germany), vol. 1, p. 12.

4. Yang, R. T. (1997) *Gas Separation by Adsorption Processes* (Imperial College Press, London).

5. Scheele, C. W. (1773) Chemische Abhandlung von der Luft und dem Feuer; see: Ostwald's Klassiker der exakten Wiss. 58 (1894).

6. Kiefer, S. and Robens, E. (2008) Some intriguing items in the history of volumetric and gravimetric adsorption measurements, *J. Thermal Analysis and Calorimetry*, **94**, pp. 613–618.

7. Kehi, D. M. (1793) Observations et Journal sur la Physique, de Chemie et d'Histoire Naturelle et des Arts, Paris: Tome XLII, p. 250.

8. Lowitz, T. (1786) Chromatographic adsorption analysis: Selected works, *Crell's Chem. Ann.,* **1**, p. 211.

9. Kistler, S. S. (1931) Coherent expanded aerogels and jellies, *Nature* (London), **127**, p. 741.

10. Hood, H. P. and Nordberg, M. E. (1934) US Patent 2,106,744.

11. Barrer, R. M. (1948) Syntheses and reactions of mordenite, *J. Chem. Soc.*, pp. 2158–2163. doi: 10.1039/JR9480002158.

12. Ostrejko, R. V. (1901) German Patent 136,792.

13. Schueth, F., Sing, K. S. W., and Weitkamp, J. (eds.). (2002) *Handbook of Porous Solids* (Wiley-VCH, Weinheim, Germany), vols. 1–5.

14. Dabrovski, A. (2001) Adsorption: From theory to practice, *Adv. Colloid Interface Sci.*, **93**, pp. 135–224.

15. Humphrey, J. L. and Keller, G. E. (1997) *Separation Process Technology* (McGraw-Hill, New York).

16. Yang, R. T. (2003) *Adsorbents: Fundamentals and Applications* (Wiley, New York).

17. Wilson, S. T., Lok, B. M., Messina, C. A., Cannan, T. R., and Flanigen, E. M. (1982) Aluminophosphate molecular sieves: A new class of microporous crystalline inorganic solids, *J. Am. Chem. Soc.*, **104**, pp. 1146–1147.

18. Martens, J. A. and Jacobs, P. (1994) Crystalline microporous phosphates: A family of versatile catalysts and adsorbents, in advanced zeolite science and application, *Stud. Sur. Sci. Catal.*, **85**, pp. 653–685.

19. Beck, J. S., Vartuli, J. C., Roth, W. J., Leonowicz, M. E., Kresge, C. T., Schmitt, K. D., Chu, C. T. W., Olson, D. H., Sheppard, E. W., McCullen, S. B., Higgins, J. B., and Schlenker, J. L. (1992) A new family of mesoporous molecular sieves prepared with liquid crystal templates, *J. Am. Chem. Soc.*, **114**, pp. 10834–10843.

20. Inagaki, S., Fukushima, Y., and Kuroda, K. (1993) Synthesis of highly ordered mesoporous materials from a layered polysilicate, *J. Chem. Soc., Chem. Commun.*, pp. 680–682. doi: 10.1039/C39930000680.

21. Schoonheydt, R. A., Pinnavaia, T., Lagaly, G., and Gangas, N. (1999) Pillared clays and pillared layered solids, *Pure Appl. Chem.*, **71**, p. 2367.

22. Iijima, S. (1991) Helical microtubules of graphitic carbon, *Nature*, **354**, pp. 56–58.

23. Harris, P. J. F., Tsang, S. C., Claridge, J. B., and Green, M. L. H. (1994) High-resolution electron microscopy studies of a microporous carbon produced by arc-evaporation, *J. Chem. Soc. Faraday Trans.*, **90**, pp. 2799–2802.

24. Pekala, R. W., Alviso, C. T., Kong, F. M., and Hulsey, S. S. (1992) Aerogels derived from multifunctional organic monomers, *J. Non-Crystalline Solids*, **145**, pp. 90–98.

25. Kaneko, K. and Ishii, C. (1992) Super-high surface area determination of microporous solids, *Colloids Surf.*, **67**, pp. 203–212.

26. Kroto, H. B., Heath, J. R., O'Brien, S. C., Curl, R. F., and Smalley, R. E. (1985) C60: Buckminsterfullerene, *Nature*, **318**, pp. 162–163.

27. Yaghi, O. M., O'Keeffe, M., Ockwig, N. W., Chae, H. K., Eddaoudi, M., and Kim, J. (2003) Reticular synthesis and the design of new materials, *Nature*, **423**, pp. 705–714.

28. Kitagawa, S., Kitaura, R., and Noro, S. (2004) Functional porous coordination polymers, *Angew. Chem. Int. Ed.*, **43**, pp. 2334–2375.

29. Wang, R., Wang, L., and Wu, J. (2014) *Adsorption Refrigeration Technology: Theory and Application* (John Wiley & Sons, Singapore). doi: 10.1002/9781118197448.

30. Tomphson, G. (1986) *The Museum Environment* (Butterworth-Heinemann, London-Boston).

31. Kaviany, M. (1995) *Principles of Heat Transfer in Porous Media*, 2nd ed. (Springer, New York).

32. Roes, A. W. M. and Swaaij, W. P. M. (1979) Hydrodynamic behaviour of a gas-solid counter-current packed column at trickle flow, *Chem. Eng. J.*, **17**, p. 81.

33. ASHRAE (American Society of Heating, Refrigeration and Air-conditioning Engineers) (1997) *Handbook-Fundamentals* (ASHRAE, Atlanta, GA, USA).

34. DallBauman, L. A. and Finn, J. E. (1999) Adsorption processes in spacecraft environmental control and life support systems. In: *Adsorption and Its Applications in Industry and Environmental Protection*, Dabrowski, A. (ed.) (Elsevier, Amsterdam), vol. 120B, p. 455.

35. Pons, M., Meunier, F., Cacciola, G., Critoph, R. E., Groll, M., Puigjaner, L., Spinner, B., and Ziegler, F. (1999) Thermodynamic based comparison of sorption systems for cooling and heat pumping, *Int. J. Refrig.*, **22**, pp. 5–17.

36. Matsushita, M. (1987) Adsorption chiller using low-temperature heat sources, *Energy Conservation*, **39**, p. 96.

37. Aristov, Yu. I. (2013) Challenging offers of material science for adsorption heat transformation: A review, *Appl. Therm. Eng.*, **50**, pp. 1610–1618.

38. Jaenchen, J., Ackermann, D., and Stach, H. (2002) Adsorption properties of aluminophosphate molecular sieves: Potential applications for low temperature heat utilization, *Proc. Int. Sorption Heat Pump Conference*, Shanghai, China, pp. 635–638.

39. Kakiuchi, H., Iwade, M., Shimooka, S., Ooshima, K., Yamazaki, M., and Takewaki, T. (2005) Water vapor adsorbent FAM-Z02 and its applicability to adsorption heat pump, *Kagaku Kogaku Ronbunshu*, **31**, pp. 273–277.

40. Barton, T. J., Bull, L. M., Klemperer, W. G., Loy, D. A., McEnaney, B., Misono, M., Monson, P. A., Pez, G., Scherer, G. W., Vartuli, J. C., and Yaghi, O. M. (1999) Tailored porous materials, *Chem. Mater.*, **11**, pp. 2633–2656.

41. Galwey, A. and Brown, M. (1999) *Thermal Decomposition of Ionic Solids* (Elsevier Science, Amsterdam).

42. Andersson, J. Y. (1986) Kinetic and mechanistic studies of reactions between water vapour and some solid sorbents. Department of Physical Chemistry, The Royal Institute of Technology, S-100 44, Stockholm, Sweden, pp. 27–44.

43. Kirk-Othmer. (1992) *Encyclopedia of Chemical Engineering*, 4th ed. (Wiley, New York), vol. 4.

44. US Patent N 1,740,351. Dehydrating substance, H. Isobe, Dec. 17, 1929.

45. Chi, C. W. and Wasan, D. T. (1969) Measuring the equilibrium pressure of supported and unsupported adsorbents, *Ind. Eng. Chem. Fundam.*, **8**, pp. 816–818.

46. Heiti, R. V. and Thodos, G. (1986) Energy release in the dehumidification of air using a bed of $CaCl_2$-impregnated celite, *Ind. Eng. Chem. Fundam.*, **25**, pp. 768–771.

47. Levitskii, E. A., Aristov, Yu. I., Tokarev, M. M., and Parmon, V. N. (1996) "Chemical Heat Accumulators": A new approach to accumulating low potential heat, *Sol. Energy Mater. Sol. Cells*, **44**, pp. 219–235.

48. Fedorov, N. F., Ivahnyuk, G. K., and Babkin, O. E. (1990) The parameters determining sorption properties of impregnated dessicants, *Rus. J. Appl. Chem.*, **63**, pp. 1275–1279.

49. Chi, C. W. and Wasan, D. T. (1970) Fixed bed adsorption drying, *AIChE J.*, **16**, pp. 23–31.

50. Aristov, Yu. I., Tokarev, M. M., Cacciola, G., and Restuccia, G. (1996) Selective water sorbents for multiple applications: 1. $CaCl_2$ confined in mesopores of the silica gel: sorption properties, *React. Kinet. Cat. Lett.*, **59**, pp. 325–334.

Chapter 2

Description of Adsorption Equilibrium and Dynamics

The term adsorption is commonly used to denote the gas concentrated on a surface in contrast to the gas entering the bulk, as in absorption [1]. As we will see in the following sections, in composites "salt in porous matrix" (CSPMs), a sorbate can be involved in both processes; therefore, further on we will refer to either adsorption or simply sorption, regardless of the real location of the sorbate, which is often difficult to unambiguously determine.

A "gas (sorptive)–solid (sorbent)" system in equilibrium can be characterized by the temperature T, the sorptive pressure P, and the sorbate uptake (loading) w. According to the Gibbs phase rule, a common adsorption system is bi-variant, i.e., it possesses two degrees of freedom. It means that the equilibrium adsorbed uptake w is a function of the two variables P and T. Mono-variant absorption systems are considered in Sec. 2.5. They have one degree of freedom and behave in a different way than the bi-variant systems.

We start this chapter with Sec. 2.1, which concerns with the adsorption theories based on the fundamental Langmuir and Gibbs approaches from which many basic adsorption equations were derived. These equations link the equilibrium adsorbed uptake w with the gas pressure P and temperature T. The next section presents

---

Nanocomposite Sorbents for Multiple Applications
Yu. I. Aristov
Copyright © 2020 Jenny Stanford Publishing Pte. Ltd.
ISBN 978-981-4267-50-2 (Hardcover), 978-981-4303-15-6 (eBook)
www.jennystanford.com

Description of Adsorption Equilibrium and Dynamics

useful empirical approaches that we use in this book for formulating requirements to the adsorbent optimal for a given application. Then we shortly introduce and discuss other basic definitions and quantities that are important for further considerations in this book, namely, the classification of adsorption isotherms (Sec. 2.3) and the isosteric heat of sorption (Sec. 2.4).

In addition to various aspects of sorption equilibrium, in Sec. 2.6 we give basic information about sorption kinetics, which is often limited by gas diffusion through the pore network for adsorption systems. Both the rigorous Fickian diffusion model and the approximate linear driving force (LDF) model are considered thereafter.

2.1 Theoretical Adsorption Equations

2.1.1 Langmuir-Type Isotherms

The Langmuir adsorption theory [2] is one of the earliest theories suggested for describing adsorption equilibrium. It appears to be so fruitful that the Langmuir-type isotherms still remain to be the most widely used in practice. The Langmuir approach is kinetic by nature and is based on the assumption that the rate of adsorption is equal to the rate of desorption. The Langmuir model considers that (a) the surface is homogeneous, (b) each surface site can be occupied by just one gas molecule, and (c) the adsorbed molecule is immovable. This results in the following basic equation:

$$w(P,T) = w_0 \frac{b(T)P}{1 + b(T)P} \tag{2.1}$$

where w_0 is the maximum gas amount adsorbed in gram per unit mass of adsorbent, $b(T) = b_0 \cdot \exp(Q/RT)$ is the affinity constant, Q is the adsorption heat, and R is the universal gas constant. This equation is successful in describing many experimental data, although it fails in the case of complex adsorbents or adsorbates for which the aforementioned assumptions are not readily satisfied. This complexity can be rationalized within the concept of a heterogeneous surface. For exponential distribution of the adsorption heat Q, the Freundlich equation can be obtained:

$$w(P,T) = KP^{1/n} \qquad (2.2)$$

where K and n are parameters with complex temperature dependence [3]. The parameter n is usually greater than unity. Since this equation gives a continuously rising uptake with an increase in pressure, it was corrected by Sips [4]:

$$w(P,T) = w_0 \frac{KP^{1/n}}{1 + KP^{1/n}} \qquad (2.3)$$

Equations (2.2) and (2.3) are not valid at low pressure as they do not go over to the linear dependence of the Henry law type. Recognizing this, Toth [5] proposed the following equation:

$$w(P,T) = w_0 \frac{KP^m}{[1 + KP^m]^{1/m}} \qquad (2.4)$$

where the parameter m is commonly less than unity. At $n = 1$ and $m = 1$, Eqs. (2.3) and (2.4) reduce to the basic Langmuir equation. Large deviation from unity indicates a strong surface heterogeneity. As the Toth equation contains three fitting parameters, it describes well many adsorption data for practical adsorbents. Note that the Freundlich, Sips, and Toth isotherms can be considered semi-empirical approaches since they do not have rigorous theoretical substantiation.

2.1.2   Gibbs Approach

The adsorption equilibrium can be described by a thermodynamic approach as developed by Gibbs [6]. The Gibbs adsorption isotherm equation

$$\left(\frac{d\pi}{d\ln P} \right)_T = \frac{n}{S} RT \qquad (2.5)$$

links the number (n) of moles adsorbed with the gas pressure P and temperature T as well as with the spreading pressure π, which plays the same role in the adsorbed phase as the pressure P in the bulk phase (S is the surface area). This general equation can be transformed into various forms more convenient for analyzing experimental data by a proper choice of the equation of state of the surface phase.

The linear Henry equation describes the equilibrium at infinite dilution:

$$\frac{n}{S} = K(T)P \qquad (2.6)$$

where the parameter $K(T) = C(T)/(RT)$ is the Henry constant.

If the adsorbed molecules have a finite size and can move along the surface, but no interaction between them is allowed, the Volmer equation can be obtained [7]:

$$b(T)P = \frac{\theta}{1-\theta}\exp\frac{\theta}{1-\theta} \qquad (2.7)$$

where $\theta = A_o/A$ is the fractional loading, A_o is the area occupied by the adsorbed molecules.

If lateral interaction between the mobile adsorbed molecules is allowed, the Hill–de Boer equation

$$b(T)P = \frac{\theta}{1-\theta}\exp\frac{\theta}{1-\theta}\exp[-c(T)\theta] \qquad (2.8)$$

can be obtained. Among the other fundamental equations derived from the generic Eq. (2.5), we mention only the Harkins–Jura equation that describes a multilayer adsorption. More details on the theoretical adsorption isotherms can be found elsewhere (see, e.g., Refs. [7–9]).

2.2 Empirical Approaches to Adsorption Equilibrium

Here we consider useful empirical approaches that will be used further on to formulate requirements for the optimal adsorbent. Indeed, in addition to numerous theoretical isotherms, many empirical equations have been proposed to describe experimental data for a variety of solids with complex pore and surface structure. We mainly focus on the Polanyi potential theory [10], which leads to very important empirical equations proposed by Dubinin et al. [11, 12].

2.2.1 Polanyi Potential Theory

This theory considers the adsorption process to be similar to condensation, and the adsorbed state to behave like a liquid. The principle of temperature invariance declares that at different temperatures T_a and T_b, equal uptake can be achieved at the gas pressures P_a and P_b, which are linked through the following formula [10]:

$$T_a \ln(h_a) = T_b \ln(h_b) \qquad (2.9)$$

where $h = P/P_0$ is a relative pressure of the adsorptive. It postulates a one-to-one correspondence between the adsorption uptake and the product $T \ln h$. Dubinin introduced a similar value $\Delta F = -RT \ln h$, called the free energy of adsorption or the adsorption potential. He showed that for a variety of microporous solids, the specific adsorption volume V (normalized to the unit mass of adsorbent, cm^3/g) can be expressed in the general form as a unique function of ΔF [11]: $V = f(\Delta F)$. If the adsorbate density does not depend on temperature, a similar relation is valid for the specific uptake (g/g):

$$w = f(\Delta F) \qquad (2.10)$$

The most famous are the Dubinin–Radushkevich equation [13]

$$V = V_0 \exp\left[-\left(\frac{\Delta F}{\beta E_0}\right)^2\right] \qquad (2.11)$$

and its generalization by Astakhov [14]:

$$V = V_0 \exp\left[-\left(\frac{\Delta F}{\beta E_0}\right)^n\right] \qquad (2.12)$$

where V_0 is the maximum sorbed volume, E_0 is the so-called characteristic energy, β is a constant that depends only on the sorptive nature, and n describes the surface heterogeneity.

The main advantage of Eqs. (2.10)–(2.12) is a one-to-one correspondence between the equilibrium uptake and single parameter $\Delta F = -RT \ln h$, instead of the usually employed two parameters (*P*, *T*). For instance, the experimental isobars and isotherms of water sorption on the composite "CaCl$_2$/silica gel KSK" (SWS-1L) measured over a temperature range of 23–140°C and a pressure range of 8–133 mbar settle on the unique characteristic sorption curve if presented as a function of the adsorption potential Δ*F* (Fig. 2.1) [15].

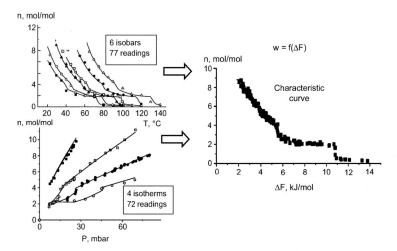

Figure 2.1 The experimental isobars and isotherms of water sorption on "CaCl$_2$/silica gel KSK" and the characteristic sorption curve. Reprinted from Ref. [15], Copyright 2012, with permission from Elsevier.

The Polanyi approach is rather universal, and its application area was generalized for many non-microporous adsorbents [16], including CSPMs [17, 18]. Therefore, we shall widely use Eq. (2.10) in this book.

2.2.2 Trouton's Rule

Another empirical relationship that we shall use in this book is known as Trouton's rule [19]. It declares that isosteric sorption lines for a given adsorbent and the equilibrium curve for a pure sorptive intersect at the temperature approaching infinity (Fig. 2.2). In the classic chemistry, Trouton's rule is formulated in a different but general way: the molar entropy of evaporation of any liquid is approximately equal to 88 J/molK [20]. The following useful relation can be obtained from simple geometrical considerations of Fig. 2.2.

$$T_2^2 = T_1 T_3 \qquad (2.13)$$

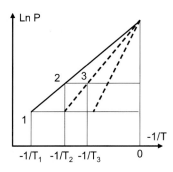

Figure 2.2 Schematics of Trouton's rule: bold line—"gas–liquid" equilibrium; dashed lines—sorption isosters.

Although Eq. (2.13) does not have solid theoretical justification, it is valid for a variety of "sorbate-sorbent" pairs, as shown in Ref. [18]. Among these pairs are water–silica gel Fuji RD, water–zeolite 13X, water–SWSs, CO_2–carbon, methanol–carbons (AC-35, TA90), methanol–hydrophobic zeolite CBV901 Y, ethanol–ACF, and ammonia–carbon PX31.

The high accuracy of this empirical approximation is demonstrated in Table 2.1, which displays a comparison between the temperature T_3 calculated from Eq. (2.13) and directly obtained from the experimental isosters of the silica gel Fuji type RD. Therefore, this equation will be used in Chapter 3 for analysis of adsorption heat transformer (AHT) thermodynamic cycles.

It was shown [18] that Eq. (2.13) is valid for a given "sorbate–sorbent" pair if the entropy change during adsorption ΔS_{ad} is equal to that during condensation ΔS_e, which means that the adsorbed state is similar to the liquid state as proposed in the Polanyi potential theory. There is nothing surprising that it is always valid if there is a one-to-one correspondence between the equilibrium adsorption uptake w and the adsorption potential ΔF [18]. Indeed, in this case, each isoster is a line of the equal potential $\Delta F = -RT°\ln(P/P_0) = $ const., and the equilibrium pressure can be expressed as $\ln(P) = $ const./RT

Description of Adsorption Equilibrium and Dynamics

+ $\ln[P_0(T)]$, where $P_0(T)$ is a function of the temperature only. At T approaching infinity, $\ln(P)$ approaches $\ln(P_0)$, and Trouton's rule is automatically fulfilled.

Table 2.1 Water–silica gel Fuji Davison RD

T_1 (°C)	T_2 (°C)	T_{3exp} (°C)	T_{3cal} (°C)	ΔT (°C)
5	37.1	73.8	72.8	−1.0
5	31.2	59.8	59.8	0.0
5	26.8	49.8	50.3	0.5
5	23.5	42.5	43.1	0.6
5	20.7	36.6	37.3	0.7
10	43.1	80.1	80.0	−0.1
10	36.8	65.8	66.2	0.4
10	32.3	55.5	56.3	0.8
10	28.7	47.9	48.7	0.8
10	25.9	41.8	42.6	0.8
20	54.2	92.3	92.4	0.1
20	47.8	77.5	78.3	0.8
20	42.7	66.1	67.1	1.0
20	38.8	58.1	58.9	0.8
20	36.0	51.8	52.8	1.0

Note: Experimental (T_{3exp}), calculated (T_{3cal}) values of temperature T_3 and the difference $\Delta T = T_{3cal} - T_{3exp}$ obtained at various T_1 and T_2 from the equilibrium isosteric chart.

Source: Reprinted from Ref. [18], Copyright 2008, with permission from Elsevier.

2.2.3 BET Equation

The Brunauer–Emmett–Teller (BET) equation

$$V = V_0 \frac{CP}{(P_0 - P)\left[1 + (C - 1)(P / P_0)\right]} \tag{2.14}$$

was developed in 1938 [21] to account for multilayer adsorption. It contains two fitting parameters: the maximum sorbed volume V_0 and the constant C, which characterizes interaction in the adsorbed layer. Despite many modifications of Eq. (2.14), it is still the most important equation for characterization of mesoporous adsorbents, e.g., for the determination of specific surface area. Equation (2.14) is invariant with respect to the relative pressure $h = P/P_0$.

2.3 Classification of Adsorbents

Pores are conventionally classified according to the IUPAC recommendations as follows: Micropores have diameters less than 2 nm; mesopores have diameters between 2 and 50 nm; and macropores have diameters greater than 50 nm [22]. Capillary condensation is typical for adsorption in mesopores, whereas strong interactions between sorbate molecules and pore walls result in filling the micropores volume [10].

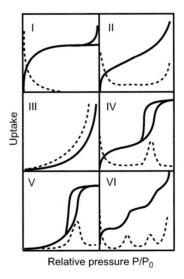

Figure 2.3 Six types of adsorption isotherms and their derivatives. Reprinted from Ref. [23], Copyright 2007, with permission from JST.

A variety of porous solids result in a spectrum of isotherm shapes that were typified in Ref. [24] as five classes I–V (Fig. 2.3) creating the base of the IUPAC classification. The five types of isotherm

Description of Adsorption Equilibrium and Dynamics

are characteristic of adsorbents that are microporous (type I), nonporous or macroporous (types II and III), or mesoporous (types IV and V). The differences between types II and III isotherms and between types IV and V isotherms arise from the relative strengths of the fluid–solid and fluid–fluid attractive interactions: Types II and IV are associated with stronger fluid–solid interactions, while types III and V are related to weaker fluid–solid interactions. The hysteresis loops usually exhibited by types IV and V isotherms are associated with capillary condensation in the mesopores. One more type (VI) of adsorption isotherms was primarily introduced to represent adsorption on nonporous or macroporous solids where stepwise multilayer adsorption occurs [22].

2.4   Isosteric Heat of Sorption

The isosteric heat of sorption ΔH_{is} is a basic property in adsorption science. It is defined from the van't Hoff equation as

$$\Delta H_{is} = R\frac{d(\ln P)}{d(1/T)}\bigg|_{w=\text{const.}} \tag{2.15}$$

The isosteric heat commonly, but not necessarily, increases with the decrease in the uptake, which reflects the fact that first portions of the sorptive stronger interact with the adsorbent surface. The isosteric heat can be obtained by direct measurement of sorption isosteres, differentiation of sorption isotherms at constant uptake, and calorimetric methods [25, 26]. The importance of minimizing the "dead volume" in apparatuses for the direct measurement of isosteres is discussed in Ref. [27]. We shall measure isobaric charts and determine the ΔH_{is} value in order to discriminate between different sorption mechanisms and compare sorption in bulk and confined states.

2.5   Mono-variant versus Bi-variant Systems

The Gibbs phase rule [6] specifies conditions of the phase equilibrium and states as follows

$$f = c - p + 2 \tag{2.16}$$

where f is the system variance, c is the number of components, and p is the number of phases. As stated above, a common adsorption system consists of two phases (solid and gas) and two components (sorbent and sorptive). According to the Gibbs phase rule, such a system is bi-variant: the equilibrium uptake depends on both gas pressure and temperature. The sorption equilibrium can be presented as a set of isotherms, isobars, or isosteres, as shown in Fig. 2.4.

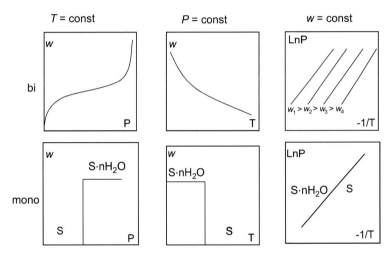

Figure 2.4 Mono-variant versus bi-variant systems.

A mono-variant system has only one degree of freedom, which results in different shapes of equilibrium sorption curves. A typical example of mono-variant system gives an interaction between a salt S and a gaseous sorptive A to give crystalline solvate according to reaction (1.1). The system consists of two components (salt and gas) and three phases (gas, salt, and solvate) and has only one degree of freedom. This means that at fixed temperature T, transition (1.1) occurs at certain pressure $P(T)$. At lower pressure, an anhydrous salt is stable, whereas at higher pressure, the salt binds the gas and wholly transforms to a salt solvate. Hence, the amount of gas bound by the salt (sorbed) is a step-like function of temperature as opposed to smooth curves of bi-variant systems (Fig. 2.4). The unique curve $P(T)$ describes the sorption equilibrium of a mono-variant system instead of an isosteric chart for a bi-variant one (Fig. 2.4).

2.6 Adsorption Kinetics

It is a common case for physical adsorption that the intrinsic rate of sorption on the surface is extremely fast, and the overall adsorption rate is controlled by mass and heat transfer. This section aims to give an overview of various gas transport mechanisms in porous media and to shortly bring in adsorption kinetics by an individual adsorbent grain. The pore structure of a real adsorbent is so complex that simplified models are strictly necessary. One of them represents a porous solid as a bundle of parallel cylindrical capillaries and considers a gas flow through a single capillary. The diffusivity D in a porous medium is related to the idealized diffusivity D_p in a straight cylindrical pore (with the diameter equal to the mean pore diameter) as follows:

$$D = \varepsilon D_p / \chi \qquad (2.17)$$

where the porosity ε takes into account the fact that gas transport cannot occur through the solid skeleton, while the tortuosity χ accumulates all other effects, e.g., irregular and twisting shape of pores. A detailed description of sorption kinetics can be found elsewhere [7, 28, 29]. Very popular for practical calculations is the LDF approach, which was suggested to simplify the numerical analysis of sorption dynamics in chromatographic columns [30].

2.6.1 Gas Transport Mechanisms

The main gas transport mechanisms in pores are molecular diffusion, Knudsen diffusion, surface diffusion as well as Poiseuille flow. When the pore diameter d is large relative to the mean free path l of gas molecules, collisions between the molecules are more frequent than collisions between the molecules and the pore wall. In this case, the diffusion mechanism is similar to that in the bulk gas and the pore diffusivity is identical to the molecular diffusivity D_m [7, 28]. In the opposite case of small pores (or low pressures), the mean free path is greater that the pore diameter and gas molecules collide mainly with the pore wall. A molecule exchanges energy with the wall and is reflected in random direction. The stochastic nature of this process results in a Fickian expression for the flux with the

Knudsen diffusivity (in m^2/c)

$$D_{kn} = 48.5\,d\sqrt{T\,/\,M} \qquad (2.18)$$

where d is in meters, the temperature T is in kelvin, and the molecular mass of gas M is in g/mol. The Knudsen diffusivity depends only on the pore size and the molecular velocity. If the mean free path is comparable with the pore diameter, a transient regime is observed when both mechanisms are important, so that the overall diffusivity through the gas phase can be written as [28]

$$\frac{1}{D_p} = \frac{1}{D_m} + \frac{1}{D_{kn}} \qquad (2.19)$$

If adsorption on the pore wall is significant and adsorbed molecules are mobile, additional gas transport occurs due to the gas diffusion along the surface. As the fluxes through the gas phase and the adsorbed phase are independent, the diffusivity can be calculated as the sum of these two contributions:

$$D = D_p + KD_s \qquad (2.20)$$

where the coefficient K accounts for the difference in molecular densities in the gas and adsorbed phases. The surface diffusion gets more important at lower temperatures.

The Poiseuille or viscous flow arises due to a pressure difference between the ends of a capillary. Although it is similar to the laminar flow of liquid, it can be formally described by the Fickian diffusivity given by

$$D_{Pois} = Pd^2/32\eta \qquad (2.21)$$

where P is the absolute pressure and η is the gas viscosity. The contribution of the viscous flow becomes important in relatively large pores and at high pressures.

2.6.2 Isothermal Adsorption Kinetics in a Single Grain

The adsorption and desorption of gas are almost always controlled by various mass or heat transfer resistances, the mutual contributions of which depend on the pore structure and thermal properties of adsorbent grain as well as on the parameters of the adsorption process. Here we consider only two simple but frequent cases of

28 | *Description of Adsorption Equilibrium and Dynamics*

mass transfer resistances at the isothermal conditions, namely, the surface resistance control and the pore diffusion control. Many other resistances and their combinations are considered in Ref. [28].

2.6.2.1 Surface resistance control

For a nonporous grain or a porous grain with very fast intragrain diffusion, the sorption rate is controlled by the gas transport to (or through) the grain external surface. The adsorbate is distributed in the grain homogeneously, and the sorption rate

$$\frac{dw}{dt} = k_s a(c - c_s) = \frac{3k_s}{R}(c - c_s)$$

(2.22)

depends on the mass transfer coefficient k_s, the specific grain surface $a = 3/R$ for a spherical grain of radius R, and the difference $(c - c_s)$ between the gas concentrations in the bulk and at the grain surface. If the concentration change is sufficiently small, the equilibrium relation $w_{eq} = K \cdot c$ is linear, where K is the local slope of the sorption isotherm, and Eq. (2.22) may be rewritten as

$$\frac{dw}{dt} = \frac{3k_s}{KR}(w_{eq} - w)$$

(2.23)

and directly integrated with the following initial conditions:

$$t = 0, c = q = 0$$
$$t > 0, c = c_\infty = w_{eq}/K$$

to give a simple exponential dependence on time

$$\frac{w}{w_\infty} = 1 - \exp(-\frac{3k_s t}{KR})$$

(2.24)

2.6.2.2 Pore diffusion control

If the intragrain diffusion controls the overall adsorption rate, the uptake rate is given by the appropriate solution of the differential mass balance equation. For uniform spherical adsorbent grains, it may be written as a Fickian diffusion (FD) equation [28]:

$$(1 - \varepsilon)\frac{\partial w}{\partial t} + \varepsilon\frac{\partial c}{\partial t} = \varepsilon D_p \left(\frac{\partial^2 c}{\partial R^2} + \frac{2}{R}\frac{\partial c}{\partial R} \right)$$

(2.25)

If the step change of concentration is small (from c_o to $c_o + \Delta c$, where $\Delta c << c_o$), then the equilibrium relationship is expected to be linear:

$$\frac{\partial w}{\partial t} = K \frac{\partial c}{\partial t} \tag{2.26}$$

with $K = K(c_0) = $ const. Now Eq. (2.25) may be written as

$$\frac{\partial c}{\partial t} = \frac{\varepsilon D_e}{\varepsilon + (1-\varepsilon)K} \left(\frac{\partial^2 c}{\partial R^2} + \frac{2}{R} \frac{\partial c}{\partial R} \right) \tag{2.27}$$

The relevant initial and boundary conditions for a step concentration change at the external surface of the particle at $t = 0$ are

$$c(R, 0) = c_0, \qquad\qquad w(R,0) = w_0,$$

$$c(R_p, \infty) = c_\infty, \qquad\qquad w(R, \infty) = w_\infty,$$

$$(\partial c / \partial R)|_{R=0} = (\partial w / \partial R)|_{R=0} = 0$$

The solution of Eq. (2.27) is well known [29]:

$$\frac{\Delta w_t}{\Delta w_\infty} = \frac{w - w_0}{w_\infty - w_0} = 1 - \frac{6}{\pi^2} \sum_{n=1}^{\infty} \frac{1}{n^2} \exp\left(-\frac{n^2\pi^2 \dfrac{\varepsilon D_p}{\varepsilon + (1-\varepsilon)K} t}{R^2} \right) \tag{2.28}$$

$$= 1 - \frac{6}{\pi^2} \sum_{n=1}^{\infty} \frac{1}{n^2} \exp\left(-\frac{n^2\pi^2 D_{ap} t}{R^2} \right)$$

where D_{ap} is the apparent water diffusivity $= \varepsilon D_p / [\varepsilon + (1 - \varepsilon)K]$. Thus, this diffusivity depends on the pore diffusivity D_p, the grain porosity ε, and the shape of the water sorption isotherm (the local slope of the isotherm $K = dw/dc$). A simplified expression for the initial region of the uptake curve may be obtained as [28]

$$\frac{m_t}{m_\infty} \approx \frac{2S}{V} \sqrt{\frac{D_{ap}\, t}{\pi}} = A\sqrt{t} \tag{2.29}$$

where S/V is the ratio of external area-to-grain volume ($S/V = 3/R$ for a spherical grain). Equation (2.29) can be used for determining the apparent diffusion time constant k_D ($= D_{ap}/R^2$), the apparent diffusivity D_{ap}, and the effective pore diffusivity D_p.

The isothermal differential step (IDS) method based on this rigorous model is a popular and effective tool for measuring the apparent diffusivity and the gas diffusivity in pores [28]. This

Description of Adsorption Equilibrium and Dynamics

method, however, has to be applied with due caution as the uptake evolution is very sensitive to thermal effects originated from the heat released during gas sorption [31].

2.6.2.3 Linear driving force model

The LDF model was originally proposed for adsorption chromatography [32]. Nowadays, it is frequently used for describing adsorption kinetics in column because it is simple and physically consistent [33]. According to the LDF model, the rate of adsorption of a sorptive by an adsorbent grain is given by

$$\frac{d\bar{w}}{dt} = K(w_{eq} - \bar{w}) \tag{2.30}$$

where $\bar{w}$ is the current average sorbate concentration in the grain (grams per unit mass of the adsorbent), w_{eq} is the equilibrium sorbate concentration, and K is the lamped mass transfer coefficient. The latter is assumed to be $K(T) = 15D_{ap}(T)/R^2$, where $D_{ap}(T)$ = $D_{ap}^0 \exp(-E_a/RT)$ is the temperature-dependent diffusivity, which is determined by the activation energy E_a. The discrepancy in the solutions to the FD and LDF models can be significant: the LDF underestimates the uptake at short times and overestimates it at long times [33]. The uptakes are equal at ca. 70% of the equilibrium conversion. Despite this, the LDF model allows a simple mathematical treatment of adsorption kinetic data, ensures satisfactory description of the packed-column breakthrough curves and the separation efficiency.

2.6.3 Kinetics of Gas–Solid Reactions

Reaction (1.1) between a solid salt S and a gaseous sorptive A, S (s) + nA (g) $\leftrightarrow$ S·nA (s), is more complicated than adsorption processes considered earlier. Its kinetics can be controlled by the chemical reaction itself (i.e., the redistribution of bonds), the changing geometry of reaction interface, and the transport of reagents to or from the reaction interface [34]. The models of solid-state reactions are based on the processes of nucleation and the growth of product nuclei near the reaction interface. Combination of rate equations for nucleation together with those for growth results in the Avrami–Erofeev (A–E) equation

$$[-\ln(1 - \alpha)]^{1/n} = k\,(t - t_o) \tag{2.31}$$

where α is the conversion degree (also called the fractional reaction or the product yield), constants n, k, and t_o are commonly empirical and have no physical meaning, except for several idealized cases. A typical feature of Eq. (2.31) is a sigmoid kinetic curve with an induction period, which is required for the generation of growth nuclei. Once these nuclei have been established, the reaction rate gradually increases during the acceleratory period to the maximum rate. After this, the rate reduces progressively until the reaction is completed.

Another important kinetic equation

$$1 - (1 - \alpha)^{1/3} = kt \tag{2.32}$$

comes from the shrinkage (or contracting) volume model, which is applied when nucleation occurs on certain crystal surfaces and the advance of reaction interfaces into the bulk of the reactant particle proceeds inward from those crystal surfaces [34]. This equation also describes propagation of the adsorption front toward the center of a spherical grain for strongly adsorbed species [28].

References

1. Sing, K. S. W., Everett, D. H., Haul, R. A.W., Moscou, L., Pierotti, R. A., Rouquerol, J., and Siemieniewska, T. (1985) Reporting physisorption data for gas/solid systems (Recommendations 1984), *Pure Appl. Chem.*, **57**, pp. 603–619.

2. Langmuir, I. (1916) The constitution and fundamental properties of solids and liquids, *J. Am. Chem. Soc.*, **38**, pp. 2221–2295.

3. Zeldowitsch, J. V. (1934) Adsorption site energy distribution, *Acta Physicochem. USSR*, **1**, pp. 961–973.

4. Sips, R. (1948) Combined form of Langmuir and Freundlich equations, *J. Chem. Phys.*, **16**, pp. 490–495.

5. Toth, J. (1971) State equations of the solid gas interface layer, *Acta Chem. Acad. Hung.*, **69**, pp. 311–317.

6. Gibbs, J. W. (1928) *Collected Works*, New York.

7. Do, D. D. (1998) *Adsorption Analysis: Equilibria and Kinetics*, Imperial College Press, p. 892.

8. Adamson, A. W. and Gast, A. P. (1997) *Physical Chemistry of Surfaces*, 6th ed., Wiley-Interscience, New York.

9. Rudzinski, W. and Everett, D. H. (1992) *Adsorption of Gases on Heterogeneous Surfaces*, Academic Press, San Diego.

10. Polanyi, M. (1932) Theories of the adsorption of gases, *Trans. Faraday Soc.,* **28**, pp. 316–333.

11. Dubinin, M. M. (1960) The potential theory of adsorption of gases and vapors for adsorbents with energetically non-uniform surface, *Chem. Rev.,* **60**, pp. 235–266.

12. Dubinin, M. M. (1975) Physical adsorption of gases and vapors in micropores, in: D. A. Cadenhead, J. F. Danielli, and M. D. Rosenberg (Eds.), *Progress in Surface and Membrane Science*, vol. 9, ISBN 0-12-571809-8, Academic Press, New York, p. 1.

13. Dubinin, M. M. and Radushkevich, L. V. (1947) The equation of the characteristic curve of the activated charcoal, *Dokl. Acad. Nauk SSSR,* **55**, pp. 331–337.

14. Dubinin, M. M. and Astakhov, V. F. (1971) Adsorption on microporous sorbents, *Sov. Chem. Bull. Chem*, pp. 5–16.

15. Aristov, Yu. I. (2012) Adsorptive transformation of heat: Principles of construction of adsorbents database, *Applied Therm. Engn.,* **42**, pp. 18–24.

16. Jaroniec, M. (1997) Fifty years of the theory of the volume filling of micropores, *Adsorption,* **3**, pp. 187–198.

17. Prokopiev, S. I. and Aristov, Yu. I. (2000) Concentrated aqueous electrolyte solutions, *J. Solution Chem.,* **29**, pp. 633–649.

18. Aristov, Yu. I., Sharonov, V. E., and Tokarev, M. M. (2008) Universal relation between the boundary temperatures of a basic cycle of sorption heat machines, *Chem. Eng. Sci.,* **63**, pp. 2907–2912.

19. Alefeld, G. and Radermacher, R. (1994) *Heat Conversion Systems*, CRC Press, Boca Raton, U.S.A.

20. Cengel, Yu. A. and Boles, M. A. (2002) *Thermodynamics: An Engineering Approach*, 4th ed., McGray-Hill Inc.

21. Brunauer, S., Emmet, P. H., and Teller, E. (1938) Adsorption of gases in multimolecular layers, *J. Amer. Chem. Soc.,* **60**, pp. 309–316.

22. Gregg, S. J. and Sing, K. S. W. (1982) *Adsorption, Surface Area and Porosity*, Academic Press, N.Y.

23. Aristov, Yu. I. (2007) Novel materials for adsorptive heat pumping and storage: Screening and nanotailoring of sorption properties, *J. Chem. Eng. Japan*, **40**, pp. 1241–1251.

24. Brunauer, S., Deming, L. S., Deming, E., and Teller, E. (1940) On a theory of the van der Waals adsorption of gases, *J. Amer. Chem. Soc.*, **62**, pp. 1723–1732.

25. Bering, B. P., Zhukovskaja, E. G., Rachmukov, B. C., and Serpinsky, V. V. (1969) *Z. Chem. (Leipzig)*, **9**, pp. 13–22.

26. Rouquerol, F., Partyka, S., and Rouquerol, J. (1972) In: Thermochimie, Colloques Internationaux du CNRS, no. 201 Editions du CNRS, Paris, p. 547.

27. Shen, D. and Buelow, M. (1998) Isosteric study of sorption thermodynamics of single gases and multi-component mixtures on microporous materials, *Microporous Mesoporous Mater.*, **22**, pp. 237–249.

28. Kaerger, J. and Ruthven, D. M. (1992) *Diffusion in Zeolites and Other Microporous Solids*, Wiley, New York.

29. Crank, J. (1956) *Mathematics of Diffusion*, Oxford University Press, London.

30. Glueckauf, E. (1955). Part 10. Formulae for diffusion into spheres and their application to chromatography, *Trans. Faraday Soc.*, **51**, pp. 1540–1551.

31. Lee, L. K. and Ruthven, D. M. (1979) Analysis of thermal effects in adsorption rate measurements, *J. Chem. Soc., Faraday Trans. I*, **75**, p. 2406.

32. Gleuckauf, E. and Coates, J. I. (1947) The influence of incomplete equilibrium on the front boundary of chromatograms and the effectiveness of separation, *J. Chem. Soc.*, pp. 1315–1321.

33. Sircar, S. and Hufton, J. R. (2000) Why does the linear driving force model for adsorption kinetics works? *Adsorption*, **6**, pp. 137–147.

34. Galwey, A. and Brown, M. (1999) *Thermal Decomposition of Ionic Solids*, Elsevier Science, Amsterdam.

Chapter 3

Optimal Adsorbent: Basic Requirements

As mentioned in Chapter 1, it is reasonable to suppose that a specific type of adsorbent material is required for adsorptive systems that perform a particular function. The selection of a proper sorbent is a complex problem even for common applications. Hence, from a practical point of view, it would be highly desirable to have a clear idea about the requirements for an optimal sorbent for each specific application. In this chapter, we will briefly consider how these requirements can be formulated and do so for several adsorptive applications. Dynamic demands for an optimal sorbent will be surveyed in Sec. 3.3.

The rational basis for the target-oriented design of sorbents should rely on fundamentals of theoretical surface chemistry and adsorption theory. The nature of sorbent–sorbate interactions is very complex, and modern theoretical models are successful in describing relatively simple interactions and chemical bonds. Therefore, the predominant scientific basis for specifying the thermodynamic requirements is actually the adsorption equilibrium and appropriate equations considered in Chapter 2. These requirements will serve as a guiding star for the synthesis of real materials with sorption properties that are close as much as possible to those predicted for an optimal adsorbent.

Nanocomposite Sorbents for Multiple Applications
Yu. I. Aristov
Copyright © 2020 Jenny Stanford Publishing Pte. Ltd.
ISBN 978-981-4267-50-2 (Hardcover), 978-981-4303-15-6 (eBook)
www.jennystanford.com

3.1 How to Formulate Thermodynamic Requirements?

The requirements can be expressed in either qualitative or quantitative manner. Such qualitative parameter could be, for instance, a desirable shape of the equilibrium sorption isotherm, which certainly results in optimal performance. For example, in long adsorption columns, a favorable, convex shape of the sorption isotherm leads to an adsorption front that spreads as it propagates [1]. As it will be demonstrated in the following sections, for several applications the most profitable are materials with a step-like adsorption isotherm. The pore shape can strongly affect the sorption hysteresis phenomena, while the "hydrophobic" or "hydrophilic" notions reflect the ability to sorb nonpolar or polar sorptives. A diversity of quantitative requirements may specify, e.g., the position of the mentioned step, pore curvature, specific surface area, pore volume, sorbent capacity at given temperature and pressure, sorption potential, adsorption heat, operating or regeneration temperature, and so on.

In both cases, the predominant scientific basis for formulating the requirements is the equilibrium sorption equations overviewed in Chapter 2, which describe the equilibrium uptake as a function of T and P. The empirical Polanyi principle of temperature invariance allows more general presentation of the requirements because it operates with just one parameter, namely, the adsorption potential ΔF. Fortunately, this principle is valid not only for microporous solids, but also for many mesoporous and composite adsorbents [2]. Another convenient invariant to present the equilibrium sorption is the relative pressure $h = P/P_0$.

3.2 Particular Adsorption Technologies

Among numerous adsorption technologies mentioned in Chapter 1, we analyze here the following processes that have been selected, first of all, in line with the historically established scientific interests of the author: gas drying, transformation of low-temperature heat, maintaining relative humidity (RH), shifting chemical equilibrium, and extraction of water from atmospheric air. For these applications,

Particular Adsorption Technologies | 37

we consider the requirements to the optimal adsorbent, which will serve as a model to be replicated at synthesis stage (Chapter 4). It will be clearly shown that these applications ask for a quite different adsorbent, ensuring optimal operation. Results of testing these "replications" are presented further in this book.

3.2.1 Gas Drying

Gas drying is one of the most important industrial adsorption technologies [1, 3]. Drying apparatus is commonly a fixed bed adsorber with periodic thermal regeneration of adsorbent. The main requirements for optimal adsorbents are large dynamic water capacity, low dew point at the adsorber outlet, and low regeneration temperature.

For the sake of simplicity, let us consider a step-like adsorption front that is a good approximation for the sufficiently long adsorber.

Let the temperature in the saturated part of the adsorber be T_{in} and the partial pressure of water vapor be P_{in}. The equilibrium water uptake $w_{in}(P_{in}, T_{in}) = w_{max}$ can be obtained directly from the equilibrium relation $w(P, T)$. If the equilibrium uptake is invariant with respect to the sorption potential ΔF, then $w_{in} = w(\Delta F_{in}) = w[RT_{in}\ln(P_o/P_{in})]$, where P_o is a saturated vapor pressure at temperature T_{in}. At the adsorption stage, the equilibrium uptake w_{in} corresponding to the inlet temperature and vapor pressure evidently has to be maximized.

At the regeneration stage, the adsorbent has to give up the adsorbed water back; hence, the residual equilibrium uptake $w_{reg} = w(P_{reg}, T_{reg})$ or $w_{reg} = w(\Delta F_{reg}) = w[RT_{reg}\ln(P_o/P_{reg})]$ has to be minimized. Thus, if the values of P and T at the adsorption and regeneration stages are fixed, an adsorbent *optimal for gas drying* should have the maximal uptake difference $\Delta w = w_{in} - w_{reg}$. Both boundary uptakes (w_{in} and w_{reg}) can be obtained from the equilibrium relation $w(P, T)$ or, more conveniently, from the characteristic curve of water sorption $w(\Delta F)$.

The latter is presented for the composite "CaCl$_2$/silica KSK" (SWS-1L) in Fig. 3.1 for making the following estimations: At the inlet relative pressure $P/P_o = 0.4$ and $T_{in} = 303$ K, $\Delta F_{in} = 2.3$ kJ/mol and the equilibrium uptake can reach $w_{in} = 0.45$ g H$_2$O per 1 g of the sorbent. It is common that for regeneration, the same air is used (the moisture content is 12.8 g/m³), but after heating up to T_{reg} = 373 K, so that P/P_o = 0.0167, ΔF_{reg} = 12.7 kJ/mol and w_{reg} = 0.02 g/g. Hence, SWS-1L exchanges $\Delta w = w_{in} - w_{reg}$ = 0.43 g/g, which gives an upper estimate of the SWS-1L sorption capacity under the fixed conditions.

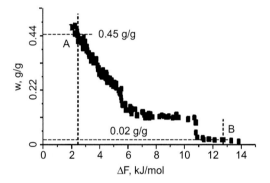

Figure 3.1 Equilibrium water uptake was a function of the adsorption potential ΔF and the relative pressure h for composite sorbent SWS-1L. Vertical lines correspond to $\Delta F_{ads}°=°2.3$ kJ/mol and $\Delta F_{reg}°=°12.7$ kJ/mol.

In fact, the dynamic sorption capacity is lower than the estimated static capacity, because the concentration front is not step-like and a length of unusable bed (LUB) > 0. Of course, a sharp concentration front (or a short LUB) is highly desirable because it results in a high sorbent productivity as well as high product purity [3]. Pore-size distribution can play a role in LUB, but not as important as the grain size and the shape of equilibrium isotherm. For a convex (favorable) isotherm, an initially dispersed adsorption front will become sharper, eventually approaching a shock transition (self-sharpening behavior). A convex (unfavorable) isotherm leads to the adsorption front that spreads as it propagates [1, 4], which results in increasing LUB and decreasing adsorbent efficiency.

Requirements of a low dew point at the adsorber outlet and a low regeneration temperature are mutually exclusive. Indeed, lower outlet dew point can be obtained for an adsorbent having higher affinity for water vapor, which means stronger adsorption sites. Accordingly, higher regeneration temperature is required to desorb water from these sites. As specified in Chapter 2, the Dubinin adsorption potential $\Delta F = -RT \ln(P/P_o)$ was introduced as a universal measure of the adsorbent's affinity for the adsorptive [5] (here P_o

is the saturated vapor pressure at temperature T). For instance, the equilibrium curve $w(\Delta F)$ in Fig. 3.1 presents a distribution of the SWS-1L adsorption sites on their affinity for water vapor: e.g., point A shows that the sites with affinity exceeding 2.3 kJ/mol are responsible for sorbing 0.45 g/g, whereas only 0.02 g/g are retained by strong sites with affinity larger than 12.7 kJ/mol (point B). The affinity of these strongest adsorption sites determines the dew point of the outlet air. If the regenerated adsorbent bed is cooled down to $T_{in} = 303$ K, the partial pressure of water vapor, P, over these strong sites can be estimated as follows: $\ln(P/P_o) = -\Delta F_{reg}/(RT_{in}) = -5.04$ and $P = 0.0065P_o(303$ K$) = 0.28$ mbar, which corresponds to a dew point of $-33°$C.

3.2.2 Heat Transformation and Storage

Fulfillment of the Paris Agreement requires the replacement of fossil fuels with renewable energy sources and rational use of heat in industry, transport, and dwellings. These new heat sources have lower temperature potential, which opens a niche for applying adsorption technologies for energy transformation and storage. Adsorptive heat transformers (AHTs) and chemical heat transformers (CHTs) are considered an alternative to common compression units [6]. A detailed description of AHT principles can be found elsewhere [7]. Here we give a short introduction to the adsorptive heat transformation necessary for further analysis.

In contrast to a classical heat engine, an AHT consumes and produces only thermal energy and operates between three thermostats—I, II, and III—with high (T_H), intermediate (T_M), and low (T_L) temperatures, respectively (Fig. 3.2). The AHT can transform thermal energy under three modes: (1) cooling, (2) heating, and (3) increasing temperature potential (Fig. 3.2). Any AHT unit consists of adsorber **A**, evaporator **E**, and condenser **C**, between which vapor V of the adsorbate is exchanged and which transfer heat to/ from thermostats I, II, and III (Fig. 3.2).

Two types of the cycles suggested for AHT essentially differ by the way of adsorbent regeneration:

- By isobaric adsorbent heating up to the temperature sufficient for the adsorbate removal;

- By reducing the adsorptive pressure over adsorbent at a constant temperature.

The first, *temperature-driven* (TD), cycles are very common and widely used to realize cooling and heating modes (1 and 2 in Fig. 3.2) [8, 9]. The second, *pressure-driven* (PD), cycles are much less spread and suggested mainly for the temperature upgrading mode (3 in Fig. 3.2) [10, 11]. In this section and Chapter 16, the TD cycles are considered the most prevalent. The PD cycles can be essentially different from the TD ones [12] and are specifically considered in Chapter 17.

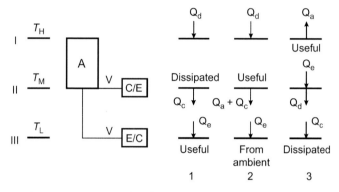

Figure 3.2 Operation principle of a chemical (adsorption) heat pump: I, II, and III thermostats at temperatures T_L, T_M, and T_H; A—adsorber, C—condenser, E—evaporator, V—vapor of adsorbate. Other symbols are defined elsewhere in the chapter. Adapted from Ref. [13], Copyright 2008, with permission from Elsevier.

A basic three-temperature (3T) TD cycle of AHT consists of two isosteres and two isobars (Fig. 3.3a). For modes 1 and 2, the stages of the AHT cycle are:
- "1–2" is an isosteric heating of the saturated adsorbent along the rich isostere (1–2) from the initial temperature T_1 (which is equal to the condenser temperature $T_c = T_M$) to the minimal desorption temperature T_2;
- "2–3" is an isobaric desorption of the adsorbate due to adsorbent heating from T_2 to the desorption temperature T_3 (equal to the temperature T_H of an external heat source) with subsequent vapor condensation in the condenser at $T_c = T_M$ with the heat release Q_c;
- "3–4" is an isosteric cooling of the adsorbent along the weak isostere (3–4) down to the initial adsorption temperature T_4;

- "4–1" is an isobaric adsorption that is driven by adsorbent cooling down to $T_1 = T_M$ and the evaporation of sorptive in the evaporator, where the useful cold Q_e is generated at temperature $T_e = T_L$. Thus, the AHT unit can simultaneously produce both cold and useful heat.

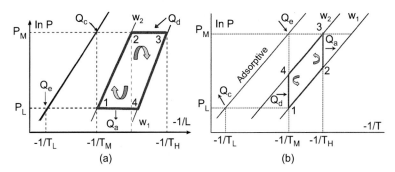

Figure 3.3 P–T diagram of the 3T adsorptive cycles with the regeneration caused by a jump of the adsorbent temperature (a) and by a drop of pressure over the adsorbent (b). Reprinted from Ref. [11], Copyright 2017, with permission from Elsevier.

The 3T TD cycle is uniquely characterized by the three temperatures: evaporator temperature $T_e = T_L$, condenser temperature $T_c = T_M$, and temperature of the external heat source, $T_g = T_H$. If all thermal masses are neglected, the efficiency of the AHT cycle, historically called coefficient of performance (COP), is equal to

$$COP_c = Q_e/Q_d \qquad (3.1)$$

$$COP_h = (Q_c + Q_a)/Q_d \qquad (3.2)$$

for cooling and heating, respectively. If the thermal masses are taken into account, both the efficiencies gradually increase when the amount of adsorbate exchanged in the cycle, $\Delta w = w_{max} - w_{min}$, rises [7]. Hence, if the three cycle temperatures (T_H, T_M, and T_L) are fixed, the optimal adsorbent should exchange the maximal mass of adsorbate between the boundary isosteres, that is, the difference $\Delta w = w(P_L, T_M) - w(P_M, T_H)$ or $\Delta w = w(\Delta F_1) - w(\Delta F_2)$ or $\Delta w = w[RT_M \ln(P_o(T_M)/P_L)] - w[RT_H \ln(P_o(T_H)/P_c)]$ has to be maximized (Fig. 3.4a). These conditions are quite similar to those formulated in Sec. 3.2.1 for gas drying.

Now let us analyze what should be an optimal adsorbent if only two temperatures (e.g., T_L and T_M) are fixed and the third one can be chosen to the best advantage. The values of T_L and T_M uniquely determine the rich isostere and the minimal desorption temperature T_2. The latter can be obtained as an intersection of the line $\ln P = \ln P_o(T_M) = $ const. with the rich isostere (Fig. 3.3a). If heat from the external heat source is supplied to the adsorber right at the minimal desorption temperature $(T_2 = T_H)$, no entropy is generated and the highest second law efficiency (equal to the Carnot efficiency!) can be reached [8, 14]. In this case, the rich and weak isosteres degenerate into a single line (not presented) as in a reversible chemical heat transformer [2, 7].

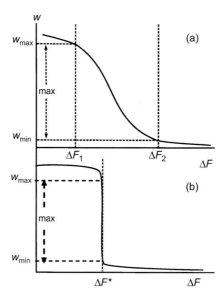

Figure 3.4 Schematic presentation of the characteristic sorption curves for optimal adsorbent: gas drying (a), heat transformation (a, b), maintaining RH (b), and shifting chemical equilibrium (b).

The optimal adsorbent should retain the maximal amount of adsorbate at temperature T_2 and pressure P_M (point 2 in Fig. 3.3a). The best gain can be made if the adsorbate is completely released by an infinitely small increase in temperature above T_2. Thus, the optimal adsorbent has, at $P = P_M$, a step-like sorption isobar, so that the function $w(T)|_{P=P_c} = 0$, when the temperature is approaching T_2 from the right side, and takes large uptake values, when it is approaching T_2 from the left side [14] (Fig. 3.4b). For "adsorbate–adsorbent" pairs that follow the Polanyi potential theory, the position of this step at fixed T_M and T_L can be estimated by using Trouton's rule (see Sec. 2.2.2) as $T_2 = T_M^2/T_L$. Hence, the optimal adsorbent should be different for various cooling applications (freezing, ice making, air conditioning, etc.). We consider this in more details in Chapter 16.

3.2.3 Maintaining Relative Humidity

The relative pressure $h = P/P_o$ or the RH $= h \cdot 100\%$ is an important parameter that has to be controlled in many technological processes, such as the production of noodles and drying of paper, polymers, wood, fruits, etc. Maintaining the RH is required for safe demonstration, conservation, and transportation of various museum exhibits, library values, and archival documents. The RH is of primary importance since it strongly affects the degradation rate of artworks. The exhibits are made of a variety of materials, ranging from stone and metal to wood, leather, paper, textiles, and each of them requires different microclimatic conditions to be preserved correctly [15, 16]. For instance, the optimal RH is 30% for photographs, 30–40% for archival books, 50–60% for books in use, 45% for textiles, etc. In any case, the RH should not exceed 70% to avoid the formation of mold and fungus. Thus, the optimal RH for the conservation of various exhibits varies from 30% to 65%. The RH between 50% and 60% is recommended for many types of artworks [15].

One way is to maintain the RH not in the whole museum space, but within restricted volumes, such as display cases, picture frames, or containers for transporting artworks (for all the cases the term "showcase" will be used further) [15, 17].

An optimum adsorbent should absorb and retain an excess of moisture when the RH value increases above the required value and should release the absorbed water back as the RH value decreases below the required level to compensate this decrease. It is evident that the adsorption isotherm of water for this optimum adsorbent at the process temperature T_p should be stepwise, and the step of sorption should occur at pressure P, which corresponds to the required RH $= [P/P_0(T_p)] \cdot 100\%$ or, in ΔF terms, at $\Delta F = -RT_p \ln[P/P_0(T_p)]$ (Fig. 3.4b). The larger the water sorption capacity, i.e., the height of the isotherm step, the longer time the adsorbent will exchange the moisture and maintain the required RH. Practical development of such an adsorbent is presented in Chapter 19 together with its testing under conditions of a real library and museum in Russia [18].

3.2.4 Shifting Chemical Equilibrium

The conversion of reactants in many important catalytic processes (e.g., in the large-scale ammonia and methanol syntheses) is limited by the chemical equilibrium of the reversible reaction. To increase conversion, it is desirable to shift the equilibrium of such a process toward the formation of products. Removal of the product in the reactor appears to be one of the promising ways to increase the conversion of reversible reactions. This can be performed by removing a product from the reaction mixture by its adsorption on appropriate material [19, 20].

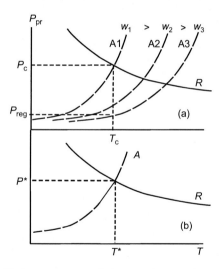

Figure 3.5 Equilibrium pressure of the reaction product as a function of temperature: R—in the course of chemical reaction, A, A1–A3—over adsorbent at various uptakes for bi-variant (a) and mono-variant (b) sorption equilibrium.

Let the temperature dependence of the equilibrium pressure P_{pr} of the gaseous reaction product B be described by curve R and that of the equilibrium pressure of gas B over an adsorbent is represented by isosteres A1–A3, which correspond to the uptakes $w_1 > w_2 > w_3$

(Fig. 3.5a). At the desired reaction temperature T_c, the adsorbent can absorb the product B if the uptake $w < w_1 = w(T_c, P_c)$. As soon as the uptake reaches w_1, the adsorbent should be regenerated, e.g., by a pressure swing down to the regeneration pressure P_{reg} (Fig. 3.5), which corresponds to the equilibrium uptake $w(T_c, P_{reg})$. The difference $\Delta w = w(T_c, P_c) - w(T_c, P_{reg})$ should be maximized in order to enhance the time before regenerations or reduce the amount of the adsorbent used.

For mono-variant equilibrium, isosteres A1–A3 degrade into single line A (Fig. 3.5b) that intersects the reaction equilibrium line R at point (P^*, T^*). The adsorbent shifts the equilibrium toward the formation of product B, if the reaction temperature $T < T^*$. For sorbent regeneration, it is sufficient to reduce the pressure just a little bit lower than P^*: $P_{reg} < P^*$.

These conditions can be equivalently formulated in terms of the adsorption potential ΔF: the optimal adsorbent should have a step-like adsorption isotherm at temperature T_c with the step positioned at $\Delta F = -RT_c \ln[P_c/P_0(T_c)]$, as schematically represented in Fig. 3.4b. The height of the step (Δw) has to be maximal to increase the operation time. This criterion specifies a general algorithm for searching the optimal adsorbent to be used for increasing the output of the gaseous product B. Appropriate examples will be considered in Chapter 20 for the catalytic synthesis of methanol.

3.2.5 Extraction of Water from the Atmospheric Air

A sustainable water supply is of vital importance for the development of humanity. The world's freshwater resources are distributed very unevenly, as is the world population [21]. The water-stressed basins are situated mainly in northern Africa, Mediterranean region, Middle East, Near East, southern Asia, northern China, Australia, the USA, Mexico, north-eastern Brazil, and the west coast of South America. At the same time, the earth's atmosphere is a huge reservoir of moisture. The total amount of water in the atmosphere $(13 \times 10^3 \text{ km}^3)$ greatly exceeds the global demand for a decade. In contrast to rainfall, which is characterized by seasonal irregularity, the atmospheric humidity is the only continuously available water resource for most desert regions. Adsorptive extraction of water from the atmosphere could significantly alleviate the problem of water shortage in arid regions.

First let us consider a general scheme for the sorptive extraction of water vapor from the atmosphere [22, 23]. The absolute air humidity in arid areas undergoes seasonal changes but is relatively stable during the day. On the contrary, the RH varies greatly throughout the day: It is minimal in the daytime and rises sharply at night when the ambient temperature falls. The general process scheme consists of two stages (Fig. 3.6). The first stage is water sorption in the night at high RH. Ambient air passes throughout a bed of solid sorbent and gets dry. The latter stage—water desorption and subsequent water condensation—is carried out in the daytime. The heat needed for the desorption of water must be supplied from an external heat source: In hot and sunny climates, solar energy can be used. Finally, the water desorbed is collected in a condenser. A reservoir buried underground at a depth exceeding 1 m can serve as the condenser. This condenser, supplied with a heat exchanger at the bottom, can maintain inside a temperature of 10–20°C even in the daytime, when the desorption and condensation stages proceed.

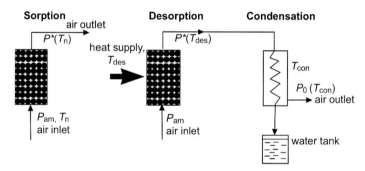

Figure 3.6 General scheme for water extraction from air. Adapted from Ref. [22], Copyright 2007, with permission from Elsevier.

The efficiency of water extraction depends on the climatic conditions and is expected to increase at a higher daytime absolute humidity and larger difference between day and night temperatures and, hence, between the day and night RH. Indeed, the mean diurnal temperature variation may reach 15–20°C; therefore, the relative pressure of water vapor may change from 0.2–0.4 to 0.5–0.8 throughout the day [24].

To analyze the process of sorptive water extraction, we use the Clausius–Clapeiron diagram (Fig. 3.7). This P–T chart presents the liquid–vapor equilibrium of pure H$_2$O (line 1) and the equilibrium between the vapor and the solid sorbent. For the sake of simplicity, we assume the latter equilibrium to be mono-variant and, thus, represented by the single line 2 (Fig. 3.7). This represents the case of a gas–solid reaction, e.g., the formation/decomposition of salt hydrate.

The water sorption occurs at the ambient vapor pressure P_{am} and the average night temperature T_n. Consequently, reaction (1.1) has to occur at the equilibrium temperature $T^*(P_{am})$ obeying the relation $T_n < T^*(P_{am}) < T_{des}$. In other words, the hydration isostere of the optimal sorbent (or the equilibrium line of the optimal hydration reaction) has to lie between the points (P_{am}, T_n) and (P_{am}, T_{des}) (Fig. 3.7). During sorption, the water partial pressure in the outlet air approaches the equilibrium pressure $P^*(T_n)$ over the salt-hydrate system, and the degree of water extraction from the inlet air flux is $\delta_{sorb} = \Delta P_{ads}/P_{am} = [P_{am} - P^*(T_n)]/P_{am}$. Evidently, this degree increases at lower pressure $P^*(T_n)$ in the outlet air flux. Thus, the affinity of optimal sorbent for water should be high enough to get low $P^*(T_n)$.

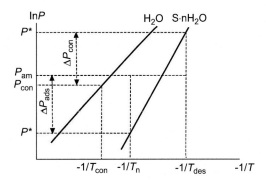

Figure 3.7 Requirements for the optimal adsorbent/salt (see the text).

At the second stage, desorption proceeds by heating the sorbent up to T_{des}, that is, the temperature level of heat used for desorption (regeneration). At this temperature, the equilibrium water pressure in the outlet air equals $P^*(T_{des})$ (Fig. 3.7). The water desorbed has to be collected in the condenser at temperature T_{con}. The degree of water collection is $\delta_{col} = \Delta P_{con}/P^*(T_{des}) = [P^*(T_{des}) - P_{con}]/P^*(T_{des})$. Evidently, this degree increases at higher pressure $P^*(T_{des})$ in the outlet air and lower condensation pressure P_{con}. Hence, at the desorption stage, the affinity of the optimal solid sorbent for water should not be too high, which would otherwise reduce the pressure $P^*(T_{des})$. Searching for appropriate compromise may be done by considering the $P-T$ diagram of a particular solid sorbent [23]. For these studies, the conditions at the sorption and desorption stages are fixed as follows (the so-called standard conditions): $P_{am} = 8$ mbar, $T_n = T_{con} = 15°C$ ($P_{con} = 17$ mbar), and $T_{des} = 80°C$, which are typical of many arid areas [24].

To summarize, the thermodynamic requirements of an adsorbent optimal for the examined applications boil down to a large change in the adsorption uptake either between the two adsorbent states, which are characterized by the fixed sets (T_1, P_1) and (T_2, P_2) (or the appropriate adsorption potentials ΔF_1 and ΔF_2), or at fixed T and P (or ΔF). Evidently, the latter is a degenerated case of the former and corresponds to the mono-variant type of sorption equilibrium (see Sec. 2.5). The two cases are schematically shown in Fig. 3.4 in terms of the adsorption potential ΔF. We hope that this approach will be disseminated to other adsorptive technologies, and appropriate demands will be formulated and used for planning the synthesis procedure of a real adsorbent adapted to this application.

Hopefully, with the further development of theoretical surface chemistry, it will be possible to calculate thermodynamic functions of the "adsorbate–adsorbent" system in a wide temperature and pressure range and, hence, consider the mentioned requirements at the theoretical level. However, nowadays the predominant scientific basis for specifying the requirements is mostly the experimental adsorption equilibrium and approaches reviewed in Chapter 2 as demonstrated above in this chapter.

3.3 Dynamic Requirements

The thermodynamic demands designate the sorption properties of the optimal adsorbent in equilibrium conditions. To reach the equilibrium, the sorption process has to be sufficiently fast, which imposes specific conditions on dynamic properties of the optimal adsorbent and process arrangement. We consider some of them in this section focusing mainly on the requirements that relate to inherent adsorbent properties (favorable isotherm shape, pore size, transport mechanisms, etc.). Organization of the adsorption process (heat and mass transfer to the adsorbent layer, adsorber configuration, residence time, etc.) is also taken into account when necessary.

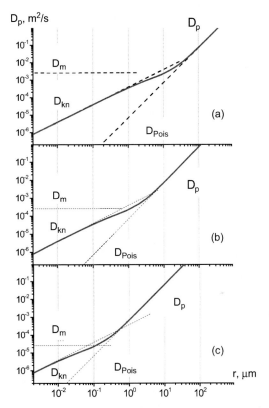

Figure 3.8 The overall diffusivity D_p of water vapor in a straight cylindrical pore of radius r at $T = 20°C$ and pressure $P = 10$ (a), 100 (b), and 1000 (c) mbar.

The dynamic response of the adsorbent has to be well correlated with the timescale of the general process. For a short timescale, the dynamic requirements of the adsorbent used are more severe. For example, an adsorbent for passive hygrostats can easily smooth out alterations of the ambient RH because their typical timescale is an hour and longer. Contrariwise, an optimal adsorbent for shifting the equilibrium of a catalytic reaction should be in a dynamic agreement with a catalyst of this reaction, which is typically very efficient

Optimal Adsorbent: Basic Requirements

and converts reagents to products at a gas residence time less than 0.1–1 s. A desirable cycle time of AHTs is considered to be 1–10 min.

As the adsorption rate is often limited by gas diffusion through the pore network of the adsorbent, a characteristic diffusional time τ_d gives a useful estimation of the mass transfer resistance. This time can be calculated as $\tau_d = R^2/D_g$ for a single grain and as $\tau_d = L^2/D_l$ for an adsorbent layer of thickness L. Here D_g and D_l are the gas diffusivities in the grain and the layer, respectively. As pointed out in Sec. 2.6.2.2, the two diffusivities characterize the gas propagation in an adsorbent grain. The effective pore diffusivity D_p determines the motion of a non-adsorbing gas molecule. The apparent diffusivity D_{ap} describes the propagation of the adsorption front in the grain and, hence, the temporal evolution of the adsorption uptake. The larger the slope of the adsorption isotherm $K = dw/dc$, the slower the adsorption front propagation in comparison with the gas molecule diffusion. Typically, $D_{ap} \approx (10^{-3} - 10^{-4})\, D_p$.

The only intrinsic parameter of the adsorbent that can affect the diffusional gas transport is the average pore diameter d. Indeed, the Knudsen diffusivity $D_{kn} \sim d$ and the Poiseuille diffusivity $D_{Pois} \sim d^2$, while the molecular diffusivity does not depend on the pore size (see Sec. 2.6). Overall, the increase in the pore size improves the gas transport.

In small pores, the Knudsen diffusivity is dominant (Fig. 3.8). The essential rise in the overall diffusivity occurs in the pores where the Poiseuille flow makes the main contribution; for example, at the water vapor pressures of $P = 10$ and 100 mbar, it happens, respectively, in pores of 40–60 and 4–6 µm size or larger (Fig. 3.8b). Thus, it is convenient to use an adsorbent that contains large pores (10–100 µm), which ensure efficient gas transport, together with micro- and mesopores (1–10 nm), which supply necessary internal surface and adsorption uptake. In particular, adsorption heat transformers operate at the water vapor pressure between 10 and 100 mbar, and transport pores of the optimal adsorbent should be larger than 50 µm. AHTs with ammonia as a working fluid operate at $P > 1$ bar, and it is sufficient that the smallest transport pores are ca. 0.5 µm in size.

For such a bimodal structure, it is important to harmonize the mass resistances in macro- and mesopores. Let us consider a porous layer composed of consolidated loosely packed primary grains that, in turn, have intrinsic porosity. Two pore subsystems, namely, the voids between the primary grains, macropores, and the channels inside these grains, mesopores (Fig. 3.9), are connected so that both diffusional resistances affect the water transport. Their relative contributions essentially depend on the pore geometry. Under isothermal conditions, the contributions are qualitatively determined by the ratio α of the diffusional rate constant $k_g = (D_g/R^2)$ in the mesopores inside the primary adsorbent grains of size $2R$ and in the voids between the grains $k_l = (D_l/L^2)$: $\alpha = (D_g/R^2)/(D_l/L^2)$.

For small primary adsorbent grains, $\alpha \gg 1$, and the adsorption equilibrium in the grains is rapidly established. The adsorption dynamics, in this case, is governed by the diffusion in the layer voids, so that an adsorption front propagating into the sorbent layer is formed (Fig. 3.9b).

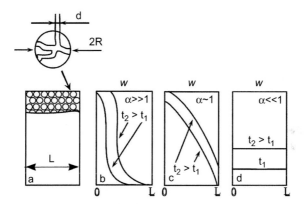

Figure 3.9 Schematic representation of the adsorbent layer consisting of the primary particles of size $2R$ (a) and of the profiles of adsorbed water for different relative importance of the diffusional resistances in macro- and mesopores (b–d). Reprinted from Ref. [26], Copyright 2010, with permission from Elsevier.

In the opposite case of large adsorbent grains ($\alpha \ll 1$), a constant water vapor concentration equal to that above the adsorbent layer is established inside the gas phase, and water content is almost uniform within the entire layer (Fig. 3.9d). In this case, the kinetics of water vapor sorption is governed by the imbibition of water by individual adsorbent grains. The formal kinetics of sorption by solids with a bi-disperse pore structure was considered in references [25, 26].

Optimal Adsorbent: Basic Requirements

The crossover between the two extreme diffusional modes is expected at $\alpha \sim 1$ (Fig. 3.9c), which can be achieved by varying the grain size R [27]. At a constant layer thickness L, the overall diffusional rate constant $k = (k_g k_l)/(k_g + k_l)$ is maximal when the inter- and intragrain diffusional resistances are equal, $k_g = k_l$. The intragrain diffusivity is usually the Knudsen one $D_g = D_{kn} = 48.5 d_g \sqrt{T/M}$ (see Eq. (2.18)), where d_g is the average size of the pores inside the adsorbent grains. The D_g value does not depend on the pressure and the grain radius R. The desired mechanism of the intergrain transport is the Poiseuille flow with the diffusivity $D_1 = D_{Pois} = P d_v^2 / 32\eta$, where d_v is the average size of the voids between the adsorbent grains. For a layer of loose grains, one can expect a linear dependence between the size of void and the size of grains $d_v = \beta R$; hence, $D_1 = P(\beta R)^2/32\eta$. The equality $k_g = k_l$ allows the estimation of the optimal size of loose adsorbent grains

$$R_{opt} = 2 \left(\frac{48.5 d_g L^2 \eta \sqrt{T/M}}{\beta^2 P} \right)^{\frac{1}{4}} \tag{3.3}$$

which ensures the most efficient gas transport through the layer. The thicker the layer and the lower the gas pressure, the larger the grains from which the layer is composed of. The coefficient β depends on the layer porosity $\beta = (2/3)[\varepsilon/(1 - \varepsilon)]$ [28]. For a dense random packing of spheres, $\varepsilon = 0.36$–0.40, and thus $\beta = 0.375$–$0.444 \approx 0.4$ will be taken as a reference value. At $P = 10$ mbar $= 103$ Pa, $T = 30°C = 303$ K, $M = 18$ g/mol (water), $d_g = 15$ nm $= 1.5 \times 10^{-8}$ m, and $L = 2 \times 10^{-3}$ m, the so-estimated optimal radius is 34 µm. At $L = 2 \times 10^{-2}$ m, $R_{opt} = 107$ µm. Even a little deviation of the grain size from the optimal one can significantly reduce the overall diffusional constant. For instance, at $R = 1.2 R_{opt}$, $k(R) = 0.47k(R_{opt})$, which means that the process becomes approximately two times slower!

In this chapter, we restrict ourselves only with these basic dynamic considerations. More complex cases are analyzed below for particular applications of composites "salt in porous matrix" (CSPMs). For instance, an adsorbent dynamically optimal for adsorptive heat transformation cycles driven by temperature drop and pressure jump is considered in Chapters 16 and 17, respectively.

3.4 Other Requirements

We have considered above only thermodynamic and dynamic requirements for adsorbents optimal for various applications. In practice, many other, more down to earth, adsorbent properties are extremely important as well, among which are low/reasonable cost, mechanical strength, thermal/hydrothermal stability, nontoxicity, incombustibility, etc. These properties differ depending on the application and have to be considered on a case-to-case basis. For instance, adsorbents of water vapor used in commercial adsorptive chillers have to be hydrothermally stable over 100,000–500,000 operation cycles with a typical duration of 1–10 min. The similar cycle used for seasonal heat storage lasts one year; accordingly, the adsorbent has to withstand less than 100 cycles.

References

1. Ruthven, D. M. (1984) *Principles of Adsorption and Adsorption Processes* (John Wiley & Sons, New York, USA).
2. Aristov, Yu. I., Sharonov, V. E., and Tokarev, M. M. (2008). Universal relation between the boundary temperatures of a basic cycle of sorption heat machines, *Chem. Eng. Sci.*, **63**, pp. 2907–2912.
3. Yang, R. T. (1997) *Gas Separation by Adsorption Processes* (Imperial College Press, London).
4. Keltsev, N. V. (1984) *Fundamentals of Adsorption Technique* (Khimiya, Moscow) (in Russian).
5. Dubinin, M. M. (1960) Theory of physical adsorption of gases and vapour and adsorption properties of adsorbents of various natures and porous structures, *Bull. Division Chem. Soc.*, pp. 1072–1078.
6. Meunier, F. (1998) Solid sorption heat powered cycles for cooling and heat pumping applications, *Appl. Therm. Eng.*, **18**, pp. 715–729.
7. Raldow, W. M. and Wentworth, W. E. (1979) Chemical heat pumps: A basic thermodynamic analysis, *Solar Energy*, **23**, pp. 75–79.
8. Aristov, Yu. I. (2008) Chemical and adsorption heat pumps: Cycle efficiency and boundary temperatures, *Theor. Found. Chem. Eng.*, **42**, pp. 873–881.
9. Li, T. X., Wang, R. Z., and Li, H. (2013) Progress in the development of solid-gas sorption refrigeration thermodynamic cycle driven by low-grade thermal energy, *Prog. Energy Combustion Sci.*, **40**, pp. 1–58.

10. Chandra, I. and Patwardhan, V. S. (1990) Theoretical studies on adsorption heat transformer using zeolite-water vapour pair, *Heat Recovery Systems CHP*, **10**, pp. 527–537.

11. Aristov, Yu. I. (2017) Adsorptive transformation of ambient heat: A new cycle, *Appl. Therm. Eng.*, **124**, pp. 521–524.

12. Aristov, Yu. I. (2019) A new adsorptive cycle "HeCol" for upgrading the ambient heat: the current state of the art, *Int. J. Refrig.*, **105**, pp. 19–32.

13. Sharonov, V. E. and Aristov, Yu. I. (2008). Chemical and adsorption heat pumps: Comments on the second law efficiency. *Chem. Eng. J.*, **136**, pp. 419–424.

14. Aristov, Yu. I. (2007) Novel materials for adsorptive heat pumping and storage: Screening and nanotailoring of sorption properties, *J. Chem. Eng. Japan*, **40**, pp. 1241–1251.

15. Tomphson, G. (1986) *The Museum Environment* (Oxford, Butterworth Heinemann).

16. Camuffo, D. (1998) *Microclimate for Cultural Heritage* (Elsevier, Amsterdam).

17. Cursiter, S. (1936–1937) Control of humidity in cases and frames, *Technical Studies Fine Arts*, **5**, pp. 109–111.

18. Glaznev, I. S., Alekseev, V. N., Salnikova, I. V., Gordeeva, G., Shilova, I. A., Elepov, B. S., and Aristov, Yu. I. (2009) ARTIC-1: A new humidity buffer for showcases, *Studies Conservation*, **54**, pp. 133–148.

19. Kuczynski, M., Oyevaar, M. H., Pieters, R. T., and Westerterp, K. R. (1987) Methanol synthesis in a countercurrent gas–solid–solid trickle flow reactor. An experimental study, *Chem. Eng. Sci.*, **42**, pp. 1887–1898.

20. Roes, A. W. M. and van Swaaij, W. P. M. (1979) Hydrodynamic behavior of a gas–solid counter–current packed column at trickle flow, *Chem. Eng. J.*, **17**, pp. 81–89.

21. Food and Agricultural Organization of the United Nations (2007) Coping with water scarcity: Challenge of the twenty-first century. http://www.fao.org/nr/water/docs/escarcity.pdf (last accessed July 16, 2017).

22. Aristov, Yu. I., Tokarev, M. M., Gordeeva, L. G., Snitnikov, V. N., and Parmon, V. N. (1999) New composite sorbents for solar-driven technology of fresh water production from the atmosphere, *Solar Energy*, **66**, pp. 165–168.

23. Gordeeva, L. and Aristov, Yu. I. (2014) Extraction of water from the atmosphere in arid areas by employing composites "a salt inside a porous matrix," in: *Transport and Reactivity of Solutions in Confined*

Hydrosystems, L. Mercury, N. Tas, and M. Zilberbrand (Eds.) (Springer Science + Business Media BV, Dodrecht), pp. 257–268.

24. Arakawa, H. (Ed.) (1969) *World Survey of Climatology. V.8. Climates of Northern and Eastern Asia.* Elsevier Scientific Publishing Company; Takahashi, K. and Arakawa, H. (Eds.) (1981) *World Survey of Climatology. V.9. Climates of Southern and Western Asia.* Elsevier Scientific Publishing Company; Griffiths, J. F. (Ed.) (1972) *World Survey of Climatology. V.10. Climates of Africa.* Elsevier Scientific Publishing Company.

25. Ruckenstein, E., Vaidyanathan, A. S., and Yuongquist, G. R. (1971) Sorption by solids with bi-disperse pore structures, *Chem. Eng. Sci.*, **26**, pp. 1305–1318.

26. Ruthven, D. M. and Loughlin, K. F. (1971) The sorption and diffusion of n-butane in Linde 5A molecular sieve, *Chem. Eng. Sci.*, **26**, pp. 1145–1154.

27. Glaznev, I. S., Koptyug, I. V., and Aristov, Yu. I. (2010). A compact layer of alumina modified by $CaCl_2$: The influence of composition and porous structure on water transport, *Micropor. Mesopor. Materials*, **131**, pp. 358–365.

28. Heifets, L. I. and Neimark, A. V. (1982) *Multiphase Processes in Porous Media,* (Moscow, Chemistry) (in Russian).

Chapter 4

Basic Synthesis Methods

Exploration of entirely new approaches to tailoring advanced porous materials is the backbone of modern materials science [1]. As opposed to this general tendency, the synthesis of composites "salt in porous matrix" (CSPMs) is concerned with the utilization of known routes to porous materials, mainly of methods for the preparation of the traditional two-component supported catalysts [2, 3]. One initial component of these catalysts is often an inorganic salt that afterward transforms into an active state by thermal decomposition, hydrolysis, reduction, etc. Methods for mixing components can be conditionally divided into three groups:

1. Impregnation with the salt;
2. Co-precipitation or gelation of components;
3. Mechanical mixing of components with further thermal dispersion.

As the first method is predominately used for the synthesis of CSPMs, we consider it in more details than the two other approaches.

The initial preparation step concerns the deposition of an active salt from an impregnating solution onto the surface of a common porous matrix (mainly oxides or carbons). Appropriate implementation of this step allows the fine regulation of texture characteristics and phase composition of the deposited salt and, thus,

---

Nanocomposite Sorbents for Multiple Applications
Yu. I. Aristov
Copyright © 2020 Jenny Stanford Publishing Pte. Ltd.
ISBN 978-981-4267-50-2 (Hardcover), 978-981-4303-15-6 (eBook)
www.jennystanford.com

a target-oriented synthesis of CSPMs with the sorption properties close to those revealed for the optimal adsorbent (Chapter 3), which serve as a guiding star for such nanotailoring.

4.1 Impregnation of the Porous Matrix with an Active Salt

The most common method of CSPM synthesis is impregnation of the ready matrix (e.g., porous silica gel, alumina, carbon, etc.) with an aqueous solution of the active salt to be delivered inside the matrix pores and dispersed there. The whole procedure involves preliminary drying of the matrix in order to remove adsorbed water, its filling with the salt solution, filtration, and final drying of the wet composite (Fig. 4.1).

Three main modifications of this method are commonly used [4]:
1. Dry (or incipient wetness, or pore volume) impregnation,
2. Wet (or adsorption, or equilibrium deposition) impregnation,
3. Combined impregnation,

which are considered in the following sections. These impregnation methods are widely used at an industrial scale for catalyst preparation.

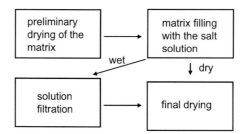

Figure 4.1 Schematics of the impregnation process.

4.1.1 Dry (Incipient Wetness, Pore Volume) Impregnation

Dry impregnation involves preparing a solution of the salt to be dispersed. The solution is then mixed with the porous matrix so that the solution volume V_s equals the pore volume V_p of the matrix [4–7]. The liquid phase soaks quite fast into the matrix grains due to capillary forces, though the salt ions often penetrate rather slowly to the grain center due to successive adsorption on and desorption from the matrix surface. Thus, to reach the uniform distribution of the salt in the grains, the wet composite has to be kept for several hours. As a result, the solution is located only inside the pores, and the filtration stage is omitted (Fig. 4.1). Accordingly, dry impregnation is easier in realization. Another technological advantage of this approach is that no salt solution exhaust is produced as the salt completely precipitates inside the grain pores.

After the matrix has imbibed the salt solution into its pore voids, the wet composite is heated to remove the solvent. When choosing the solvent, one should keep in mind that

- The salt has to be sufficiently soluble in this solvent to ensure that enough salt is contained in the volume of solution sucked into the matrix pores;
- The solvent is able to wet the matrix surface.

In the majority of cases, water is used as a solvent. For impregnating hydrophobic matrices, e.g., porous carbons, salt solutions in methanol or ethanol are utilized as they better wet the carbon surface.

The amount of the salt inserted inside the pores, C_s (g salt/g matrix), can be calculated from the specific volume V_p (cm^3/g matrix) of the matrix's pores and the salt concentration c_s (g salt/cm^3) in the solution by $C_s = c_s V_p$. For instance, for a saturated CaCl$_2$ aqueous solution, c_s = 0.56 and 0.75 g/cm^3 at T = 10 and 30°C, respectively [8]. Since the KSK silica gel has the specific pore volume of 1.0 cm^3/g, the salt content C_s amounts to 0.56 and 0.75 (g salt)/(g silica) if the pores are filled with those solutions or 0.36 and 0.43 (g salt/g composite). Thus, dry impregnation permits the deposition of the desired amount of the active salt on the surface of the host matrix.

For many CSPM applications, a large salt content is needed to enhance the equilibrium uptake (Chapter 2). In this case, salts with good solubility (at least, 0.2–0.3 g/cm^3) and matrices with large pore volume (more than 0.4–0.6 cm^3/g) are preferable. Other important parameters of the matrix are its specific surface area, hydrophobicity/hydrophilicity, adsorption or ion-exchange capacity, acidity/basicity, etc.

Two main mechanisms of salt deposition on the surface of the pores are adsorption and precipitation/crystallization. The adsorption of salt (or interfacial deposition) may contribute significantly for matrices with high adsorption/ion-exchange capacity and/or dilute solutions. In this case, the distribution of salt on the matrix surface is homogeneous and resembles a salt layer (Fig. 4.2). The precipitation/crystallization (or bulk deposition) of salt is the dominant mechanism for weakly adsorbing surfaces and highly concentrated solutions. In this case, when the wet composite dries, the solvate evaporates and the salt particles precipitate on the matrix surface. The texture and distribution of the salt inside the matrix are affected by many factors such as the salt's concentration, ionic strength, and pH in the impregnating solution, its viscosity, the drying scenario, etc. [4, 9]. Nanocrystallites or amorphous nanoparticles of the salt, weakly bound to the matrix surface, are formed (Fig. 4.2). Formation of these two salt phases and the inherent difference in their sorption behavior are considered in Sec. 4.4.1.2.

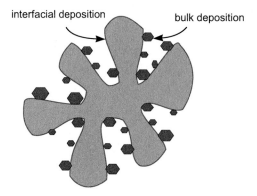

Figure 4.2 Schematics of salt deposition for dry and wet impregnations.

Generally, typical active salts for CSPMs are halides, nitrates, sulfates of alkali and alkali-earth metals (Chapter 5) that are weakly adsorbed on the surface of the matrix (silica gel, carbon, etc.). For this reason, precipitation is the dominant deposition mechanism, so that the salt tends to form crystalline or amorphous particles inside the pores, the size of which is restricted by the pore walls. In dry impregnation, formation of salt particles on the external surface of the grain was not observed [10–12].

Appropriate implementation of dry impregnation allows fine regulation of the texture and phase composition of the deposited salt and, thus, management of the sorption behavior of the final composite. The main tools to be used for the target-oriented synthesis of CSPMs are considered in Sec. 4.4.

4.1.2 Wet (Adsorption, Equilibrium Deposition) Impregnation

When the matrix grains are immersed in a dilute salt solution, the volume of which exceeds significantly the pore volume of the matrix, $V_s > V_p$, the method is called wet impregnation [4]. It is highly desirable that the impregnation occurs at a fixed pH and ionic strength.

After equilibration for several hours under stirring, the excess solution has to be driven off by filtration [4, 13, 14]. Commonly, the remaining concentration of the salt in the spent solution is small; therefore, almost all the salt is deposited on the matrix surface due to the adsorption process. This process includes diffusion of the salt ions from the mother waters toward the internal surface of the matrix, electrostatic interaction between the ions and the surface, proper adsorption of ions with the formation of ion pairs, hydrogen bonds, inner-sphere surface complexes as well as surface-induced hydrolysis. Three basic models, used for describing the charging mechanism of a porous oxide surface, were suggested in Refs. [15–17] (see the summary in Ref. [4]). Under certain conditions, the matrix surface may act as a reagent that results in partial surface dissolution and the formation of a mixed solid phase, which includes the oxide and the salt [4, 7, 18, 19].

As a rule, the interfacial deposition results in composites with a very high (up to atomic) dispersion of the supported salt. Nevertheless, a little portion of the solution can remain on the external surface of the grains after filtration of the excess of the salt solution. Therefore, when the composite dries, the salt can partially precipitate on the outside of the grains, forming large crystals. In order to remove them from the grain surface, Gong et al. [20] suggested additional treatment of the composite in a special temperature/humidity chamber. "The salt on the external surface of the composite absorbs water and transforms into the salt solution,

Basic Synthesis Methods

then falls down through the strainer. The composite sorbent is taken out when there is no salt solution formed any more and then dried" [20].

If the matrix grains are large (D_{gr} > 2–3 mm), wet impregnation can be carried out under evacuation in order to avoid the crush of the matrix grains due to the disjoining action of capillary forces [20].

4.1.3 Combined Impregnation

Combined impregnation is intermediate between dry and wet impregnations [4]: It does not involve the filtration step, and the deposition of the salt on the matrix surface takes place during a slow evaporation of the solvent out of the impregnating solution during the drying step. In this case, the salt deposition can occur according to both mechanisms mentioned above, namely, bulk and interfacial depositions. Their relative contribution depends on the impregnation conditions (concentration and pH of the impregnating solution, temperature, evaporation rate, etc.). It can be varied in a wide range in order to make use of advantages and escape disadvantages of each procedure.

4.2 Co-precipitation

In both dry and wet impregnations, the host porous matrix is a feed material that is commercially available or synthesized in a separate route. This ready matrix is impregnated with an active salt as described earlier. A sol–gel method offers an alternative approach: direct insertion of the active salt right during the sol–gel synthesis of the entire composite. This is expected to result in a homogeneous and fine dispersion of the salt across the composite and its stronger interaction with the surrounding "matrix." Such a method was proved to be effective in giving highly active catalysts and biocatalysts [21, 22]. Refs. [23–26] report on the preparation of composite sorbents of water vapor, $CaCl_2/SiO_2$ and $LiBr/SiO_2$, by the sol–gel route. This method gives CSPMs with aero- and xero-gel structures that possess an extremely large pore volume and sorption capacity. Here we give a brief description of the procedure and summarize the main results on sorption and structural properties of the CSPMs obtained.

Four composites (1–4) were prepared by the sol–gel route described in Fig. 4.3, and one composite (5) was prepared for comparison by impregnating pieces of silica aerogel with 14 wt.% aqueous $CaCl_2$ solution. Samples 1 and 3 were prepared to achieve a target $CaCl_2$ content of 17 wt.% (dry base), while samples 2 and 4 were prepared for a target of 30 wt.%. The alcogel samples were synthesized from tetraethoxysilane (TEOS), ethanol, H_2O, HCl, NH_4OH, $CaCl_2$ following the two-step preparation procedure described by Brinker et al. [27]. More details of the synthesis procedure can be found elsewhere [23].

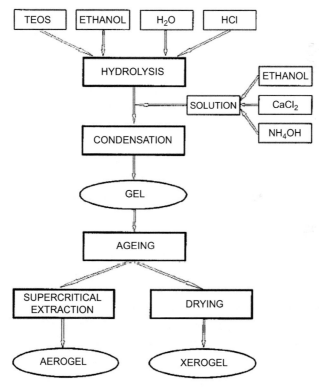

Figure 4.3 Schematics of the preparation procedure of the $CaCl_2$/aerogel composites. Reprinted with permission from Ref. [23]. Copyright 1997 American Chemical Society.

The mechanism of composite formation includes two stages:
1. Creation of completely or partially hydrolyzed monomers, dimers, and oligomers of SiO_4-tetrahedrons;

2. Polymerization and further hydrolysis, which are accompanied by the interaction of salt ions with numerous silanol groups due to electrostatic, ion–ion, ion–dipole, and ion-exchange forces.

These processes lead to the formation of surface complexes that are likely to promote specific salt precipitation and, at the drying stage, give salt clusters with higher dispersion and stronger binding than for CSPMs prepared by dry impregnation. No X-ray diffraction (XRD) pattern was detected for the sol–gel CSPM composites.

Their sorption properties and comparison with common CSPMs are reported in Refs. [23, 25]. Here we briefly summarize that the co-precipitation method results in CSPMs with enhanced affinity for water vapor and adsorption capacity exceeding 1 g/g (!) at higher relative pressures, which are markedly larger than those for CSPMs similar in composition yet prepared by the impregnation of silica gels or aerogels [23, 25]. The hybrid materials afforded by the sol–gel route do not dissolve at higher relative pressures of water vapor, in contrast to common CSPMs produced by impregnation, which could be an effective tool to avoid leaking of the salt solution out of the sorbent pores. The disadvantages of the sol–gel procedure include high cost of the necessary equipment and reagents as well as low mechanical strength of the composites obtained.

4.3 Mechanical Mixing and Thermal Dispersion

In this approach, powders of a metal salt and a porous material are mixed at a predetermined ratio that depends on the required salt content in the final composite. After the finely divided powders of the salt and the matrix are thoroughly mixed, the mixture is heated to a temperature that can be selected between the Tammann temperature and the melting point of the salt. It is assumed that at the Tammann temperature, the crystal lattice begins to become sufficiently mobile to ensure that the salt spreads over the matrix surface (Fig. 4.4). By convention, this temperature is considered to be approximately 70% of the salt's melting point (in absolute temperature) [28]. If the dispersion temperature is too low, the process may be unacceptably long. Too high dispersion temperature may result in the salt's oxidation, decomposition, or reaction with the matrix, which would lead to gradual degradation of the composite.

This method is especially useful if the active salt is insoluble in water and other common solvents so that the impregnation methods cannot be used.

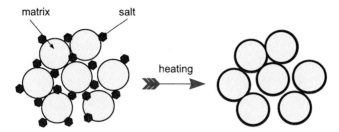

Figure 4.4 Schematics of the thermal dispersion method.

If the amount of salt in the composite is selected to obtain a monolayer of salt on the matrix surface, this version of the method is called thermal monolayer dispersion [29]. Russell and Stokes [30] showed the first evidence for the monolayer dispersion of MoO_3 on γ-Al_2O_3. Monolayer dispersion of many ionic metal oxides and salts, particularly halides, has been accomplished by Xie and Tang [31]. The matrices involved were γ-Al_2O_3, silica gel, TiO_2, activated carbon, and zeolites. A typical example of preparation of the $AgNO_3$/SiO_2 composite was presented in Ref. [29]. That composite was successfully used as a new sorbent for olefin/paraffin separation by adsorption via π-complexation.

The thermal monolayer dispersion technique requires the use of fine powder and thorough mixing. Thus, pelletizing is needed when sorbents in the pellet form are to be used. Again, an advantage of this technique is that insoluble salts can be directly dispersed, as in the case with CuCl.

The method of thermal dispersion is also widely used for the synthesis of new composite solid electrolytes ("ionic salt–porous oxide"), which can be considered a new class of ionic conductors with high ionic conductivity that occurs via interfaces [32, 33]. The main thermodynamic reason for the formation of non-autonomous interface phases relates to the adhesion energy. If it is sufficiently high, an ionic salt tends to spread along the oxide surface and to form a nanolayer if the oxide is nanocrystalline or nanoporous. The thermodynamic driving force for the spreading of a mobile phase

66 | *Basic Synthesis Methods*

S (salt) on the surface of an immobile phase M (matrix) can be expressed by the equation [34–36]

$$\Delta G_s = \alpha_{S\text{-}M} + \alpha_S - \alpha_M < 0 \qquad (4.2)$$

where $\alpha_{S\text{-}M}$, α_S, and α_M are the interface free energy and the free surface energies of components S and M, respectively. Unfortunately, the value of interface energy $\alpha_{S\text{-}M}$ is unknown in almost all the cases. The phenomenological analysis shows that $\alpha_{S\text{-}M}$ is given by the following equation [37]:

$$\alpha_{S\text{-}M} = \alpha_S + \alpha_M - \alpha_{adh} = \alpha_S + \alpha_M - \left(U_{int} - U_{strain}\right) \qquad (4.3)$$

where the adhesion energy α_{adh} includes two terms: the interaction energy U_{int} due to the chemical interaction at the interface, and the elastic energy U_{strain} associated with a possible mismatch between the lattice parameters of contacting phases. Finally, one comes to the following criterion of the self-spreading of salt:

$$\Delta G_S = 2\alpha_S - \left(U_{int} - U_{strain}\right) < 0 \qquad (4.4)$$

which illustrates the important impact of U_{int} and U_{strain} on the ΔG_S value and the driving force of spreading. The formed salt nanolayer is commonly amorphous or has enhanced concentration of structural defects, such as point vacancies and dislocations [33]. At low salt content, a salt monolayer or atomic clusters of the salt are formed, whose properties can significantly differ from those of the bulk salt.

4.4 Dry Impregnation: Tools for CSPM Nanotailoring

Since dry impregnation appears to be the dominant procedure for synthesizing CSPMs, we consider below the main synthetic tools available for a target-oriented synthesis of composites with predetermined properties. According to the nanotailoring concept (Chapter 1), these properties should be specifically adapted to the requirements of various practical applications (see Chapter 3). Among the parameters to be properly selected are

- Chemical nature of the active salt,
- Porous structure of the host matrix,
- Salt content, and
- Synthesis conditions.

4.4.1 Active Salt

The active substance in any CSPM is an inorganic salt, so its choice is of prime importance. Indeed, Fig. 4.5 displays experimental isobars of water sorption on a commercial silica gel KSK both pure and modified by CaCl$_2$ and LiBr. First, the modification leads to a significantly larger amount of water sorbed in comparison with that for pure silica. Second, the shape of isobar depends on the salt's nature: At the same temperature, the LiBr/KSK composite binds larger mass of water, which is especially pronounced at $T > 60$–$100°C$. Here we briefly discuss how the salt's nature and state affect the sorption propeties of CSPM, while a comprehensive analysis will be made in Chapter 5.

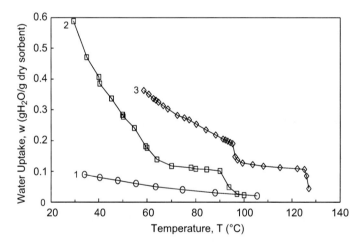

Figure 4.5 Comparison of the water sorption isobars of pure silica KSK (1), composites CaCl$_2$/KSK (2), and LiBr/KSK (3); P = 25 mbar. Reprinted by permission from Springer Nature from Ref. [11], Copyright 1996.

4.4.1.1 Chemical nature of the salt

Fortunately, a plenty of salts can form crystalline solvates with water, methanol, and ammonia according to reaction (1.1). As these systems in bulk are mono-variant (see Sec. 2.5), the equilibrium temperature of the reaction, T_r, depends only on the vapor pressure P. Hence the isotherm (isobar) of this process is stepwise, which is desirable for many adsorption applications (Chapter 3). At fixed pressure P, the step position (T_r) is a unique function of the pressure. This would allow selection of the salt for which the sorption step is positioned at a desirable temperature.

Another important requirement for an optimal salt (or optimal reaction) is that the salt must exchange a large mass of sorbate in the course of reaction (1.1). This mass related to the dry salt mass is maximal for light salts, such as lithium chloride. It is worthy to remind that 1 mol of lithium chloride can bind 3 mol of methanol with the formation of LiCl·3CH$_3$OH (see reaction (1.3)), giving the mass of methanol exchanged as large as 2.26 g/g.

4.4.1.2 Crystalline versus amorphous salt phases

As reported earlier, the mode of deposition (bulk versus interfacial) may strongly affect the texture of the deposited salt; for instance, interfacial deposition results in CSPMs with a very high dispersion of the supported salt (Fig. 4.2). One might expect that this textural diversity results in perceptibly different sorption properties of the final composites.

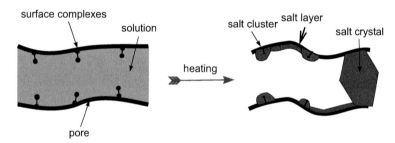

Figure 4.6 Formation of bulk crystalline and surface amorphous phases of the confined salt.

Our analysis of various CSPMs based on CaCl$_2$, CuSO$_4$, MgSO$_4$, and Na$_2$SO$_4$ by potentiometry, titration, differential dissolution, XRD, and differential scanning calorimetry (DSC) techniques [37, 38] suggests the following deposition pattern of the salt MX$_n$ (Fig. 4.6). During the impregnation stage, the salt co-ions M^{n+} get adsorbed on the silica gel surface, which results in the formation of surface complexes according to the following solid–liquid processes [17, 38]:

$$\equiv\text{Si-OH} \Leftrightarrow \equiv\text{Si-O}^- + \text{H}^+ \quad (4.5)$$

$$\equiv\text{Si-O}^- + \text{M}^{n+} \Leftrightarrow \equiv\text{Si-OM}^{(n-1)+} \quad (4.6)$$

$$2\equiv\text{Si-OH} + \text{M}^{n+} \Leftrightarrow 2(\equiv\text{Si-O})\text{M}^{(n-2)+} \tag{4.7}$$

$$\equiv\text{Si-OH} + \text{M}^{n+} + \text{H}_2\text{O} \Leftrightarrow \equiv\text{Si-OM}^{(n-1)+}\text{OH}^- + 2\text{H}^+ \tag{4.8}$$

After drying, the surface complexes $\equiv\text{Si-OM}^{(n-1)+}$, $2(\equiv\text{Si-O})\text{M}^{(n-2)+}$, and $\equiv\text{Si-OM}^{(n-1)+}\text{OH}^-$ yield the surface X-ray amorphous phase of the salt. The double electric layer near and these complexes on the silica gel surface may affect the process of nucleation and growth of the solid phase of the salt during thermal drying of the composite and, consequently, the phase composition and the morphology of the anhydrous confined salt. The double electric layer near the silica gel surface [4] promotes interfacial deposition of the disordered surface layer of the salt (Fig. 4.6). The surface complexes can act as centers of the crystal nucleation and result in the formation of a large number of salt clusters of atomic size instead of the growth of salt crystals with the size close to the pore dimension.

Table 4.1 Relative vapor pressure η_r corresponding to various transitions between salt hydrates, both bulk and confined to silica pores of 15 nm in size (silica gel KSK). Δm is the mass of water sorbed and m is the initial mass of salt (hydrate)

Salt	Transition	η_r (bulk)	η_r (confined)	$\Delta m/m$, g/g
Na_2SO_4	$1 \Rightarrow 7$	0.80	0.5–0.6	0.79
	$7 \Rightarrow 10$	0.91	—	0.205
Na_2HPO_4	$0 \Rightarrow 2$	0.37–0.39	—	0.255
	$2 \Rightarrow 7$	0.52–0.57	—	0.505
$\text{Ca(NO}_3)_2$	$0 \Rightarrow 2$	0.21–0.23	0.15–0.19	0.22
LiNO_3	$3 \Rightarrow 0$	0.20–0.26	0.06–0.12	0.78
MgSO_4	$0 \Rightarrow 1$	0.03	0.001–0.003	0.17
	$1 \Rightarrow 2$	0.05–0.06	0.003–0.01	0.15
	$2 \Rightarrow 4$	0.15–0.17	—	0.23
	$4 \Rightarrow 6$	0.24–0.30	—	0.19
CaCl_2	$0 \Rightarrow 1/3$	< 0.01	—	0.054
	$1/3 \Rightarrow 2$	0.04–0.05	0.03–0.04	0.265
	$2 \Rightarrow 4$	0.16–0.18	0.12–0.14	0.245
	$0 \Rightarrow 6$	—	—	0.97
LiBr	$0 \Rightarrow 1$	0.01	0.01	0.21
	$1 \Rightarrow 2$	0.04–0.05	0.02–0.03	0.17
LiCl	$0 \Rightarrow 1$	0.09–0.12	0.04–0.05	0.435
	$1 \Rightarrow 2$	0.12–0.13	0.06–0.09	0.30
	$0 \Rightarrow 3$	—	—	1.30

The differential dissolution method [40] reveals that the fraction of metal cations linked to the surface complexes (the "linked" metal) increases at smaller salt concentration and higher pH of the impregnating solution [37, 38]. This fraction also depends on the nature of salt (CuSO$_4$ > CaCl$_2$ ≈ MgSO$_4$). The major part of the confined salt (the "free" salt) commonly precipitates/crystallizes on the matrix surface to form stoichiometric nanocrystals (Fig. 4.6).

These two forms of the deposited salt appear to have quite different sorption properties [37, 38]. For instance, the water uptake as a function of water relative pressure for composite CaCl$_2$ (33.7 wt.%)/(silica gel KSK) has a plateau corresponding to the water uptake n = 2 mol/mol (Fig. 4.7a), which indicates the formation of the crystalline hydrate CaCl$_2$·2H$_2$O in the silica pores, which is similar to the *mono-variant* bulk system "CaCl$_2$–H$_2$O" [11, 41]. This hydrate is stable in the range of relative pressure P/P_o = 0.03–0.10. At P/P_o < 0.03, the crystalline dihydrate decomposes. At P/P_o ≥ 0.10, the water uptake grows monotonically with the increase in pressure, which indicates the formation of the CaCl$_2$ aqueous solution in the silica pores (for more details, see Chapter 5).

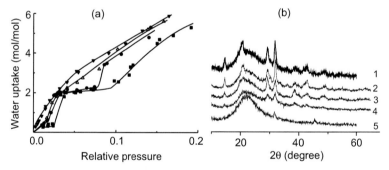

Figure 4.7 (a) Water uptake of CaCl$_2$ (33.7%)/SiO$_2$ (■), CaCl$_2$ (23.3%)/SiO$_2$, prepared at pH of solution 5.5 (●), 8 (△), and CaCl$_2$ (9.2%)/SiO$_2$ (▼) as a function of the relative pressure of water vapor; (b) XRD patterns of the composites CaCl$_2$ (33.7 wt.%)/SiO$_2$ (1), CaCl$_2$ (23.3 wt.%)/SiO$_2$, prepared at a pH of solution 2.0 (2), 5.5 (3), 8.0 (4), and CaCl$_2$ (9.2 wt.%)/SiO$_2$ (5) saturated with water up to n = 2. Figure (a) reprinted by permission from Springer Nature from Ref. [39], Copyright 2007, and Figure (b) reprinted from Ref. [38], Copyright 2006, with permission from Elsevier.

The water uptake in the composite CaCl$_2$ (9.2 wt.%)/SiO$_2$ gradually increases with the relative pressure (Fig. 4.7a). Such a *bi-variant* behavior is typical for salt solutions. No plateau corresponding to the formation of stable crystalline hydrates is observed. The isotherm of the composite CaCl$_2$ (23.3 wt.%)/SiO$_2$ is of an intermediate type (Fig. 4.7a). The plateau at $n = 2$ indicates the formation of dihydrate CaCl$_2$·2H$_2$O, but the P/P_o range where this hydrate is stable is significantly narrower (0.03–0.07) than for CaCl$_2$ (33.7 wt.%)/(silica gel). The same composite prepared at pH = 8 has no plateau at $n = 2$, thus demonstrating a crossover to bi-variant behavior. Similar changes in the type of sorption equilibrium with a decrease in salt content were observed for other CSPMs consisting of CuSO$_4$, MgSO$_4$, and Na$_2$SO$_4$ confined to the silica pores [38].

These two types of the sorption behavior correlate well with two types of the phase composition of the studied sorbents (Fig. 4.7b). Indeed, in the composite CaCl$_2$ (33.7 wt.%)/(silica gel), the hydrated salt forms a crystalline phase CaCl$_2$·2H$_2$O in the silica pores. This phase is also observed in CaCl$_2$ (9.2 wt.%)/SiO$_2$, but its integral intensity is very low. It means that the main part of the salt in this composite is X-ray amorphous. The intensity of reflections of CaCl$_2$ (23.3 wt.%)/SiO$_2$ appears to depend on the pH of the solution. Thus, the amounts of crystalline CaCl$_2$·2H$_2$O in the samples prepared at pH = 2.0, 5.5, and 8.0 estimated from the integral intensity of the reflection at $2\Theta = 14.6°$ are 28, 17, and 9 wt.%, which correspond to 21, 13, and 7 wt.% of anhydrous CaCl$_2$, respectively (Fig. 4.7b). Hence, the shift in the ion-exchange equilibrium between the silica surface and the salt solution toward the formation of surface complexes causes a decrease in the amount of the crystalline phase of the salt.

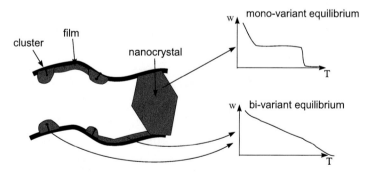

Figure 4.8 Interrelation between the texture of the deposited salt and the type of its sorption equilibrium.

To sum up, during impregnation of the silica gel with an aqueous salt solution, the ion-exchange interaction results in the formation of various surface complexes. During the drying stage, it stabilizes the dispersed salt in two phases: a volume crystalline phase and a surface X-ray amorphous phase. The fractions of the crystalline and amorphous phases depend on the salt content and the preparation conditions (e.g., pH of the salt solution) and strongly affect the sorption equilibrium between the composite and the sorbate (Fig. 4.8):

- For the volume crystalline phase of the confined salt, the equilibrium of solvation reaction (1.1) is mono-variant, and the sorption isotherm is S-shaped with plateaux and steps at certain temperatures;
- The surface X-ray amorphous phase demonstrates the bi-variant equilibrium with smooth sorption isotherms.

Thus, the stage in which an active salt from an impregnating solution deposits on the surface of a porous matrix is very important, and its appropriate implementation allows fine regulation of the texture characteristics and the phase composition of the deposited salt. The intelligent combination of interfacial and bulk depositions would allow a target-oriented synthesis of CSPMs with sorption properties close to those of an optimal adsorbent (see Chapter 3).

4.4.2 Host Matrix

Choice of the host matrix is also very important. The initial idea was to disperse an active salt in the pores of a host matrix in order to avoid or at least abate several shortages typical for bulk salts (see Sec. 1.6). Here we briefly consider how the matrix may affect the properties of the confined salt, and how these effects can be utilized for nanotailoring CSPMs with predetermined sorption properties.

It is well known that dividing a bulk substance into tiny pieces may strongly change its properties such as the melting temperature and enthalpy [42, 43] (see Chapter 11). From the thermodynamic point of view, the confinement of a salt in the restricted pore geometry leads to enormous increase in its surface energy, which may cause variation in the salt's sorption properties. In the following, we briefly consider this "pore size" effect. Additionally, the confined salt may be affected by the strong potential of the pore wall, ion adsorption, wetting phenomenon, and other types of "guest–host" interaction (see Sec. 4.4.2.2). A comprehensive study of the matrix effect on the sorption properties of CSPMs is presented in Chapter 6. Melting and solidification of the hydrated salt in the pores are considered in Chapter 11; its specific heat capacity is discussed in Chapter 10, and dynamics of the hydrated water is explained in Chapter 12.

4.4.2.1 Pore size effect

The "pore size" effect on the sorption properties of CSPMs was systematically studied by confining Ca(NO$_3$)$_2$ to silica pores of various sizes [44]. A gradual shift to higher hydration temperature was found for the transition between anhydrous Ca(NO$_3$)$_2$ and its dihydrate (synthesis reaction)—from 45–48°C for the bulk salt to 53–58°C and 83–105°C for the salt in the silica pore of 12 and 3.5 nm size, respectively (Fig. 4.9). Thus, for a selected host matrix, the proper choice of its pore size provides a way to change the hydration temperature by some tens of degree Celsius. For Ca(NO$_3$)$_2$ dihydrate, a hysteresis between the synthesis and decomposition reactions is found. Indeed, to be decomposed, the bulk hydrate has to be heated up to 140°C, that is, by 90–95°C higher than the synthesis temperature. Evidently, this strong overheating does not allow usage of the bulk salt for heat storage. Confining the salt inside the silica pores reduces the hysteresis, which becomes almost negligible in the 3.5 nm pores (Fig. 4.9). The sorption properties of the salt in such small pores significantly differ from those of the bulk salt, e.g., the sorption equilibrium becomes purely bi-variant.

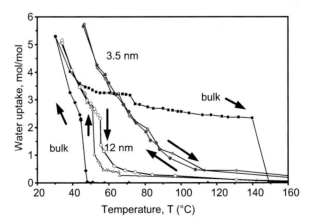

Figure 4.9 Isobars of sorption/desorption for Ca(NO$_3$)$_2$: bulk and confined to pores of 12 nm and 3.5 nm [45].

Similar effects are found in the sorption of methanol by confined salts (see Chapter 7). Thus, the confinement of salts to sufficiently small pores may be used for the intent variation of the salt's sorption properties due to the "pore size" effect. As demonstrated further in this book, these effects become appreciable if the size of pores is 10–50 nm and smaller.

4.4.2.2 Guest–host effect

Chemical interaction of the deposited salt with the matrix surface may be another reason for changing the salt's sorption properties. As discussed earlier, this is a case of interfacial deposition when the salt cations adsorb on the matrix surface. This results in the formation of surface complexes according to processes (4.5)–(4.8) and subsequent formation of the surface X-ray amorphous phase of the salt. It may be expected that the relative contribution of the "guest–host" effect increases for matrices with a large "specific surface area/specific pore volume" ratio, at low salt concentration and high pH of the impregnating solution as well as at high calcination temperature. The latter favors an efficient thermal dispersion of the salt on the matrix surface (see Sec. 4.4.3.3).

Another expectation is that the "guest–host" effect levels individual properties of the salt, thus promoting crossover from a mono-variant sorption equilibrium typical of the authentic salt to a bi-variant one as confirmed by Figs. 4.7, 4.8, and 4.10.

4.4.3 Synthesis Conditions

As demonstrated earlier, appropriate realization of the impregnation and drying stages of CSPM synthesis may result in the fine-tuning of the salt's texture and phase composition of the deposited salt.

4.4.3.1 Solution concentration

The salt's concentration in the impregnating solution determines the salt content in the final composite. A lower concentration of the impregnating solution is advantageous for the interfacial deposition of the salt and, hence, for the formation of the X-ray amorphous salt with a bi-variant sorption equilibrium. Indeed, a gradual transformation of the sorption isotherms from mono- to bi-variant type is observed for the CaCl$_2$/SiO$_2$ composites when the salt content decreases from 33.7 to 9.2 wt.% (Fig. 4.7a).

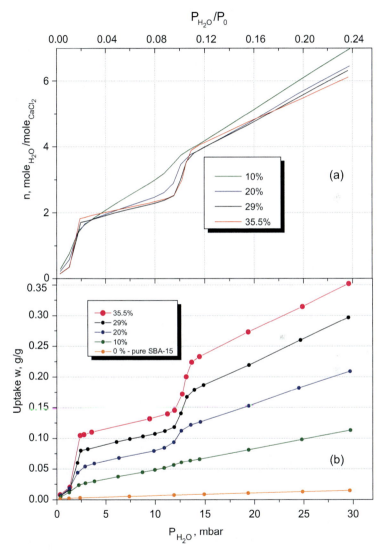

Figure 4.10 The isotherms of water sorption on CaCl$_2$/SBA-15 with various salt content plotted as $n(P_{H_2O})$ (a) and $w(P_{H_2O})$ (b). $T = 50°C$.

To highlight the effect of the salt concentration and separate it from the "pore size" effect, it is convenient to use as a host matrix porous materials with mono-dimensional pores, such as SBA-15 or MCM-41 [46, 47]. The main effect of the increasing $CaCl_2$ concentration in the impregnation solution and, hence, the salt content in the $CaCl_2$/SBA-15 composites is a gradual enhancement of the water uptake (Fig. 4.10b). The uptake reaches 0.33 g/g at $C =$ 35 mass.%, which is much larger than the sorption of water by the pure matrix (SBA-15).

At increasing salt content, the steep (S-shaped) isotherm segments become more and more pronounced; hence a bi-variant sorption equilibrium observed at the low salt content (10 wt.%) approaches a mono-variant one typical of the bulk salt. Again, the salt's sorption properties at high salt contents significantly differ from those at small contents.

If present the sorption as a function of vapor pressure $P(H_2O)$, it is seen that the hydration pressure reduces from 12.0–13.5 to 11.0–12.5 mbar, when the salt content decreases from 29–35 to 20 mass.% (Fig. 4.10a). This effect is similar to the increase in the $Ca(NO_3)_2$ hydration temperature in smaller silica pores (see Sec. 4.4.2.1), and may have the same origin. Indeed, the size of salt particles inside the matrix pores can depend on the salt content in the composite. It is reasonable to expect that a larger content could lead to bigger salt particles and, hence, to higher hydration pressure. Hence, the variation of the salt concentration in the impregnating solution gives an efficient tool for controllable CSPM design.

4.4.3.2 Solution pH

We have already mentioned that the pH of the impregnating solution may have an important impact on the phase composition of the deposited salt (see Fig. 4.7 and Refs. [38, 48]). This effect is additionally illustrated in Fig. 4.11, which shows that the increase in pH makes the step transitions less sharp and proceeding over a wider range of relative pressure. This is in good agreement with the general pattern of the interfacial and bulk depositions (see Sec. 4.4.1.2 and Refs. [37, 38, 48]). Thus, the sorption properties of CSPMs can be efficiently and intently monitored by varying the pH of the impregnating solution.

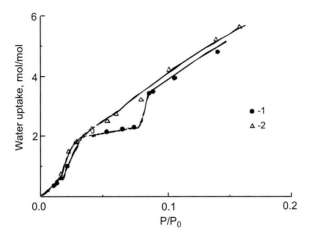

Figure 4.11 Temperature-independent curves of water sorption by composite CaCl$_2$ (23.3 wt.%)/(silica KSK) prepared at pH of salt solution 5.5 (Δ) and 8.0 (●). T_{calc} = 200°C. Adapted from Ref. [38], Copyright 2006, with permission from Elsevier.

4.4.3.3 Calcination temperature

The main effect is smoothing of the sorption isotherms at higher T_{calc} so that the equilibrium becomes bi-variant (Fig. 4.12a). This smoothing is detected at T_{calc} = 370°C, while no change is observed at T_{calc} = 170–200°C and lower. As the melting temperature of CaCl$_2$ in the bulk is 772°C (=1045 K), its Tammann temperature can be estimated as 458°C (=732 K). Probably, the thermal dispersion of CaCl$_2$ is also possible at somewhat lower T_{calc} = 370°C, and its final effect may depend on the calcination time. The latter effect has not been studied yet.

We have prepared a set of composites CaCl$_2$ (23.3%)/SiO$_2$(KSK) calcined at various temperatures between 170 and 470°C and then hydrated to form a hexahydrate CaCl$_2$·6H$_2$O. DSC thermograms in Fig. 4.12b present melting peaks of this hydrate. At T_{calc} ≤ 270°C, the melting peak remains invariable. At higher calcination temperature, it gradually decreases, which may be caused by growing amorphization of the salt. No melting peak is observed for the composite calcined at 470°C, close to the salt Tammann temperature, which means the salt is entirely amorphous. Thus, for drying temperature above 270°C, the thermal dispersion of the salt over the silica gel surface should be taken into account, which promotes the formation of the X-ray amorphous phase.

78 | Basic Synthesis Methods

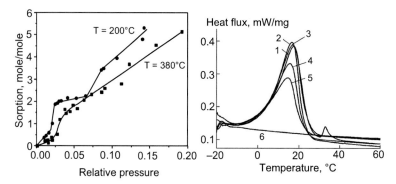

Figure 4.12 (a) Isotherms of water sorption by composite CaCl$_2$ (23.3%)/SiO$_2$(KSK) calcined at 200 and 380°C; (b) DSC thermograms of this composite calcined at 170 (1), 220 (2), 270 (3), 320 (4), 370 (5), and 470°C (6) and hydrated up to n = 6. Reprinted by permission from Springer Nature from Ref. [37], Copyright 2005.

This chapter addressed the practical tools that are available for designing CSPMs with predetermined sorption properties. It demonstrated that these properties can be monitored by a proper choice of the chemical nature and content of the confined salt, the porous structure of the host matrix, and the synthesis conditions. All these tools can be used to adjust the real composite sorbent to an optimal one, a mental representation of which is given in the previous chapter for several adsorption technologies.

References

1. Schueth, F., Sing, K. S. W., and Weitkamp, J. (Eds.) (2002) *Handbook of Porous Solids* (Wiley-VCH, Weinheim, Germany), Vols. 1–5.
2. Regalbuto, J. (Ed.) (2007) *Catalyst Preparation. Science and Engineering* (CRC Press, Taylor & Francis Group, Boca Raton).
3. Dzisko, V. A., Karnaukhov, A. P., and Tarasova, D. V. (1978) *Physico-Chemical Foundations of Oxide Catalysts Synthesis* (Nauka, Novosibirsk) (in Russian).
4. Bourikas, K., Kordulis, C., and Lycourghiotis, A. (2006) The role of the liquid–solid interface in the preparation of supported catalysts, *Catal. Rev.*, **48**, pp. 363–444.
5. Okada, K., Nakanome, M., Kameshima, Y., Isobe, T., and Nakajima, A. (2010) Water vapour adsorption of CaCl$_2$-impregnated activated carbon, *Mater. Res. Bull.*, **45**, pp. 1549–1553.

6. Fortier, H., Westreich, P., Selig, S., Zelenietz, C., and Dahn, J. R. (2008) Ammonia, cyclohexane, nitrogen and water adsorption capacities of an activated carbon impregnated with increasing amounts of $ZnCl_2$, and designed to chemisorb gaseous NH_3 from an air stream, *J. Colloid Interface Sci.*, **320**, pp. 423–435.

7. Bandosz, T. J. and Petit, C. (2009) On the reactive adsorption of ammonia on activated carbons modified by impregnation with inorganic compounds, *J. Colloid Interface Sci.*, **338**, pp. 329–345.

8. Conde, M. R. (2004) Properties of aqueous solutions of lithium and calcium chlorides: Formulations for use in air conditioning equipment design, *Int. J. Thermal Sci.*, **43**, pp. 367–382.

9. Neimark, A., Heifez, L. I., and Fenelonov, V. B. (1981) IEC. *Product Res. Devel.*, **20**, pp. 439–450.

10. Gordeeva, L. G., Glaznev, I. S., Savchenko, E. V., Malakhov, V. V., and Aristov, Yu. I. (2006) Impact of phase composition on water adsorption on inorganic hybrids "salt/silica," *J. Colloid Interface Sci.*, **301**, pp. 685–691.

11. Aristov, Yu. I., Tokarev, M. M., Cacciola, G., and Restuccia, G. (1996) Selective water sorbents for multiple applications: 1. $CaCl_2$ confined in mesopores of the silica gel: Sorption properties, *React. Kinet. Cat. Lett.*, **59**, pp. 325–334.

12. Simonova, I. A., Freni, A., Restuccia, G., and Aristov, Yu. I. (2009) Water sorption on the composite "silica modified by calcium nitrate": Sorption equilibrium, *Micropor. Mesopor. Mater.*, **122**, pp. 223–228.

13. Daou, K., Wang, R. Z., and Xia, Z. Z. (2006) Development of a new synthesized adsorbent for refrigeration and air conditioning applications, *Appl. Therm. Eng.*, **26**, pp. 56–65.

14. Wu, H., Wang, S., and Zhu, D. (2007) Effects of impregnating variables on dynamic sorption characteristics and storage properties of composite sorbent for solar heat storage, *Sol. Energy*, **81**, pp. 864–871.

15. Parks, G. A. (1965) The isoelectric points of solid oxides, solid hydroxides, and aqueous hydroxo complex systems, *Chem. Rev.*, **65**, pp. 177–198.

16. Van Riemsdijk, W. H., De Wit, J. C. M., Koopal, L. K., and Bolt, G. H. (1987) Metal ion adsorption on heterogeneous surfaces: Adsorption models, *J. Colloid Interface Sci.*, **116**, pp. 511–522.

17. Hiemstra, T., Van Riemsdijk, W. H., and Bolt, G. H. (1989) Multisite proton adsorption modeling at the solid/solution interface of (Hydr) oxides: A new approach. I. Model description and evaluation of intrinsic reaction constants, *J. Colloid Interface Sci.*, **133**, pp. 91–104.

18. Van de Water, L. G. A., Bergwerff, J. A., Nijhuis, T. A., de Jong, K. P., and Weckhuysen, B. M. (2005) UV-Vis microspectroscopy: Probing the initial stages of supported metal oxide catalyst preparation, *J. Am. Chem. Soc.*, **127**, pp. 5024–5025.

19. Uvarov, N. F., Isupov, V. P., Sharma, V., and Shukla, A. K. (1992) Effect of morphology and particle size on the ionic conductivities of composite solid electrolytes, *Solid State Ionics*, **51**, pp. 41–52.

20. Gong, L. X., Wang, R. Z., and Xia, Z. Z. (2010) Adsorption equilibrium of water on composite adsorbent employing lithium chloride in silica gel, *J. Chem. Eng. Data*, **55**, pp. 2920–2923.

21. Schneider, M. and Baiker, A. (1995) Aerogels in catalysis, *Catal. Rev. Sci. Eng.*, **37**, pp. 515–556.

22. Avnir, D., Braun, S., Lev, O., and Ottolenghi, M. (1994) Enzymes and other proteins entrapped in sol-gel materials, *Chem. Mater.*, **6**, pp. 1605–1614.

23. Mrowiec-Bialon, J., Jarzebski, A. B., Lachowski, A. L., Malinowski, J. J., and Aristov, Yu. I. (1997) Effective inorganic hybrid adsorbents of water vapor by the sol–gel method, *Chem. Mater.*, **9**, pp. 2486–2490.

24. Gordeeva, L. G., Mrowiec-Bialon, J., Jarzebski, A. B., Lachowski, A. I., Malinowski, J. J., and Aristov, Yu. I. (1999) Selective water sorbents for multiple applications: 8. Sorption properties of $CaCl_2$–SiO_2 sol–gel composites, *React. Kinet. Catal. Lett.*, **66**, pp. 113–120.

25. Mrowiec-Bialon, J., Lachowski, A. L., Jarzebski, A. B., Gordeeva, L. G., and Aristov, Yu. I. (1999) SiO_2-LiBr nanocomposite sol-gel adsorbents of water vapour: Preparation and properties, *J. Colloid Interface Sci.*, **218**, pp. 500–503.

26. Tohidi, S. H., Novinrooz, A. J., and Nabipour, A. (2012) Effect of inorganic hybrid LiBr on the silica matrix xerogels, *Int. J. Eng. Trans. B: Appl.*, **25**, pp. 51–55.

27. Brinker, C. J., Keefer, K. D., Scaefer, D. W., Assink, R. A., Kay, B. D., Ashley, C. (1984) Sol–gel transition in simple silicates II, *J. Non-Cryst. Solids*, **63**, pp. 45–59.

28. Tammann, G. (1929) *Lehrbuch der Metallkunde*, 4th edn, Berlin, Verlag Voss.

29. Padin, J. and Yang, R. T. (2000) New sorbents for olefin/parafinn separations by adsorption via π-complexation: Synthesis and effects of substrates, *Chem. Eng. Sci.*, **55**, pp. 2607–2616.

30. Russell, A. S. and Stokes, J. J. (1946) Role of surface area in dehydrocyclization catalysis, *Ind. Eng. Chem.*, **38**, pp. 1071–1074.

31. Xie, Y. C. and Tang, Y. Q. (1990) Spontaneous monolayer dispersion of oxides and salts onto surfaces of supports: Application to heterogeneous catalysis, *Adv. Catalysis*, **37**, pp. 1–43.

32. Liang, C. C. (1973) Conduction characteristics of the lithium iodide-aluminum oxide solid electrolytes, *J. Electrochem. Soc.*, **120**, pp. 1289–1292.

33. Uvarov, N. F. (2011) Composite solid electrolytes: Recent advances and design strategies, *J. Solid State Electrochem.*, **15**, pp. 367–389.

34. Adamson, A. W. and Gast, A. P. (1997) *Physical Chemistry of Surfaces* (Wiley, New York), p. 108.

35. Knozinger, H. and Taglauer, E. (1993) Towards supported oxide catalysts via solid–solid wetting, in: *Catalysis*, Vol. 10, J. J. Spivey and S. K. Agarwal (Eds.) (The Royal Society of Chemistry, Cambridge), pp. 1–40.

36. Neiman, A. Y., Uvarov, N. F., and Pestereva, N. N. (2007) Solid state surface and interface spreading: An experimental study, *Solid State Ionics*, **177**, pp. 3361–3369.

37. Gordeeva, L. G., Gubar, A. V., Plyasova, L. M., Malakhov, V. V., and Aristov, Yu. I. (2005) Composite water sorbents of the salt in silica gels pore type: The effect of the interaction between the salt and the silica gel surface on the chemical and phase compositions and sorption properties, *Kinetics and Catalysis*, **46**, pp. 736–742.

38. Gordeeva, L. G., Glaznev, I. S., Savchenko, E. V., Malakhov, V. V., and Aristov, Yu. I. (2006) Impact of phase composition on water adsorption on inorganic hybrids "salt/silica," *J. Colloid Interface Sci.*, **301**, pp. 685–691.

39. Aristov, Y. I. (2007) New family of solid sorbents for adsorptive cooling: Material scientist approach. *J. Engin. Thermophys.*, **16**, pp. 63–72.

40. Malakhov, V. V. (2000) Stoichiography as applied to studying composition and real structure of catalysts, *J. Molec. Catal.*, **158**, pp. 143–148.

41. Werich Pietsch, E. H. (Ed.) (1957) *Gmelins Handbuch der Anorganishen Chemie, Calcium T. B–L.2* (Verlag Chemie GmbH) (in German).

42. Gibbs, J. W. (1928) *The Collected Works of J. Willard Gibbs* (Longmans Green, New York).

43. Defay, R., Prigogine, I., Bellemans, A., and Everett, D. H. (1966) *Surface Tension and Adsorption* (Wiley, New York).

44. Simonova, I. A. and Aristov, Yu. I. (2005) Sorption properties of calcium nitrate dispersed in silica gel: The effect of pore size, *Rus. J. Phys. Chem.*, **79**, pp. 1307–1311.

45. Aristov, Y. I. (2007) Novel materials for adsorptive heat pumping and storage: Screening and nanotailoring of sorption properties. *J. Chem. Eng. Japan*, **40**, pp. 1242–1251.

46. Ponomarenko, I. V., Glaznev, I. S., Gubar, A. V., Aristov, Yu. I., and Kirik, S. D. (2010) Synthesis and water sorption properties of a new composite "$CaCl_2$ confined to SBA-15 pores," *Micropor. Mesopor. Mat.*, **129**, pp. 243–250.

47. Tokarev, M. M., Gordeeva, L. G., Romannikov, V. N., Glaznev, I. S., and Aristov, Yu. I. (2002) New composite sorbent "$CaCl_2$ in mesopores of MCM-41" for sorption cooling/heating, *Int. J. Thermal Sci.*, **41**, pp. 470–474.

48. Gordeeva, L. G., Glaznev, I. S., Malakhov, V. V., and Aristov, Yu. I. (2003) Influence of calcium chloride interaction with silica surface on phase composition and sorption properties of dispersed salt, *Russ. J. Phys. Chem.*, **77**, pp. 2019–2023.

Chapter 5

Composite Sorbents of Water Vapor: Effect of a Confined Salt

In this chapter, we consider composites "salt in porous matrix" (CSPMs) for which the main function is adsorption of water vapor. We first discuss the ability of various inorganic salts to form solid complexes with water molecules and consider thermodynamic aspects of their formation in bulk. Then we consider what happens if a salt is introduced into the pores of a host matrix and analyze whether sorption properties of the "salt–matrix" composite are a linear addition of salt and matrix properties, or any nonlinear effects take place. In the former case, the composite sorption properties can be easily predicted on the basis of the salt and matrix properties. The latter case can lead to interference of the salt and matrix properties to give synergetic effects and provide additional tools for designing a desirable sorbent. In both cases, an enormous number of new composite sorbents can be obtained via the target-oriented approach (see Sec. 1.5).

5.1 Salts and Hydrates in Bulk

Many porous solids that are used as host matrices are good adsorbents of water vapor themselves, such as silica gel and alumina,

---

Nanocomposite Sorbents for Multiple Applications
Yu. I. Aristov
Copyright © 2020 Jenny Stanford Publishing Pte. Ltd.
ISBN 978-981-4267-50-2 (Hardcover), 978-981-4303-15-6 (eBook)
www.jennystanford.com

Composite Sorbents of Water Vapor

which are able to adsorb 0.3–0.7 g/g. Therefore, an introduced salt has to be rather active to make a valuable contribution to the overall sorption capacity of the composite.

5.1.1 Water-Sorbing Ability of Bulk Salts

Many salts interact with water molecules giving "salt–water" complexes known as hydrates. For instance, one molecule of $Al_2(SO_4)_3$ can absorb 18 molecules of water according to the reaction

$$Al_2(SO_4)_3 + 18H_2O = Al_2(SO_4)_3 \cdot 18H_2O \tag{5.1}$$

Some other well-known reactions are as follows:

$$Na_2HPO_4 \ (s) + 12H_2O \ (g) = Na_2HPO_4 \cdot 12H_2O \ (s) \tag{5.2}$$

$$CaCl_2 \ (s) + 6H_2O \ (g) = CaCl_2 \cdot 6H_2O \ (s) \tag{5.3}$$

$$MgSO_4 \ (s) + 7H_2O \ (g) = MgSO_4 \cdot 7H_2O \ (s) \tag{5.4}$$

$$Na_2CO_3 \ (s) + 10H_2O \ (g) = Na_2CO_3 \cdot 10H_2O \ (s) \tag{5.5}$$

$$LiNO_3 \ (s) + 3H_2O \ (g) = LiNO_3 \cdot 3H_2O \ (s) \tag{5.6}$$

Being related to 1 g of the salt, the water content corresponds to 0.95, 1.52, 0.97, 1.05, 1.70, and 0.78 g of H_2O, respectively. Thus, it is expected that inorganic salts confined to pores of traditional adsorbents can significantly enhance the water sorption capacity and this increment depends on the vapor relative pressure P/P_0 (see Fig. 4.5 and Table 4.1).

Prior to proceed any further, let us briefly remind the main thermodynamic requirement to the composite sorbent and, hence, to *the active salt* optimal for a particular application (Sec. 3.2). It is a large change in the sorption uptake *either between* the two equilibrium states, which are characterized by the fixed sets (T_1, P_1) and (T_2, P_2), or *at* particular T and P. The latter case corresponds to a stepwise sorption isotherm (isobar) with the large uptake step positioned at a desirable pressure (temperature).

5.1.2 Mono-variant Equilibrium for Bulk Salts/Hydrates

"Water–salt" systems described by Eqs. (1.1) and (5.1)–(5.6) are mono-variant as discussed in Sec. 2.5. At fixed equilibrium temperature T_r, hydrate formation takes place in a step-like way at the vapor pressure P_r given by the integral van't Hoff equation

$$\mathrm{Ln} P_r = -\Delta G°/RT \quad \text{or} \quad \mathrm{Ln} P_r = -\Delta H°/RT_r + \Delta S°/R \qquad (5.7)$$

where $\Delta G°$, $\Delta H°$, and $\Delta S°$ correspond to the variation in the standard Gibb's free energy, enthalpy, and entropy, respectively (Fig. 5.1). The isobars are also stepwise, and the threshold temperature can be calculated using Eq. (5.7), too.

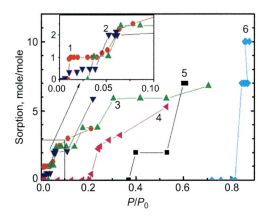

Figure 5.1 Equilibrium amount of water in various hydrated salts (in bulk) versus P/P_0: LiBr (1), $CaCl_2$ (2), $MgSO_4$ (3), $Ca(NO_3)_2$ (4), Na_2HPO_4 (5), and Na_2SO_4 (6). Reprinted from Ref. [1], Copyright 2007, with permission from JST.

As there is a large difference in the values of $\Delta H°$ and $\Delta S°$ for a plenty of known massive salts/hydrates, one might expect that the reaction temperature T_r can vary in a very wide range for different salts. It gives a valuable opportunity to select an appropriate salt that, at a given P, undergoes hydration/dehydration transition at a needed temperature T_r in accordance with the requirements discussed in Chapter 3. Table 4.1 and Fig. 5.1 demonstrate that for various bulk salts, the temperature of step-like hydrates formation is, indeed, distributed over a wide P/P_0 range [1].

Thus, the first step of tailoring an optimal CSPM is the search for a proper salt/hydrate that transforms at a temperature close to the desirable one. Then the chosen salt is to be placed inside the pores of a host matrix to avoid the problems discussed in Sec. 1.6: suppression of reaction rate, noticeable hysteresis, salt swelling/shrinkage, and threat of corrosion.

5.2 Various Salts Confined to the Mesoporous Silica Gel KSK

Here we consider the selected salts (CaCl$_2$, LiCl, LiBr, MgSO$_4$, Ca(NO$_3$)$_2$, Na$_2$SO$_4$) confined inside the pores of the same matrix, that is, a commercial silica gel KSK (Reakhim, Russia). It belongs to the family of low-density silica gels that are widely used in many adsorption technologies [2]. They are relatively cheap, available, and produced in large scale [3]. The KSK silica is similar to other mesoporous industrial silica gels such as BASF KC-Trockenperlen AF125, Davisil 62 (643–646), etc. The pore size distribution (PSD) of KSK silica shows mesopores with sizes between 5 and 30 nm, which gives an average of 15 nm (Fig. 5.2). The matrix has a specific surface area S_{sp} = 350 m^2/g and quite large pore volume V_{sp} = 1.0 cm^3/g, which allows an essential salt quantity to be inserted into.

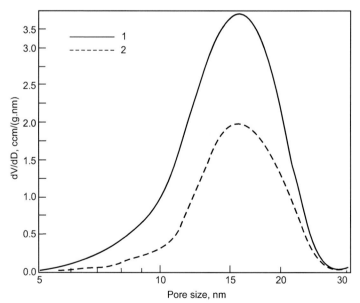

Figure 5.2 Pore size distributions of the KSK silica gel before (1) and after (2) impregnation with CaCl$_2$. Reprinted by permission from Springer Nature from Ref. [32], Copyright 1996.

5.2.1 Effect of Salt's Nature on Water Sorption Equilibrium

Figure 5.3 presents the water sorption isotherms and isobars for various composites "salt/KSK" plotted as a function of the relative pressure P/P_0 of water vapor. It shows that

- The addition of salt significantly increases the water sorption capacity as compared with the pure silica KSK (by a factor of 2–10). The isotherm shape is determined by the nature of the inserted salt, which is, therefore, the main water sorbing component;
- Isotherm sections with steep increment in the water uptake are observed, which correspond to mono-variant formations of the appropriate hydrates. For instance, the "step" on the $CaCl_2$/KSK isotherm at $P/P_0 = 0.02$–0.03 appears due to the formation of the salt dihydrate $CaCl_2 \cdot 2H_2O$;
- The isotherm "steps" are smoother than those for the appropriate bulk salt and shifted to smaller P/P_0 (when compared with Fig. 5.1 for bulk salts/hydrates);
- The step position can be managed in a wide range through the choice of salt (e.g., by some 50–70°C, Fig. 5.3b). The step amplitude (height) depends on the salt content (see Sec. 5.2.2).

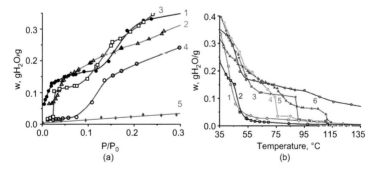

Figure 5.3 (a) Isotherms of water sorption on the composites "salt/KSK" [4]. The salts are $MgSO_4$ (1), LiBr (2), $CaCl_2$ (3), $Ca(NO_3)_2$ (4), pure KSK (5). (b) Isotherms of water sorption on the composites "salt/KSK." The salts are $LiNO_3$ (1), $Ca(NO_3)_2$ (2), $CaCl_2$ (3), LiCl (4), LiBr (5), and $MgSO_4$ (6). $P(H_2O) = 17$ mbar. Figure (a) reprinted from Ref. [4], Copyright 2009, with permission from Elsevier.

Thus, the nature of the inserted salt is the most important factor that influences the water sorption properties of the composite;

therefore, the choice of salt is a matter of great consequence for a target-oriented synthesis of CSPMs with desirable properties. This choice is mainly leant on the sorption properties of the bulk salt, which are collected in many handbooks (see Refs. [5–8]), reviews (e.g., Ref. [9]), and numerous original papers. When a salt with the proper hydration temperature is selected, several other factors (pore size, synthesis conditions, etc.) can also be used for fine-tuning the sorption properties of the CSPM.

5.2.2 Effect of the Salt Content on Water Sorption Equilibrium

One such factor is the salt content C in the composite, which can be varied by managing the salt concentration in the impregnating solution (see Sec. 4.4.3.1). The first effect of increasing the salt content is a gradual enhancement of the water uptake (see, e.g., Figs. 4.7a, 4.10b and Refs. [10–14]). It confirms that the salt is the main water-sorbing component. At large C values, the equilibrium uptake can even be proportional to the salt content as found for the LiBr/KSK composites with C = 32 and 57 wt.% [15] and the CaCl$_2$/silica with C = 17, 26, and 33 wt.% [14]. In this special case, the adsorption of water by the matrix can be neglected at all. The opposite situation is observed for the "salt/zeolite" composites for which the water sorption increases insignificantly [16] or even decreases [14]. The decrease occurs because the hygroscopic salt blocks some of the microporosity of the zeolite.

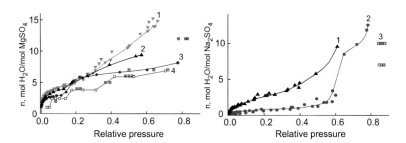

Figure 5.4 Water sorption isotherms for MgSO$_4$/KSK (left) and Na$_2$SO$_4$/KSK (right) composites. MgSO$_4$ content: 13.5 (1), 24.4 (2), 38.0 wt.% (3), and the bulk salt (4). Na$_2$SO$_4$ content: 10.3 (1), 27.3 (2), and the bulk salt (3).

When the salt content reduces, a gradual transformation of the sorption isotherms from mono- to bi-variant type is observed [17, 18]. This is first found for the $CaCl_2$/KSK composites when the salt content decreases from 33.7 to 9.2 wt.% (Fig. 4.7a) [17]. Even more pronounced transformation is shown in Fig. 4.10b for the same salt located in the mono-sized pores of SBA-15 [19]. Two tendencies are revealed when the $CaCl_2$ content decreases: the isotherms (a) become smoother and (b) shift to lower relative pressure. The same trends can be found for many other salts in the pores of KSK silica, such as $MgSO_4$, Na_2SO_4 (Fig. 5.4), and $CuSO_4$ (not presented) [15–18]. Thus, the sorption properties of the salt at large contents significantly differ from those at small contents. Both the tendencies are also revealed for changing the water sorption isotherms when the salt is placed in smaller silica pores (see Sec. 4.4.2.1 and Chapter 6) and may have the same origin. Indeed, the size of salt particles inside the matrix pores can depend on the salt content in the composite. It is reasonable to expect that a larger content could lead to bigger salt particles and, hence, to higher hydration pressure. Hence, the variation in the salt content C gives an efficient tool for controllable CSPM design.

5.2.3 Water Sorption Properties of the Salt/Matrix Composites: Non-additivity

As the composite isotherms smoothen and shift at small salt contents, it indicates that the sorption properties of the "salt–matrix" composite are not a linear addition of the properties of the salt and the matrix. One may expect that nonlinear effects take place.

This issue is specially examined in Ref. [4]. Isobars of water sorption on silica KSK, bulk calcium nitrate, and the composite $Ca(NO_3)_2$(45 wt.%)/KSK (SWS-8L) were experimentally measured at 17.0 mbar (Fig. 5.5a). For the pure silica gel, sorption increases gradually with the rise in temperature up to 0.04 g of H_2O per 1 g of silica. In contrast, the sorption of water by the bulk salt proceeds stepwise with the formation of $Ca(NO_3)_2 \cdot 2H_2O$ in the course of the salt hydration. Because of the mono-variant equilibrium in the bulk $Ca(NO_3)_2$–H_2O system, at a fixed water vapor pressure P, hydration occurs at a certain threshold temperature $T_r(P)$. At $P = 17.0$ mbar, this temperature was measured to be $T_r = 46$–$47°C$ (Fig. 5.5a). On the other hand, this transition temperature can be calculated from the van't Hoff equation, since the standard enthalpy $\Delta H^0_{f,298}$ and entropy S^0_{298} of formation are known for the bulk salt and its dihydrate [20]: T_r (17 mbar) = $48°C$ in good agreement with the measured value.

The foreseen sorption by the composite was calculated as the amount of water sorbed by the bulk salt and the silica taken with appropriate weight coefficients:

$$w_\Sigma = C \cdot w_{Ca(NO3)2} + (1-C) \cdot w_{SiO2} \qquad (5.8)$$

where $w_{Ca(NO3)2}$ and w_{SiO2} are the water uptakes for $Ca(NO_3)_2$ and SiO_2, respectively (curve 4 in Fig. 5.5a). Comparison of this calculated curve with the experimental isobar for SWS-8L shows that the sorption properties of the composite cannot be presented as a simple addition of Eq. (5.8). Indeed, if the salt is confined to the silica pores, the temperature of hydration increases up to 48–52°C (Fig. 5.5a), probably due to the dispersion of calcium nitrate to a nanosize scale. This augmentation of hydration temperature conforms with the shift in isotherm to smaller P/P_0 values described earlier. Both findings are due to the stronger bonding between water molecules and the salt in the confined state. Hence, a dispersed salt is a better desiccant agent than the bulk salt.

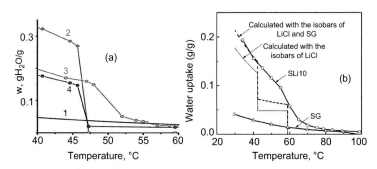

Figure 5.5 (a) Isobars of water sorption on the pure silica KSK (1), bulk calcium nitrate (2), composite SWS-8L (3), measured at P = 17 mbar. Curve 4 is calculated as the linear addition by Eq. (5.8). (b) Calculated and experimental isobars of water sorption at 8.8 mbar for the composite LiCl/silica (SLi10). Figure (a) reprinted from Ref. [4], Copyright 2009, and Figure (b) reprinted from Ref. [21], Copyright 2014, with permission from Elsevier.

A similar analysis performed for LiCl/silica (Fig. 5.5b) [21] and Na_2SO_4/silica [22] composites also reveals the non-additivity of water sorption on these composites and an interference of the properties of the salt and the matrix. This synergetic effect provides additional tools for designing a desired sorbent.

5.2.4 Synthesis/Decomposition Hysteresis

It is well known that there can be a "pressure–temperature" region near equilibrium where the gas–solid reaction (1.1) in bulk is inhibited [23]. This inhibition region may result in noticeable hysteresis between synthesis and decomposition reactions or, in other words, adsorption/desorption hysteresis. This is one of the main restrictions on the direct application of bulk salts for sorption of gases/vapors. The suggested solution is insertion of a salt inside the pores of a common adsorbent (see Chapter 1 and Refs. [24–27]).

Indeed at $P(H_2O) = 17$ mbar, synthesis of $Ca(NO_3)_2 \cdot 2H_2O$ from the *bulk* salt occurs at 46–47°C in one step (Fig. 5.6) [28]. Its decomposition runs at much higher temperatures in two steps: the monohydrate forms at 80–90°C and anhydrous salt at 100°C. It is not convenient for many practical applications. For instance, the heat stored during the dihydrate decomposition at 80–100°C is released during adsorption at 46–47°C, thus having a dramatically lower

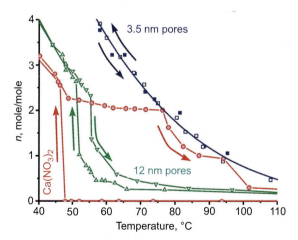

Figure 5.6 Isobars of water sorption (↑) and desorption (↓) on bulk $Ca(NO_3)_2$ (red) and $Ca(NO_3)_2$ inside the pores of 12 nm size (composite SWS(Ca)-12, green) and 3 nm size (SWS(Ca)-3, blue) [28]. $P(H_2O) = 17$ mbar.

value. Several reasons for such strong hysteresis, typical for many other salts/hydrates, are suggested in the literature [23, 29, 30]:

- Very slow formation of critical nuclei, which results in pronounced induction period during both synthesis and decomposition [30, 31];
- Strong structural distortions and stresses caused by different molecular volumes and crystal lattices of the bulk salt and its hydrate [23].

If $Ca(NO_3)_2$ is confined to the silica mesopores (12 nm), the synthesis and decomposition branches approach each other as the hydration temperature increases and the dehydration one significantly decreases (Fig. 5.6). The difference between them reduces to ca. 5°C, which is quite acceptable for most practical applications. No hysteresis is recorded when the salt is confined to the silica pores of 3.5 nm size. The degree at which the hysteresis is reduced depends on the nature of salt as well: It disappears completely for $CaCl_2$ dispersed in the silica mesopores (15 nm) [32] but only partially for $MgSO_4$ in the same pores [33].

The tentative reason for this desirable change could be more defect/less crystalline structure of the confined salt nanoparticles, which facilitates

- The formation of the critical nuclei. Indeed, the confined salt was found to be X-ray amorphous, at least, partially [17, 18];
- Relaxation of mechanical stresses;

and brings both synthesis and decomposition reactions closer to their thermodynamic equilibrium.

5.2.5 Dynamics of Hydrate Synthesis/Decomposition

Many issues concerning the hysteresis and discussed right above can also be responsible for slow dynamics of synthesis/decomposition reactions observed for many bulk salts/hydrates [30, 34]. A typical example is a long induction period and an S-shape of the kinetic curves for both hydrate synthesis and decomposition [30, 31]. One more kinetic obstacle is the formation of solid hydrate layer on the surface of the salt. Further hydration reaction requires the diffusion of vapor molecules through this increasing layer, which could be a very slow process [34].

Indeed, the dynamics of Ca(NO$_3$)$_2$ hydration in bulk is the S-shaped curve with the induction period of 100–200 s and complete hydration by 10,000 s (Fig. 5.7a) [28]. Hydration of this salt confined to the mesopores (15 nm) of silica KSK is much faster and ends in 2000–3000 s. The initial part of the kinetic curve is a linear function of $t^{0.5}$, and the curve can be described by the Fickian diffusion model [35]. This allows determination of the efficient water diffusivity in the composite pores, $D_e = (1.2 \pm 0.5)*10^{-6}$ m^2/s, which is close to the Knudsen diffusivity corrected by the porosity and tortuosity of the composite [28]. A similar acceleration was found for the hydration of Na$_2$SO$_4$ confined to the pores of silica KSM (Fig. 5.7b) [22]. These data prove that for the confined salt, the water sorption rate is controlled by vapor diffusion inside the matrix pores, as it takes place for common adsorbents, rather than by the salt hydration itself. Thus, dispersing the salt into tiny pieces inside the matrix pores allows great improvement in the unfavorable dynamics of its hydration/dehydration.

Overcoming of the problems connected with the salt swelling and deliquescence during its hydration will be considered in Chapter 6.

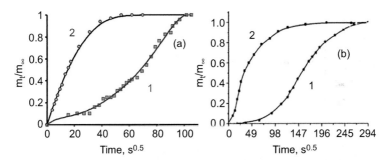

Figure 5.7 Kinetic hydration curves for Ca(NO$_3$)$_2$ (a) and Na$_2$SO$_4$ (b) in the bulk (1) and confined (2) states. Figure (a) reprinted from Ref. [4], Copyright 2009, and Figure (b) reprinted from Ref. [22], Copyright 2014, with permission from Elsevier.

5.3 Mechanisms of Water Sorption on CSPMs

Calcium chloride confined to various mesoporous matrices is one of the most extensively studied systems. It was first synthesized and studied at the Boreskov Institute of Catalysis, and its equilibrium with water vapor has been reported in Refs. [32, 36]. In 7–10 years, its numerous analogs have been prepared and studied in many laboratories all over the world [10, 11, 13, 14, 19, 37]. That is precisely why the composite "CaCl$_2$ in the mesopores of KSK silica" (SWS-1L) is considered in the following sections to ascertain the main mechanisms of water sorption on CSPMs [38].

5.3.1 Isobars of Water Sorption on SWS-1L

The isobars of water sorption on SWS-1L are measured with a CAHN C2000 thermal balance in the temperature range of 293–423 K at vapor partial pressure P_{H_2O} ranging from 8 to 133 mbar (see Ref. [32] for details). The equilibrium water uptake is calculated as $w = m(P_{H_2O},T)/m_0$, where $m(P_{H_2O},T)$ is the mass of water sorbed at

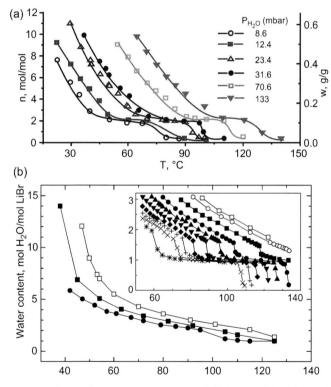

Figure 5.8 Isobars of water sorption on CaCl$_2$/(silica KSK) [32] (a) and LiBr/(silica KSK) [39] (b): C = 32 wt. %, $P(H_2O)$ = 36.0 (●), 53.0 (■), 81.0 (□) mbar. Inset: C = 57 wt. %, $P(H_2O)$ = 7.5 (*), 10.0 (x), 13.0 (+), 15.0 (♦), 19.5 (□), 24.5 (□), 32.5 (●), 50.0 (■), 68.0 (○), 78.0 (□) mbar.

(P_{H_2O}, T) and m_0 is the mass of dry sorbent. As the water sorption by the salt is dominant (see Sec. 5.2.1), it is convenient to present the uptake as the equilibrium number n of sorbed water molecules with respect to one molecule of calcium chloride: $n(P_{H_2O}, T) = [m(P_{H_2O}, T)/\mu(H_2O)]/[m(salt)/\mu(salt)]$, where μ is the molecular weight.

All the sorption isobars are found to be similar but shifted and partially extended along the temperature axis. Each curve has a plateau corresponding to $n = 2$, which indicates the formation of the solid crystalline hydrate $CaCl_2 \cdot 2H_2O$. Thus, the major part of water is sorbed by the salt rather than by the host matrix. This hydrate is directly detected by an X-ray diffraction [37]. The salt dihydrate is rather stable and undergoes *mono-variant* decomposition at high temperature to form the crystalline hydrates $CaCl_2 \cdot H_2O$, $CaCl_2 \cdot 0.33H_2O$, and, finally, the anhydrous salt. At low temperature, the amount of sorbed water increases and reaches 0.6–0.7 g/g, which is much higher than that for pure (non-modified) silica (Fig. 4.5). At temperatures lower than that of the left boundary of the plateau, water sorption depends on both temperature and partial vapor pressure. Hence, the equilibrium becomes *bi-variant*, typical for water vapor absorption by the salt solution [5]. Interestingly, almost complete removal of the sorbed water (except last 0.1 g/g corresponding to the plateau) can be performed by heating the sorbent up to only 50–100°C (Fig. 5.8a), which makes the material promising for storing low-grade heat.

Crystalline calcium chloride tetrahydrate ($CaCl_2 \cdot 4H_2O$) does not manifest itself as a plateau on the isobars, because inside the pores, it coexists with its melt and is observed as a subtle shoulder (Fig. 5.8a). The melting temperature of $CaCl_2 \cdot 4H_2O$ is 38°C in bulk and decreases down to ca. 30°C if the salt is confined to the mesopores [19]. For hexa-hydrate $CaCl_2 \cdot 6H_2O$, these temperatures are 29.8 and 21.4°C [19], so that it is always melted inside the mesopores. Thus, the confinement of $CaCl_2$ to silica pores can strongly influence the formation of salt hydrates: Only high-temperature hydrates $CaCl_2 \cdot nH_2O$ ($n = 0.33, 1, 2$) are crystalline, whereas low-temperature hydrates ($n = 4$ and 6) partially or completely melt and possess the bi-variant equilibrium with water vapors.

5.3.2 Isosteric Chart of Water Sorption on SWS-1L

Water sorption isosters on SWS-1L at $n = 1-10$ (Fig. 5.9a) are straight lines in the $\ln(P_{H_2O})$ versus $1/T$ presentation

$$\ln(P_{H_2O}) = A(w) + B(w)/T \qquad (5.9)$$

The slope depends on the water content w and gives the isosteric heat of water sorption $\Delta H_{is}(w) = B(w) \cdot R$ (Table 5.1), where R is the universal gas constant. At $n > 2$, $\Delta H_{is} = 43.9 \pm 1.3$ kJ/mol, which is close to the heat of water evaporation from aqueous $CaCl_2$ solutions [40]. A significant increase in ΔH_{is} at $n \leq 2$ is caused by the formation of solid hydrates where water molecules are bound stronger than in the solution. Very similar heat release, 60.7 kJ/mol at $n = 2$ and 46.7 kJ/mol at $n = 10$, is found for water sorption on $CaCl_2$ confined to the porous celite [26]. The isosteric chart in Fig. 5.9a can be used for analyzing the SWS-1L applications, e.g., plotting thermodynamic cycles of various heat transformation units [41].

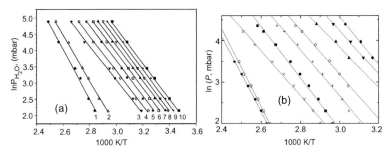

Figure 5.9 Isosters of water sorption on $CaCl_2$/(silica KSK) [32] (a) and LiBr/(silica KSK) [39]. The n value is listed near the lines. Figures reprinted by permission from Springer Nature: (a) from Ref. [32], Copyright 1996, and (b) from Ref. [39], Copyright 1998.

Table 5.1 Isosteric [32] and calorimetric [43] heats (kJ/mol) of water sorption on SWS-1L at various water contents

n (mol/mol)	0.05	0.33	1	2	3	4	5	6	7	8
w (g/g)	—	0.018	0.055	0.11	0.165	0.22	0.275	0.33	0.385	0.44
ΔH_{cal}	81.5	72.7	71.4	64.8	54.4	52.8	52.7	51.9	52.3	—
ΔH_{is}	—	—	63.1	62.3* (45.2†)	42.2	43.9	45.6	44.7	44.3	45.6
$\Delta H^{\#}$	—	75.46	73.7	62.50	—	—	—	—	—	—

*calculated from the right border of the plateau at $n = 2$
†calculated from the left border of the plateau at $n = 2$
#calculated from the solution enthalpy of the bulk hydrates [42]
Source: Reprinted by permission from Springer Nature from Ref. [32], Copyright 1996.

This isoteric heat can be compared with the calorimetric heat ΔH_{cal} of water sorption on SWS-1L presented in Fig. 5.10 [43]. First, the ΔH_{cal} value is larger than the ΔH_{is} value at any water uptake, which is the common case [44]. At very small uptake ($n < 0.15$), ΔH_{cal} (= 80 ± 2 kJ/mol) is close to the heat of water adsorption on the pure silica gel KSK, which is also presented in the figure. Hence, the very first portions of water vapor are likely adsorbed by active centers on the KSK surface. At $0.15 < n < 2.0$, ΔH_{cal} is almost constant and can be associated with the formation of crystalline hydrates $CaCl_2 \cdot nH_2O$ (n = 0.33, 1, and 2). The measured ΔH_{cal} values are close to those calculated from the solution enthalpy of the bulk hydrates [42]. At $2 < n < 4$, the calorimetric heat gradually decreases down to 52 ± 2 kJ/mol, approaching the heat of vapor absorption by an aqueous $CaCl_2$ solution [40].

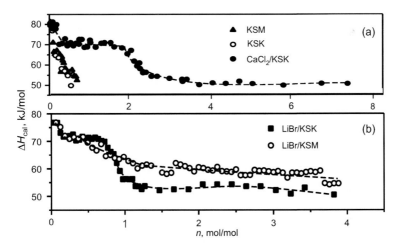

Figure 5.10 Calorimetric heat of water sorption on $CaCl_2$/KSK, meso- and microporous silica gels—KSK and KSM (a), LiBr/KSK and LiBr/KSM (b) as function of the water uptake n. Reprinted from Ref. [43], with permission from Pleiades Publishing, Ltd., Copyright 2001.

5.3.3 Universal Description of Water Sorption on SWS-1L

Dubinin introduced the adsorption potential $\Delta F = -RT \ln h = -RT \ln(P/P_o)$ and showed that for a variety of microporous solids, there is a one-to-one correspondence between the equilibrium uptake and the single parameter ΔF, instead of the usually employed two parameters (P, T) [45] (see Eqs. (2.10)–(2.12)). This approach is effectively applied to describe the water sorption on SWS-1L and other SWSs [46–48]. The experimental *isobars* of water sorption on SWS-1L (Fig. 5.8a) settle on the unique characteristic sorption curve if presented as a function of the adsorption potential ΔF (Fig. 5.11a) [46]. Moreover, this curve also describes the experimental *isotherms* of water sorption on SWS-1L (see Ref. [46] and Fig. 2.1).

Inside the KSK pores, dihydrate $CaCl_2·2H_2O$ is stable at ΔF = 8.0–10.5 kJ/mol, monohydrate $CaCl_2·H_2O$ is stable in the narrow ΔF range around 10.7 kJ/mol, and $CaCl_2·0.33H_2O$ is stable at ΔF > 11 kJ/mol (Fig. 5.11a). The $CaCl_2$ hydrated state with n = 4 reveals itself as an inflection at $\Delta F \approx 5.5$ kJ/mol. At ΔF < 5 kJ/mol, the function $n(\Delta F)$ coincides with the appropriate dependence for the aqueous $CaCl_2$ solution.

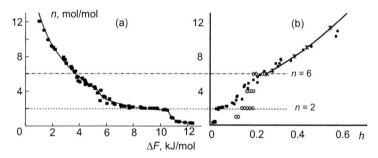

Figure 5.11 Universal isotherms of water sorption on SWS-1L (■) presented as functions of ΔF (a) and h (b) together with the literature data for the bulk $CaCl_2$ crystalline hydrates (o) and aqueous solutions (∗ and solid line in (b) [40]). Figure (b) reprinted by permission from Springer Nature from Ref. [32], Copyright 1996.

Another universal presentation of water sorption on SWS-1L gives the function $n(h)$, where h is the relative vapor pressure. All experimental data satisfactorily follow the same curve (Fig. 5.11b), giving a one-to-one correspondence between n and h. One can see once again the plateaus corresponding to $n \approx 0.3$ and 2 as well as a shoulder at $n \approx 4$. At h > 0.15–0.2, the isotherm is a smooth curve, which essentially coincides with the proper curve for a bulk aqueous solution of calcium chloride [40]. It means that the solution confinement to mesopores of the KSK silica gel does not change its water sorption properties compared to the bulk solution. It indicates that neither a confinement of the salt solution into the silica KSK mesopores nor the interaction between the solution and the silica changes the thermodynamic (sorption) properties of the solution.

It looks reasonable if one takes into consideration that thermodynamic properties of strong electrolytes mainly depend on the electrostatic interaction between cations and anions of a dissociated salt. Any ion is considered to be surrounded by an envelope of other ions [49]. For the $CaCl_2$ solutions inside the KSK pores, the diameter of the ion "atmosphere" is 0.2–0.6 nm. It is much lower than the average pore diameter of the silica gel used (15 nm). For this reason, the pore walls should not influence the ion distribution in the solution and, hence, should not affect the chemical potential of the confined solution and its thermodynamic properties.

Contrary to the liquid solutions, the sorption properties of the solid $CaCl_2$ crystalline hydrates radically change due to the impregnation into the KSK mesopores. As a result, several hydrates ($n = 4$ and 6) cannot be stabilized in the pores in a crystalline state, whereas hydrates with low-water content ($n \leq 2$) form at the relative vapor pressures that are much lower than for the bulk hydrates (Fig. 5.11). For instance, the formation of $CaCl_2 \cdot 2H_2O$ in the pores is observed at a relative pressure of about 0.02–0.03, which is much lower than at the bulk state ($h \approx 0.14$). As a result, the confined salt has a much stronger affinity for water vapor. It opens up opportunities for applying the confined $CaCl_2$ (SWS-1L) for fine dehumidification of gas mixtures [50]. It is demonstrated in the next chapter that the salt sorption properties depend on the porous structure of a host matrix to which the salt is confined. In Chapter 6, we discuss a link between the size of confined salt nanocrystals and the salt's dehydration temperature, which can be due to the contribution of the salt/hydrate surface energy.

5.3.4 Mechanisms of Water Sorption on SWS-1L

The above collection of equilibrium properties of the "water vapor–SWS-1L" system allows the following pattern of the successive water sorption to be suggested (Fig. 5.12) [38]:

- Heterogeneous adsorption of water vapor on active centers of the silica gel takes place at very low uptakes ($w < 0.01$ g/g). The heat of adsorption is very high ($\Delta H_{cal} = 76–84$ kJ/mol). Heterogeneous adsorption on the matrix surface is responsible for only 2–4% of the total sorption and can be neglected in the first approximation (not shown in Fig. 5.12). However, just these centers ensure the low dew point over SWS-1L during air drying [50].

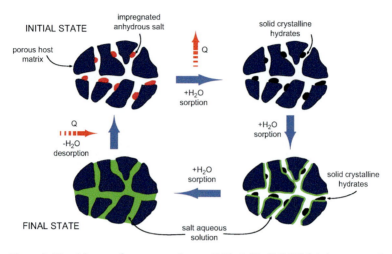

Figure 5.12 Scheme of water sorption on SWS-1L (CaCl$_2$(33%)/KSK). Reprinted by permission from Springer Nature from Ref. [38], Copyright 2007.

- The chemical reaction between the salt and water vapors with the formation of solid crystalline hydrates CaCl$_2 \cdot n$H$_2$O with n = 0.33, 1, and 2. This occurs at h < 0.15, ΔF > 8.0 kJ/mol, and w < 0.11 g/g. The heat of reaction (65–75 kJ/mol) is released and can be productively used. The chemical reaction (or solid absorption) is responsible for 15–20% of the total water sorption. Relatively high temperature (80–120°C) is necessary to remove this water (Fig. 5.8).
- Hydrates CaCl$_2 \cdot n$H$_2$O with n = 4 and 6 melt inside the KSK pores so that the solid hydrated salt coexists with the salt solution. The volume of the solution gradually increases at higher h (or lower ΔF) until it occupies the whole pore space. Such volume filling of pores leads to the advanced sorption capacity of SWS-1L (up to 0.65 g/g). This phenomenon is similar to the volume filling of micropores in activated carbons and other molecular sieves. The difference is that the driving force for water sorption on SWS-1L is an interaction of water and salt molecules (in hydrates) or ions (in solution) rather than interaction with the adsorbent walls. The water sorbed

by the solution (ca. 80% of the total sorption) can be removed by just moderate heating up to 40–80°C, depending on the vapor pressure. Thus, SWS-1L can retain a huge quantity of water and readily gives it back, which is valuable for many applications.

Similar sorption mechanisms, in a varying degree, occur in any CSPM regardless of the salt and the matrix: the salt is the main sorbing element that can significantly increase the matrix sorption capacity due to the chemical reaction with the water vapor as well as its absorption by the salt concentrated solution. Although the matrix plays a secondary role, it can have a significant impact on the composite sorption properties as analyzed in the next chapter.

References

1. Aristov, Yu. I. (2007) Novel materials for adsorptive heat pumping and storage: Screening and nanotailoring of sorption properties, *J. Chem. Eng. Japan*, **40**, pp. 1242–1251.

2. Iler, R. (1978) *The Chemistry of Silica* (Wiley, New York).

3. Yang, R. T. (2003) Gas separation by adsorption processes, in: *Adsorbents: Fundamentals and Applications* (John Wiley & Sons, New York).

4. Simonova, I. A., Freni, A., Restuccia, G., and Aristov, Yu. I. (2009) Water sorption on the composite "silica modified by calcium nitrate": Sorption equilibrium, *Micropor. Mesopor. Mater.*, **122**, pp. 223–228.

5. Kirk-Othmer. (1992) *Encyclopedia of Chemical Technology*, 4th ed. (Wiley, New York).

6. Erich Pietsch, E. H. (Ed.) (1957) *Gmelins Handbuch der Anorganischen Chemie, Calcium Teil B–Lieferung 2* (Verlag Chemie GmbH) (in German).

7. Kirgintsev, A. N., Trushnikova, L. N., and Lavrentieva, V. G. (1972) *Solubility of Inorganic Substances in Water: Handbook* (Khimia, Leningrad, USSR) (in Russian).

8. Haynes, W. M. (2016) *CRC Handbook of Chemistry and Physics*, 97th ed. (CRC Press, Boca Raton, FL).

9. Conde, M. R. (2004) Properties of aqueous solutions of lithium and calcium chlorides: Formulations for use in air conditioning equipment design, *Int. J. Thermal Sci.*, **43**, pp. 367–382.

10. Okada, K., Nakanome, M., Kameshima, Y., Isobe, T., and Nakajima, A. (2010) Water vapor adsorption of $CaCl_2$-impregnated activated carbon, *Mater. Res. Bull.*, **45**, pp. 1549–1553.

11. Zhang, X. J. and Qiu, L. M. (2007) Moisture transport and adsorption on silica gel–calcium chloride composite adsorbents, *Energy Convers. Manag.*, **48**, pp. 320–326.

12. Gordeeva, L. G. and Aristov, Yu. I. (2012) Composites "salt inside porous matrix" for adsorption heat transformation: A current state of the art and new trends, *Int. J. Low Carbon Techn.*, **7**, pp. 288–302.

13. Daou, K., Wang, R. Z., and Xia, Z. Z. (2006) Development of a new synthesized adsorbent for refrigeration and air conditioning applications, *Appl. Therm. Eng.*, **26**, pp. 56–65.

14. Cortés, F. B., Chejne, F., Carrasco-Marín, F., Pérez-Cadenas, A. F., and Moreno-Castilla, C. (2012) Water sorption on silica- and zeolite-supported hygroscopic salts for cooling system applications, *Energy Convers. Manag.*, **53**, pp. 219–223.

15. Gordeeva, L. G. (2013) Composite materials "salt in porous matrix": Design of adsorbents with predetermined properties, PhD thesis, Boreskov Institute of Catalysis, 347p (in Russian).

16. Sadeghlu, A., Yari, M., Mahmoudi, S. M. S., and Dizaji, H. B. (2014) Performance evaluation of zeolite 13X/$CaCl_2$ two-bed adsorption refrigeration system, *Int, J. Therm. Sci.*, **80**, pp. 76–82.

17. Gordeeva, L. G., Glaznev, I. S., Savchenko, E. V., Malakhov, V. V., and Aristov, Yu. I. (2006) Impact of phase composition on water adsorption on inorganic hybrids "salt/silica," *J. Colloid Interface Sci.*, **301**, pp. 685–691.

18. Gordeeva, L. G., Gubar, A. V., Plyasova, L. M., Malakhov, V. V., and Aristov, Yu. I. (2005) Composite water sorbents of the salt in silica gels pore type: The effect of the interaction between the salt and the silica gel surface on the chemical and phase compositions and sorption properties, *Kinetika i Kataliz (Russ. J. Kinetics Catalysis)*, **46**, pp. 736–742.

19. Ponomarenko, I. V., Glaznev, I. S., Gubar, A. V., Aristov, Yu. I., and Kirik, S. D. (2010) Synthesis and water sorption properties of a new composite "$CaCl_2$ confined to SBA-15 pores," *Micropor. Mesopor. Mater.*, **129**, pp. 243–250.

20. Barin, I. (1989) *Thermochemical Data of Pure Substances, Part 1* (VCH Verlagsgesellschaft mbH, Weinheim, Germany), p. 302.

21. Yu, N., Wang, R. Z., Lu, Z. S., and Wang, L. W. (2014) Development and

characterization of silica gel–LiCl composite sorbents for thermal energy storage, *Chem. Eng. Sci.*, **111**, pp. 73–84.

22. Sukhyy, K. M., Belyanovskaya, E. A., Kozlov, Y. N., Kolomiyets, E. V., and Sukhyy, M. P. (2014) Structure and adsorption properties of the composites 'silica gel–sodium sulphate,' obtained by sol–gel method, *Appl. Therm. Eng.*, **64**, 408–412.

23. Andersson, J. Y. (1986) Kinetic and mechanistic studies of reactions between water vapour and some solid sorbents. Department of Physical Chemistry, The Royal Institute of Technology, S-100 44, Stockholm, Sweden, pp. 27–44.

24. Isobe, H. (1929) Dehydrating substance, US Patent N 1,740,351.

25. Chi, C. W. and Wasan, D. T. (1969) Measuring the equilibrium pressure of supported and unsupported adsorbents, *Ind. Eng. Chem. Fundam.*, **8**, pp. 816–818.

26. Heiti, R. V. and Thodos, G. (1986) Energy release in the dehumidification of air using a bed of $CaCl_2$-impregnated celite, *Ind. Eng. Chem. Fundam.*, **25**, pp. 768–771.

27. Levitskii, E. A., Aristov, Yu. I., Tokarev, M. M., and Parmon, V. N. (1996) "Chemical Heat Accumulators": A new approach to accumulating low potential heat, *Sol. Energy Mater. Sol. Cells*, **44**, pp. 219–235.

28. Simonova, I. A., and Aristov, Yu. I. (2007) Dehydration of salt hydrates for storage of low temperature heat: Selection of promising reactions and nanotailoring of innovative materials, *Alternative Energy Ecology*, **10**, pp. 62–69.

29. Garner, W. E. and Souton, W. (1935) Nucleus formation on crystals of nickel sulphate heptahydrate, *J. Chem. Soc.*, pp. 1705–1709.

30. Lyakhov, N. Z. and Boldyryev, V. V. (1972) Mechanism and kinetics of dehydration of crystalline hydrates, *Russ. Chem. Rev.*, **XLI**, pp. 1960–1977.

31. Rae, W. N. (1916) A period of induction in the dehydration of some crystalline hydrates, *J. Chem. Soc.*, **109**, pp. 1229–1236.

32. Aristov, Yu. I., Tokarev, M. M., Cacciola, G., and Restuccia, G. (1996) Selective water sorbents for multiple applications: 1. $CaCl_2$ confined in mesopores of the silica gel: Sorption properties, *React. Kinet. Cat. Lett.*, **59**, pp. 325–334.

33. Ovoschnikov, D. S. (2009) Master's Thesis, Boreskov Institute of Catalysis.

34. Galwey, A. and Brown, M. (1999) *Thermal Decomposition of Ionic Solids* (Elsevier Science, Amsterdam).

35. Kaerger, J. and Ruthven, D. M. (1992) *Diffusion in Zeolites and Other Microporous Solids* (John Wiley and Sons, New York), 433p.

36. Aristov, Yu. I., Tokarev, M. M., Di Marco, G., Cacciola, G., Rectuccia, G., and Parmon, V. N. (1997) Properties of the system "calcium chloride–water" confined in pores of the silica gel: Equilibria "gas-condensed state" and "melting-solidification," *Rus. J. Phys. Chem.*, **71**, pp. 253–258.

37. Gordeeva, L. G., Glaznev, I. S., Malakhov, V. V., and Aristov, Yu. I. (2003) Influence of calcium chloride interaction with silica surface on phase composition and sorption properties of dispersed salt, *Russ. J. Phys. Chem.*, **77**, pp. 1843–1847.

38. Aristov, Yu. I. (2007) New family of materials for adsorption cooling: Material scientist approach, *J. Eng. Thermophys.*, **16**, pp. 63–72.

39. Gordeeva, L. G., Resticcia, G., Cacciola, G., and Aristov, Yu. I. (1998) Selective water sorbents for multiple applications: 5. LiBr confined in mesopores of silica gel: Sorption properties, *React. Kinet. Cat. Lett.*, **63**, pp. 81–88.

40. Gurvich, B. M., Karimov, R. R., and Mezheritskii, S. M. (1986) Heat conductivity of $CaCl_2$ aqueous solutions, *Rus. J. Appl. Chem.*, **59**, pp. 2692–2696.

41. Aristov, Yu. I., Restuccia, G., Cacciola, G., and Parmon, V. N. (2002) A family of new working materials for solid sorption air conditioning systems, *Appl. Therm. Eng.*, **22**, pp. 191–204.

42. Sinke, G. C., Mossner, E. H., and Curnutt, J. L. (1985) Enthalpies of solution and solubilities of calcium chloride and its lower hydrates, *J. Chem. Thermodyn.*, **17**, pp. 893–899.

43. Pankrat'ev, Yu. D., Tokarev, M. M., and Aristov, Yu. I. (2001) Heats of water sorption on silica gel containing $CaCl_2$ and LiBr, *Russ. J. Phys. Chem.*, **75**, pp. 806–810.

44. Rouquerol, F., Partyka, S., and Rouquerol, J. (1972) In: Thermochimie, Colloques Internationaux du CNRS, no. 201, Editions du CNRS, Paris, p. 547.

45. Dubinin, M. M. (1960) The potential theory of adsorption of gases and vapors for adsorbents with energetically non-uniform surface, *Chem. Rev.*, **60**, pp. 235–266.

46. Aristov, Yu. I. (2012) Adsorptive transformation of heat: Principles of construction of adsorbents database, *Appl. Therm. Eng.*, **42**, pp. 18–24.

47. Prokopiev, S. I. and Aristov, Yu. I. (2000) Concentrated aqueous electrolyte solutions, *J. Solution Chem.*, **29**, pp. 633–649.

48. Aristov, Yu. I., Sharonov, V. E., and Tokarev, M. M. (2008) Universal relation between the boundary temperatures of a basic cycle of sorption heat machines, *Chem. Eng. Sci.*, **63**, pp. 2907–2912.

49. Debye, P. and Huckel, E. (1923) Zur Theorie der Elektrolyte I. Gefrierpunktsernie drigung und verwandte Erscheinungen, *Phys. Z.*, **24**, 185.

50. Aristov, Yu. I. (2004) Selective water sorbents for air drying: From the lab to the industry, *Cat. Industry*, **6**, pp. 36–41.

Chapter 6

Composite Sorbents of Water Vapor: Effect of a Host Matrix

In this book, we demonstrate that the sorption capacity of composites "salt in porous matrix" (CSPMs) can be much superior to that of the host matrix, and this enhancement is caused basically by the interaction of a sorptive with the confined salt. In most cases, the salt can, therefore, be regarded as an active component of the composite. In Chapter 5, we analyzed how the salt's nature affects the CSPM sorption properties and, in the most part, considered various salts in mesoporous matrices, such as a silica gel KSK. In this chapter, we comprehensively analyze how changes in the salt's sorption properties caused by its confinement to a host matrix depend on the matrix. We basically investigate one salt (calcium chloride, $CaCl_2$) in a variety of host matrices with different chemical nature and pore structure. The first $CaCl_2$-based composite ("$CaCl_2$ in the mesopores of KSK silica," SWS-1L) was synthesized and studied at the Boreskov Institute of Catalysis (BIC) [1, 2]. After that, various composites "$CaCl_2$/matrix" have been synthesized and studied in many laboratories all over the world. Several typical composites are listed in Table 6.1, which is well apart from completeness.

Nanocomposite Sorbents for Multiple Applications
Yu. I. Aristov
Copyright © 2020 Jenny Stanford Publishing Pte. Ltd.
ISBN 978-981-4267-50-2 (Hardcover), 978-981-4303-15-6 (eBook)
www.jennystanford.com

Table 6.1 CaCl$_2$–based composites, host matrices used and their texture characteristics

Composite	Host matrix type	Pore volume (cm^3/g)	Specific area (m^2/g)	Pore size (nm)	Ref.
CaCl$_2$/AC	Activated carbon	1.3	1820	0.9	[3]
CaCl$_2$/zeolite 13X	Zeolite 13X	0.30	730	1.0	[4]
CaCl$_2$/FSM16	FSM16	0.85	990	2.8	[5]
CaCl$_2$/silica	Silica SPCW	0.35	600–850	1.5–3.0	[6]
CaCl$_2$/silica	Silica KSM	0.30	600	3.5	[7]
CaCl$_2$/MCM-41	MCM-41	1.1	1050	4.1	[8]
CaCl$_2$/xerogel	xerogel SiO$_2$	0.45	275	5	[9]
CaCl$_2$/alumina	A1	0.70	180	6–8	[10]
CaCl$_2$/carbon	Sibunit	0.90	450	7	[10]
CaCl$_2$/WSS	Wakkanai shale	0.90	450	7	[11]
CaCl$_2$/silica	Silica gel	1.26	312	8.2	[4]
CaCl$_2$/SBA-15(1)	SBA-15exp	0.75	520	6.5	[12]
CaCl$_2$/SBA-15(2)	SBA-15exp	0.66	483	8.1	[13]
CaCl$_2$/SBA-15(3)	SBA-15exp	0.79	486	11.8	[13]
CaCl$_2$/silica	Silica KSK	1.0	350	15	[1]
CaCl$_2$/silica	Silica gel Solvay	—	240	—	[14]
CaCl$_2$/MWCNT	Carbon nanotubes	3.0	270	8, 40–50	[15]
CaCl$_2$/MWCNT	Carbon nanotubes	> 1.3	140–170	10–20	[16]
CaCl$_2$/opal	Synthetic opal	0.28	46	50	[17]
CaCl$_2$/attapulgite	Attapulgite	—	98	64	[18]
CaCl$_2$/vermiculite	KVK	3.5	1.5	1000	[19]
CaCl$_2$/vermiculite	Micafil	4.1	15	3700	[20]

Physical effects caused by dividing the salt into tiny pieces can be different in host matrices with large (vermiculite, carbon nanotubes, attapulgite), middle (silica KSK, SBA-15, alumina), and small (activated carbons (ACs), silica gels KSM and SPCW, zeolites, xerogel SiO_2, MCM-41) pores.

For a precise study of the effect of pore size, it would be convenient to insert a guest salt into host matrices with a controllable and narrow pore size distribution (PSD). One can expect that this may lead to more narrow size distribution of the salt nanoparticles inside the pores. This exciting opportunity has been realized by using ordered silicates of SBA-15 type with mono-dispersed pores of 7–12 nm size. Porous oxides with hydrophobic (carbon Sibunit) and hydrophilic (porous oxides) surfaces as well as Brønsted (silica gel) and Lewis (alumina) surface acidity will be used to study chemical effects originated from interaction between the salt and the matrix.

Other highly hygroscopic salts discussed in this chapter are calcium nitrate $(Ca(NO_3)_2)$ and lithium nitrate $(LiNO_3)$, which were confined to a set of silica gels with different pore size [21, 22]. This investigation was aimed at elucidating the effect of the matrix pore size on the sorption properties of the confined nitrates.

Finally, we analyze a general pattern on how the hydration/dehydration temperature is affected by the porous structure of the matrix and a link between this temperature and the contribution of the surface energy to the total Gibbs energy of the "salt–hydrate" system.

6.1 "Calcium Chloride–Water" System in Bulk

We first briefly describe the physicochemical properties of the bulk "$CaCl_2$–water" system, which are necessary for analyzing experimental data.

As mentioned in Chapter 1, calcium chloride was the first salt suggested for use in composites for gas dehumidification, gas masks, and heat storage [23–25]. $CaCl_2$ is a cheap and easily available salt. Its properties make it useful in a large number of applications such as de-icing pavements and driveways, road bed stabilization and dust control, accelerating the set time of concrete, refrigeration, desiccation, and many other uses [26]. Properties of a bulk "calcium chloride–water" system (Table 6.2) have been comprehensively studied and described, for instance, in Refs. [26–32]. The salt is highly hygroscopic and liberates large amounts of heat during the formation of solid hydrates and aqueous solution.

Table 6.2 Properties of $CaCl_2$ hydrates and eutectic solution (29.6 wt.%)

n	0	0.33	1	2	4	6	eut
Molecular weight (g/mol)	110.99	116.99	129.0	147.02	183.05	219.1	—
Density (g/cm³)	2.16	—	2.24	1.85	1.83	1.71	1.28
Melting point (°C)	772	—	187	176	45.1	30.1	−49.8
Heat of formation, 25°C (kJ/mol)	−795.4	—	−1109	−1403	−2010	−2608	—
Heat of solution in H_2O (kJ/mol)	−81.85	—	−52.16	−44.05	−10.8	18.8	—
Heat of fusion (kJ/g)	28.5	—	17.3	12.9	30.6	43.4	—
Specific heat, J/(g K)	0.67	—	0.84	1.17	1.35	1.66	—

Source: Reprinted from Ref. [26], with permission from John Wiley and Sons, Copyright 2014.

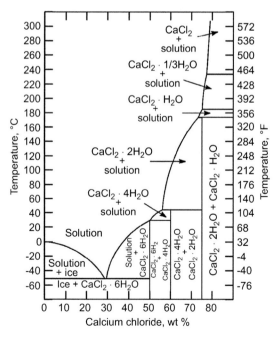

Figure 6.1 The phase diagram of the bulk "$CaCl_2$–water" system. Reprinted from Ref. [26], with permission from John Wiley and Sons, Copyright 2014.

Although calcium chloride is extremely soluble in water, the formation of the solid phase is possible under certain "temperature–concentration" conditions that are defined by the solubility diagram shown in Fig. 6.1. Five bulk hydrates $CaCl_2 \cdot nH_2O$ with n = 0.33, 1, 2, 4, and 6 have been identified and described (see Table 6.2) [26, 30]. Tetrahydrate ($CaCl_2 \cdot 4H_2O$) was found in three crystallographic forms: α, β, and γ [28, 33]. Crystallographic data have been reported for all known hydrates. X-ray diffraction (XRD)/scattering study was performed for the hydrate melts and concentrated solutions [33].

The equilibrium pressure over the salt hydrates (except for $CaCl_2 \cdot 0.33H_2O$) and solutions was measured as early as in 19th century [28, 32] and confirmed many times after that. Accurate data are available on solubility boundary, density, surface tension, dynamic viscosity, thermal conductivity, specific thermal capacity, and differential enthalpy of dilution [32]. Thus, the "calcium chloride–water" system is very convenient to make a comparison of properties in bulk and confined states as its bulk characteristics are well known. The first useful comparison is made in Fig. 6.2. It demonstrates that the hydration from $CaCl_2 \cdot 0.33H_2O$ to $CaCl_2 \cdot 2H_2O$ occurs in the bulk at a lower temperature (63°C) than in the silica mesopores (85°C), hence the confined salt differs from *the bulk one, gaining more capability of sorbing water vapor or becoming more hydrophilic.*

As $CaCl_2$ deliquescence is observed at RH > 0.2 [28] with the formation of a corrosive solution, it is convenient to isolate the salt in a porous matrix, which resulted in more technological uses, e.g., in gas drying and heat storage [23–25, 34–36]. Variation in the salt's properties due to its confinement to various porous solids is the main subject of this chapter. Composites based on calcium chloride have been and will be considered in other chapters of this book as well.

All the host matrices are conditionally divided into three groups depending on the average pore size: with large (>20–50 nm), middle (7–20 nm), and small (<7 nm) pores. We expected that this correlates with changes in the sorption properties of the confined salt: In the large pores, it is similar to the bulk salt, but in the small pores, it is very different, and in the middle pores, the salt exhibits intermediate properties.

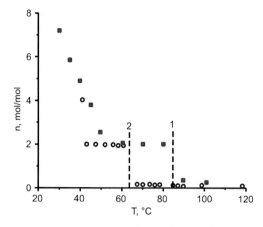

Figure 6.2 Isobars of water sorption by CaCl$_2$ in bulk (o) and confined (■) to the mesopores (D_{av} = 15 nm) of silica KSK (SWS-1L) at P_{H_2O} = 12.3 mbar. Dashed lines indicate transitions from n = 1/3 to n = 2: in the pores (1) and in the bulk (2).

6.2 CaCl$_2$ in Matrices with Large Pores

6.2.1 CaCl$_2$/Vermiculite (SWS-1V)

Vermiculite is the mineralogical name given to hydrated laminar magnesium–aluminum–ironsilicate, which resembles mica in appearance [37]. When subjected to heat, it has the property of exfoliating or expanding, due to the generation of interlaminar steam.

The expanded vermiculite KVK is found to have a wide PSD within 10–40,000 nm range with the preferable pore size of about 1000 nm [19]. According to TEM data, the expanded vermiculite consists of irregular particles of micrometer size with a flat surface. CaCl$_2$ was inserted inside the vermiculite pores by dry impregnation. As the vermiculite pore volume is rather large (1.8 cm^3/g), the salt content amounted to 57 wt.% (composite SWS-1V).

CaCl₂ in Matrices with Large Pores | 113

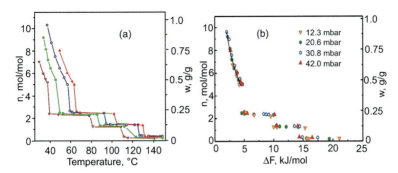

Figure 6.3 Isobars of water desorption for SWS-1V presented as a function of temperature (a) and adsorption potential (b): ▽ — 12.3 mbar, ● — 20.6 mbar, ○ — 30.8 mbar, ▲ — 42.0 mbar. Solid line in (b) illustrates the equilibrium for the salt aqueous solution in bulk [40]. Figure (a) reprinted by permission from Springer Nature from Ref. [19], Copyright 2000.

Isobars of water desorption for this composite are presented in Fig. 6.3 as dependences on the temperature and on the adsorption potential ΔF at fixed vapor pressure P_{H_2O} ranging from 12.3 to 42 mbar. Each curve has two pronounced plateaus corresponding to $n \approx 1$ and $n \approx 2$, which are due to the formation of the salt mono- and dihydrates:

$$CaCl_2 + H_2O = CaCl_2 \cdot H_2O \quad (6.1)$$

$$CaCl_2 \cdot H_2O + H_2O = CaCl_2 \cdot 2H_2O \quad (6.2)$$

whose sorption equilibrium is mono-variant (see Sec. 2.5). Further hydration step should be

$$CaCl_2 \cdot 2H_2O + 2H_2O = CaCl_2 \cdot 4H_2O \quad (6.3)$$

However, no plateau corresponding to the salt tetrahydrate CaCl₂·4H₂O is observed and the uptake gradually increases at a lower temperature. It depends on both temperature and pressure, and the "CaCl₂–water" system becomes bi-variant (Fig. 6.3). Crystalline hydrate CaCl₂·4H₂O is not detected by the XRD method probably because its melting temperature T_m in bulk is 45.1°C [26]; hence it is liquid under conditions of most of these experiments. Calcium chloride hexahydrate is not found in the confined system all the more (T_m = 30.1°C [26]). The dehydration route represents a set of successive transformations of the salt solution to the di-, monohydrate, and anhydrous CaCl$_2$ (sol→2→1→0) (Fig. 6.3). The lowest hydrate CaCl$_2$·0.33H$_2$O is not formed during the desorption run.

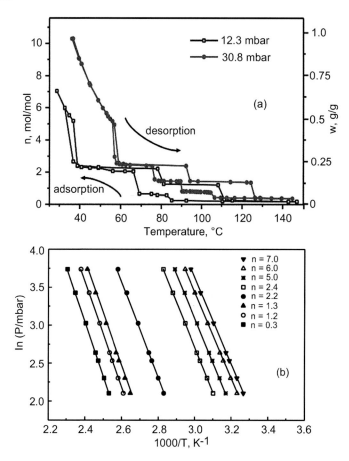

Figure 6.4 Isobars (a) and isosters (b) of water sorption on SWS-1V. Reprinted by permission from Springer Nature from Ref. [19], Copyright 2000.

The hydration route differs from the dehydration one in the following ways (Fig. 6.4a):
  a. A plateau at $n \approx 0.5$–0.6 ($w \approx 5$–6 wt.%) was observed, which indicates the formation of the lowest hydrate CaCl$_2$·0.33H$_2$O. The sorption excess over $n = 1/3$ is attributed to a strong

water adsorption by the vermiculite so that 2–3 wt.% of water are retained by the matrix even at T = 100–150°C (Fig. 6.3).

b. The successive formation of the salt hydrates with n = 1/3, 1, 2 and the salt solution is observed at higher pressures (20.6 and 30.8 mbar). At P_{H_2O} = 12.6 mbar, the monohydrate does not form, and the reduced route (0→1/3→2→sol) is observed. The monohydrate $CaCl_2 \cdot H_2O$ is known to be the least stable hydrate of this salt as it exists in a quite narrow (P, T) range (Fig. 6.1). The reduced order of transformations was also observed for the bulk salt at P_{H_2O} = 8.4 mbar, while at P_{H_2O} > 12.3 mbar, the full route (0→1/3→1→2→sol) was detected. Hence, the hydration routes of the confined salt repeat the bulk ones; however, the confined monohydrate is less stable and its formation in the vermiculite pores is observed at higher vapor pressures.

A large hysteresis is found for the synthesis and decomposition of mono- and dihydrates (Fig. 6.4a): A temperature difference of 15–35°C indicates a strong inhibition of the formation reaction. This is quite typical for gas–solid hydration reactions in the bulk [38, 39].

Table 6.3 Isosteric heat of water desorption ΔH_{is} (kJ/mol) as a function of the water content for the $CaCl_2$/vermiculite composite

n	0.33	1	1.2	1.3	2	2.2	2.4	4	6	7
ΔH_{is}	78.0	—	76.3	46.6	46.5	46.9	39.9	—	41.9	39.4

Source: Reprinted by permission from Springer Nature from Ref. [19], Copyright 2000.

Isosters of water sorption on $CaCl_2$/vermiculite composite are straight lines $\ln(P_{H_2O}) = A(w) + B(w)/T$ (Fig. 6.4b) [19]. The slope depends on the water content w and gives the isosteric water sorption heat $\Delta H_{is}(w) = B(w) \cdot R$, where R is the gas constant (Table 6.3). For $n > 2$, the ΔH_{is} value (43.9 ± 1.3 kJ/mol) appears to be close to the heat of water evaporation from aqueous $CaCl_2$ solutions [40]. A significant increase in ΔH_{is} at $n \leq 2$ is caused by the formation of solid hydrates where water molecules are bound stronger.

If water sorption is presented as a function of the adsorption potential ΔF, all experimental data satisfactorily follow the same curve, which can be considered a temperature-independent curve of the water sorption (Fig. 6.3b) [19]. Hence, there is a one-to-one correspondence between the sorbed water amount and the ΔF value. One can observe the plateaus corresponding to $n \approx 0.5$, 1, and 2 as well as a shoulder at $n \approx 5$. At $\Delta F > 5$ kJ/mol, the temperature-independent curve is smooth and essentially coincides with the proper curve for the bulk aqueous solution of calcium chloride [40]. It means that the solution confinement to the mesopores of the KVK vermiculite does not change its water sorption properties compared to the bulk solution.

It can be concluded that the sorption properties of the $CaCl_2$ hydrates confined to the large vermiculite pores are similar to the bulk ones, both concerning the transition temperature and the synthesis–decomposition hysteresis. Thus, sorption by the composite can be considered a linear addition of the sorption by the inserted salt and the host matrix, the first one being predominant. S-shaped hydration/dehydration kinetic curves [19], which will be reported in Chapter 13, give one more indication of the "bulk-like" behavior of calcium chloride confined to the large vermiculite pores. Confinement of the salt to these pores prevents agglomeration of the salt and assists in accommodating the swelling salt as well as retaining the salt solution formed at increasing vapor pressure.

6.2.2 CaCl$_2$/MWCNT

Multi-wall carbon nanotube (MWCNT) is a new promising material for advanced adsorption technologies due to its extremely large and variable pore volume and excellent thermal conductivity [41]. It was first suggested as a host matrix for composite sorbents of water and ammonia in Refs. [15, 16, 42]. In Ref. [15], the MWCNT is composed of twisted tube tangles with size exceeding 1 μm. The MWCNT is characterized by extremely high porosity: the specific pore volume is $V_p = 3.0$ cm^3/g, surface area is $S_{sp} = 270$ m^2/g (Table 6.4), and two kinds of pores. The former corresponds to the internal space of the tubes of 7.8 nm diameter. The latter are formed by voids of 40–50 nm between the tubes and have the dominant contribution to the pore volume.

CaCl₂ was introduced into the pores by a wet impregnation method: A dry MWCNT sample was immersed into an alcohol-aqueous salt solution for 1 h [15]. After that, the solution excess was removed by filtration with a water-jet air pump. The sample was rinsed out with ethanol and then dried at 160°C for 12 h. Due to the large pore volume, the salt content C_{salt} in the synthesized composite reached 53 wt.%. The XRD analysis showed that the confined salt is well crystallized. The size of coherent scattering regions equals 20–35 nm; hence the salt is located in the voids between nanotubes rather than inside these tubes.

Table 6.4 The composition and textural characteristics of the neat MWCNT and synthesized composite

Sample	Salt content C_{salt} (wt.%)	S_{sp} (m²/g)	V_p (cm³/g)	V_{calc} (cm³/g)
Neat MWCNT	0	270	3.0	3.0
CaCl₂/MWCNT	53	75	0.9	1.2

Isotherms of water sorption on the composite CaCl₂/MWCNT are stepwise curves with sharp steps corresponding to the formation of CaCl₂·nH₂O hydrates and long plateaus at n = 1 and 2 (Fig. 6.5a), position of which agrees with the literature data for the bulk salt. What is different from the bulk salt/hydrates is that the adsorption–desorption hysteresis in the composite is minor. The water sorption capacity reaches 1 g/g of composite, which is promising for adsorptive storage of low-temperature heat [15]. Very long plateaus at $n ≈ 1$ and 2 (Fig. 6.5b) are also revealed for three similar composites CaCl₂/MWCNT [17]. This indicates the "bulk-like" behavior of the salt confined to the matrix with large pores. Very steep ("bulk-like") isobars were reported for methanol sorption on the (LiCl, LiBr)/MWCNT composites [15] and for ammonia sorption on the CaCl₂/MWCNT [42].

Mixing CaCl₂ with MWCNT prevents agglomeration of the salt and accommodates the swelling salt upon hydration. The hydrophobic matrix retains the salt solution formed at high vapor pressure worse than hydrophilic matrices, like vermiculite. The retention is, however, sufficient to avoid the run-out of solution. On the other hand, water desorption from the hydrophobic pores can be easier [16].

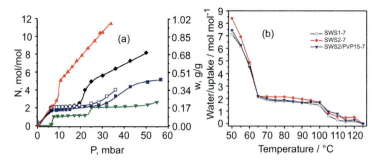

Figure 6.5 The CaCl$_2$/MWCNT composites: (a) isotherms of water sorption [15] at T = 37 (▲), 55 (♦), 60 (□, ■), and 70°C (▼) and (b) isobars of water sorption (Open symbols – desorption). Figure (a) reprinted from Ref. [15], Copyright 2016, with permission from Elsevier, and Figure (b) reprinted from Ref. [16], Copyright 2015, with permission from RSC Publishing.

6.2.3 CaCl$_2$/Attapulgite

Stepwise transitions and long plateaus between them were also observed on the isotherms of water sorption on the CaCl$_2$/attapulgite composite (not presented here, see in Ref. [18]). These are due to the formation of salt hydrates: the first isotherm step of the CaCl$_2$ dihydrate, and the second one of CaCl$_2$·4H$_2$O, which was confirmed by XRD measurements.

6.3 CaCl$_2$ in Matrices with Middle Size Pores

SWS-1L is the most studied composite based on calcium chloride confined to middle size pores as comprehensively considered in Chapter 5. Here we use those data for comparing SWS-1L with other CSPMs based on a mesoporous matrix (SBA-15, carbon Sibunit, and alumina oxide). The SBA-based composites exemplify materials with almost mono-sized mesopores, which allow the effect of pore size to be accurately studied. The carbon and alumina matrices have a very similar pore structure; however, with the pore walls being, respectively, hydrophobic and hydrophilic, this helps to elucidate the effect of the surface chemical nature.

6.3.1 CaCl$_2$/SBA-15

For a precise study of the pore size effect on the sorption properties of CSPMs, it would be convenient to insert a guest salt into the host matrices of the same chemical nature but different pore structure with a controllable and narrow PSD. One can expect that the latter may lead to a more narrow size distribution of the salt nanoparticles inside the pores. This exciting opportunity can be realized by using ordered mesoporous silicates of the SBA-15 type with monodispersed pores ranged, in principle, between 5 and 30 nm size [43].

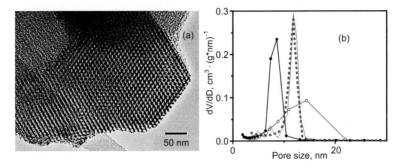

Figure 6.6 (a) TEM pictures of SBA-15; (b) pore size distribution (BJH, adsorption) of the SBA powder (■) and pressed tablet (11.8 nm) (dashed line), SBA (8.1 nm) (●) and KSK (○). Figure (a) reprinted from Ref. [44], Copyright 2013, and Figure (b) reprinted from from Ref. [13], Copyright 2011, with permission from Elsevier.

Table 6.5 Texture properties of s74, s89, and silica KSK as well as the salt content C in the appropriate composites

Material	D_{av} (nm)	V_p (cm^3/g)	S_{BET} (m^2/g)	C (wt.%)	CSD size (nm)
s74	8.1	0.66	483	28.2	10
s89	11.8	0.69	259	29.5	10
KSK	15.0	1.00	350	33.7	20

Two composites CaCl$_2$/SBA-15 (s74 and s89) with very narrow PSDs and average pore sizes of D_{av} = 8.1 and 11.8 nm (Fig. 6.6b, Table 6.5) were synthesized and studied keeping in mind their application for adsorption chillers [13]. The commercial silica gel KSK contains mesopores having sizes distributed over a wide range of 3 to 22 nm, which gives 15 nm in average (Fig. 6.6b). For the three matrices, the specific pore volume was 0.7–1.0 cm^3/g; hence the pore space was sufficient for inserting a large salt quantity (28–34 wt.%.) as well as for accepting a large water uptake (up to 0.5–0.7 g/g). The XRD study reveals that the size of coherent scattering domains (CSDs) is 10–20 nm, which confirms the localization of the salt inside the pores.

This gradual tuning of pore size allows a systematic investigation of its effect on the sorption properties of CaCl$_2$/silicate composites. Isotherms of water sorption on SBA composites have two segments with a steep increase in the uptake (Fig. 6.7a). The first one is observed at the relative pressure $P/P_0 \approx 0.02$ and corresponds to the formation of dihydrate CaCl$_2$·2H$_2$O with mono-variant equilibrium, so that the transformation occurs at almost fixed pressure. The dihydrate is formed due to the strong water bonding and can be characterized by the water affinity corresponding to the large adsorption potential $\Delta F \approx 10$ kJ/mol. The dihydrate forms in the bulk at $P/P_0 \approx 0.071$ (Fig. 6.7a), so that the affinity of the bulk salt is much smaller ($\Delta F \approx 6.5$ kJ/mol). As a result, the dew point over the confined CaCl$_2$ is much lower than over the bulk one: −28°C versus −15°C.

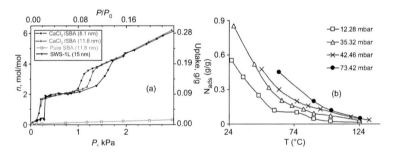

Figure 6.7 (a) Isotherms of water vapor sorption on the composites CaCl$_2$/SBA with various pore sizes, SWS-1L, pure SBA, and bulk CaCl$_2$. T = 50°C. [13]). (b) Water sorption isobars at different pressures on SCa33 (8 nm). Figure (a) re-plotted from Ref. [13], Copyright 2011, and Figure (b) reprinted from Ref. [4], Copyright 2012, with permission from Elsevier.

The second rise in sorption is observed at higher P/P_0 and attributed to further salt hydration with the formation of an aqueous salt solution. Interestingly, this hydration pressure is certainly lower in smaller pores, namely, 10–11 and 12–13 mbar for s74 and s89 as hosts. Bulk CaCl$_2$ dihydrate transforms to a tetrahydrate at a higher pressure of 20.8 mbar. For comparison, SWS-1L [7] is hydrated at intermediate pressures of 14 to 17 mbar (Fig. 6.7), which correlates with its larger pore size of 15 nm (Table 6.5). Its sorption isotherm is smoother than that for the SBA-based composites, that is, in line with its wider PSD. The composite $CaCl_2$/SBA (11.8 nm) sorbs at least one order of magnitude more water vapor than the pure SBA (11.8 nm) (Fig. 6.7a). This gives one more proof that the salt is the main sorbing component of CSPMs.

Thus, one can conclude that using host matrices with uniform pore dimension provides an effective tool for controllable tuning of the solvation temperature of CSPMs. Another tuning opportunity is associated with the variation in the salt content inside the pores. Influence of the salt content is another tool to manage the adsorption equilibrium in the pores, as seen in Fig. 4.10.

Many other commercial silica gels were used to synthesize "$CaCl_2$/silica" composites. When the pore size decreases from 15 to 5–7 nm, the water sorption isobars/isotherms become smoother, although they still partially preserve features characteristic of a bulk "$CaCl_2$–water" system, namely, long plateaus and steep transitions between them (Fig. 6.2). The smoothing is clearly seen in Figs. 6.7b and 5.8a: these samples differ only in pore size (15 nm versus 8 nm). In smaller pores, the plateau corresponding to $n = 2$ was observed only at low vapor pressure. Very smooth curves of water sorption were found for $CaCl_2$ in small silica pores (3–6 nm) (see Sec. 6.4).

6.3.2 CaCl₂/Alumina

Aluminas are a family of aluminum oxides (Al_2O_3), which can take many forms. As a general term, "alumina" refers to the lower temperature transition alumina forms. They are commonly used as adsorbents, desiccants, and catalysts [45]. Being produced in large quantity, alumina is available and relatively cheap. This porous matrix is more stable, both mechanically and hydrothermally, than silica gels.

The composite $CaCl_2$/alumina (SWS-1A) was prepared by dry impregnation of a commercial alumina A1 (Russia) with an aqueous $CaCl_2$ solution at 25°C [10]. The matrix has a wide PSD with an average pore size of ca. 8 nm. The salt content in the dry composite was 29 wt.%. The sample was dried at 150°C for 10 h. The size of $CaCl_2$ crystallites estimated by XRD was 6 nm, which confirms the localization of salt inside the matrix pores.

Isobars of water sorption on SWS-1A are smooth curves having no plateaus, steps, or inflections (Fig. 6.8a), which would resemble the mono-variant equilibrium in bulk CaCl$_2$. The maximal uptake reaches 0.5 g/g, and gradually reduces with rising temperature, thus showing purely bi-variant behavior. At 60–90°C, 80% of the sorbed water is removed, whereas for complete desorption, heating up to 100–140°C is needed. At small uptakes ($n < 4$ or $P/P_0 < 0.2$), water bonding in the composite is stronger than in the bulk salt, as clearly seen from the SWS-1A sorption isotherm plotted as $n(P/P_0)$ (Fig. 6.9a).

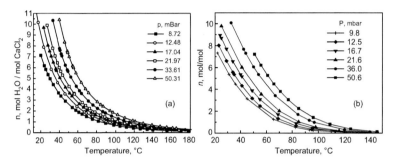

Figure 6.8 Isobars of water sorption on SWS-1A (a) and SWS-1C (b).

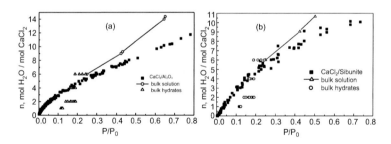

Figure 6.9 Isotherms of water sorption on SWS-1A (a) and SWS-1C (b).

This isotherm does not coincide with the isotherm for the bulk "CaCl$_2$–water" system either at high uptakes: at $n > 6$ (or $P/P_0 > 0.2$), the SWS-1A isotherm lies below the isotherm for the bulk solution [40]. It means that confinement of the solution to the alumina pores decreases its ability to bind water. Indeed, at a fixed solution concentration, the relative vapor pressure above the solution in micropores is higher than above the bulk solution. This looks surprising because the vapor pressure of wetting liquids decreases in micropores due to the capillary effect [46]. It is reasonable to suppose that the observed increment in vapor pressure is caused by a significant change in the thermodynamic properties of the solution due to its confinement to the matrix. Possible reasons of this effect are discussed in Sec. 6.6.2.

Based on the experimentally measured isobars, a family of sorption isosters is plotted. They are straight lines in a $\ln(P_{H_2O})$ versus $1/T$ presentation (Fig. 6.10a): $\ln(P_{H_2O}) = A(n) + B(n)/T$. The coefficient $B(n)$ determines the isosteric heat of the water sorption: $\Delta H_{is}(n) = B(n) \cdot R$ (Table 6.6). This heat tends to slightly decrease with the increase in water content and, at large n, approaches the evaporation heat of a bulk aqueous CaCl$_2$ solution [32, 40].

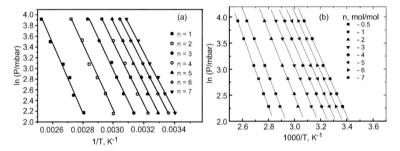

Figure 6.10 Isosters of water sorption on SWS-1A (a) and SWS-1C (b).

Thus, the confinement of calcium chloride to the alumina pores strongly influences its sorption properties in the whole range of water uptakes (or P/P_o).

6.3.3 CaCl$_2$/Carbon Sibunit

This commercial carbon has mesopores with an average size of about 7 nm and a pore volume of 0.63 cm^3/g, which are very similar to alumina A1 considered earlier. The composite CaCl$_2$/Sibunit (SWS-1C) was prepared by dry impregnation of the matrix with an aqueous CaCl$_2$ solution [10]. The salt content was 34 wt.%. The size of CaCl$_2$ crystallites estimated from XRD is 5.5 nm, which corresponds well to the average pore size of the matrix.

Similar to SWS-1A, isobars of water sorption on SWS-1C are smooth curves without plateaus, steps, or inflections (Fig. 6.8b), so

Composite Sorbents of Water Vapor

that the equilibrium is bi-variant. They differ from the bulk "CaCl$_2$–water" system in the whole range of experimental conditions as seen from the universal sorption isotherm plotted in the n versus P/P_o presentation (Fig. 6.9b). It lies above the points for bulk CaCl$_2$·H$_2$O and CaCl$_2$·2H$_2$O, and below bulk CaCl$_2$·6H$_2$O and the salt solution. The sorption isosters are straight lines $\ln(P_{H_2O}) = A(n) + B(n)/T$ (Fig. 6.10b), and the isosteric heat of water sorption is almost identical with that for SWS-1A (Table 6.6).

Table 6.6 Isosteric heat of water sorption ΔH_{is} on SWS-1A and SWS-1C at different water uptake n

n (mol/mol)		0.5	1	2	3	4	5	6	7
ΔH_{is} (kJ/mol)	SWS-1A	—	53.2	50.9	49.3	48.2	48.7	46.4	46.7
	SWS-1C	57.6	53.4	51.8	47.9	47.0	47.0	47.1	46.2

Thus, composite SWS-1C sorbs water vapor in a way that resembles in many aspects SWS-1A. On the one hand, it can be expected because the two host matrices have similar average pore size and specific volume (Table 6.1). On the other hand, one can expect essential differences because the Sibunit surface is hydrophobic, whereas the alumina surface is highly hydrophilic. The chemical nature of the host matrix seems to be of secondary importance for water sorption, probably because the interactions of water molecules with the salt in hydrates and solution are overwhelmingly stronger than with the host surface.

6.4 CaCl$_2$ in Matrices with Small Pores

Most of the commercial adsorbents have very small pores of a few nanometers [47]. Among them are ACs [47], regular density (RD) silica gels [48], zeolites [49], aluminophosphates [50], etc. These matrices are less used as hosts for CSPMs because in small pores, the interaction of sorptive molecules with the wall is very strong, so that the affinity is high and the salt addition is not as efficient as in larger pores. Nevertheless, to make the picture complete, let us briefly consider several examples.

6.4.1 CaCl$_2$/RD Silica KSM

Synthetic *silica gels* are famous commercial products formed by aggregation of primary silica particles, whose size and packing determine the specific surface area, pore size, and volume [48]. The low-density silica gel KSK was used as a host for SWS-1L, as considered in Sec. 5.3. RD silica gels have much smaller pores (2–3 nm) as well as pore volume (0.3–0.4 cm^3/g) and higher surface area (600–800 m^2/g). Their hydrated surface is highly hydrophilic, which determines their wide use as drying agents.

RD silica gel of the KSM type (Reakhim, former USSR) has the BET area S_{sp} = 600 m^2/g, pore volume V_{pore} = 0.3 cm^3/g, and average pore diameter D_{av} = 3.5 nm. Its porous structure is similar to that of other famous RD silicas, such as Fuji silica types RD and A, Davisil 30, BASF KC-Trockenperlen N, Engelhard Sorbead R, etc. The silica pores were filled with a 40 wt.% aqueous solution of CaCl$_2$. Then it was dried at 200°C until their weight remained constant. The calcium chloride content in the anhydrous composite was 21.7 wt.% (SWS-1S).

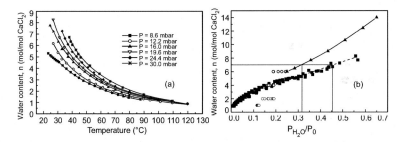

Figure 6.11 Isobars (a) and isotherm (b) of water sorption on SWS-1S. Reprinted by permission from Springer Nature from Ref. [7], Copyright 1996.

Isobars of water sorption on SWS-1S (Fig. 6.11a) are quite different from those measured for SWS-1L [1]. No plateaus, steps, and inflections resembling the solid hydrate formation are recorded. Water sorption decreases monotonously with the rise in temperature, showing a bi-variant type of sorption equilibrium even at a salt concentration that corresponds to 1 molecule of sorbed water per 1 CaCl$_2$ molecule or 86 wt.%. Contrary to a mono-variant gas–solid equilibrium, the bi-variant equilibrium is typical for liquid salt solutions. Note that in the bulk CaCl$_2$/H$_2$O system, the melting temperature of 86 wt.% solution is at least > 350°C (Fig. 6.1).

Table 6.7 Isosteric heat of water sorption, ΔH_{is}, as a function of water content

n	2	3	4	5	6	7
w (g/g)	0.07	0.105	0.140	0.175	0.210	0.245
ΔH_{is} (kJ/mol)	69.0	62.3	56.0	50.2	46.0	43.9

Source: Reprinted by permission from Springer Nature from Ref. [7], Copyright 1996.

Isosters of water sorption on SWS-1S (Fig. 6.12) can be satisfactorily approximated by straight lines $\ln(P_{H_2O}) = A(n) + B(n)/T$. Their slopes give the isosteric sorption heat $\Delta H_{is}(n)$ displayed in Table 6.7. This value tends to decrease with the rise in water content and approaches the evaporation heat of bulk aqueous $CaCl_2$ solution [25].

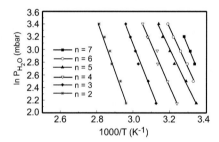

Figure 6.12 Isosters of water sorption on SWS-1S. Reprinted by permission from Springer Nature from Ref. [7], Copyright 1996.

The isotherm of water sorption plotted as a function of the relative vapor pressure P/P_o (Fig. 6.11b) allows further comparison of the SWS-1S sorption properties with those of the bulk $CaCl_2$ hydrates and solution. This isotherm coincides with the isotherm for the bulk system neither at low nor at high h values. At low h, the SWS-1S isotherm lies far above the proper points for the bulk hydrates with n = 1 and 2. At h > 0.2, the disperse system sorbs water worse than the bulk one. It means that the confinement of the salt solution to the small silica pores decreases its ability to bind water. Indeed, at a fixed solution concentration (see the straight line n = 7 in Fig. 6.11b), the relative vapor pressure over the confined solution is higher than that over the bulk one (0.45 versus 0.32). This looks surprising because the vapor pressure of wetting liquids commonly decreases in micropores due to the capillary effect [46]. It is reasonable to suppose that the observed increase in vapor pressure is caused by a significant change in the thermodynamic properties of the solution due to its confinement to micropores (see Sec. 6.6).

6.4.2 CaCl₂/Activated Carbon

ACs are the most commercialized adsorbent material with worldwide annual sales of more than \$1 billion (1997) [51]. They are characterized by large surface area (800–1500 m^2/g) and micropores of 0.5–2 nm size mixed with some meso- and macropores. The carbon surface is hydrophobic; therefore, water sorption isobars are of type V with almost zero adsorption at low relative pressure P/P_0 (IUPAC, Fig. 2.3). Therefore, the increment in sorption due to the addition of salt is especially impressive at low P/P_0, as seen in Fig. 6.13a for $CaCl_2$/AC composite reported in Ref. [3].

The neat AC mainly has 1–2 nm micropores. The N_2 adsorption/desorption hysteresis indicated the presence of mesoporosity. The authors prepared four composites impregnated with 30, 40, 50, and 70 wt.% $CaCl_2$. The isotherm of water vapor adsorption on neat AC indicates a very small adsorption degree up to $P/P_0 = 0.4$ due to the hydrophobic surface character. A steep increase in adsorption at $P/P_0 > 0.4$ is due to the clustering of water molecules in the AC micropores [52]. The desorption isotherm shows a very steep decrease at $P/P_0 = 0.55$ with a distinct hysteresis typical of bottleneck pores. The water adsorption isotherm of the AC impregnated with 30 wt.% $CaCl_2$ significantly differs from that of the neat AC especially at $P/P_0 < 0.4$–0.5 (Fig. 6.13a). Thus, impregnation of this AC by $CaCl_2$ is very effective in increasing the water adsorption at low P/P_0 values. The composite isotherms are smooth (bi-variant) except for the composite with a very large salt content of 70 wt.%. The latter behaves as $CaCl_2$ in large pores exhibiting plateaus at $w \approx 0.20$–0.25 and 0.4–0.5 g/g, which approximately corresponds to $n = 2$ and 4 with stepwise transition between them. This probably indicates that at large content, the salt is located inside the AC macropores or even out of the pores. Indeed, the SEM micrographs showed that the starting AC contains μ-sized pores [3]. Unfortunately, an XRD analysis was not performed to evaluate the size of CSDs of the salt and to identify its localization.

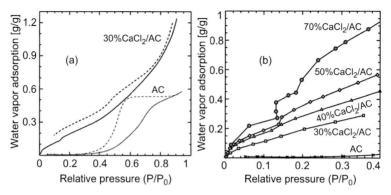

Figure 6.13 (a) Water vapor adsorption (solid lines) and desorption (dashed lines) isotherms of neat AC and 30 wt.% CaCl$_2$-impregnated AC; (b) adsorption isotherms at C = 0, 30, 40, 50, and 70 wt.%. Reprinted from Ref. [3], Copyright 2010, with permission from Elsevier.

6.4.3 CaCl$_2$/Zeolite 13X

Microporous zeolite 13X (S_{sp} = 730 m^2/g, V_p = 0.30 cm^3/g, D_{av} = 1.0 nm) was used as support for CaCl$_2$ to get a composite with 17 wt.% salt content (ZCa17) [4]. The water sorption isobars on 13X and ZCa17 demonstrate greater water uptake on the neat zeolite (Fig. 6.14). The authors ascribed this to partial blockage of the 13X microporosity by the introduced salt, which results in reducing both S_{sp} and V_p for ZCa17 by a factor of 2. The average size of CaCl$_2$ crystallites measured by XRD (72 nm) is significantly larger than the 13X pores. Hence, the salt is probably located out of the 13X pores, on its external surface. Since the salt is introduced inside the pores by the incipient wetness method, it probably escapes from the pores during drying of the wet material at 120°C. This can be due to the fact that the formation of salt particles of 1 nm size or less is highly disadvantageous from the thermodynamic point of view because of enormous contribution of their surface energy.

Interestingly, the water sorbed on ZCa17 can be removed at lower temperature (180–200°C) than from pure 13X (>250°C). Hence, it is primarily retained by the salt, rather than by the matrix micropores despite stronger interaction with the zeolite walls.

To conclude, this comparison clearly demonstrates the main drawbacks of microporous solids as support for a salt:

- The blocking effect of the impregnating salt;
- The salt escape from the pores and localization on the matrix external surface.

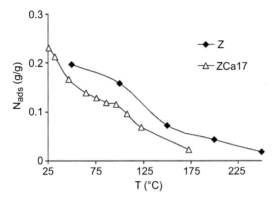

Figure 6.14 Isobars of water sorption on the neat zeolite 13X (Z) and 13X impregnated with CaCl$_2$ (ZCa17). Reprinted from Ref. [4], Copyright 2012, with permission from Elsevier.

6.5 Calcium and Lithium Nitrates in the Silica Pores of Various Size

Other hygroscopic salts for which the pore size effect was systematically studied are calcium and lithium nitrates (Ca(NO$_3$)$_2$, LiNO$_3$). These salts are well studied, and their data on solubility, equilibrium vapor pressure, and melting temperature are available in the literature [29, 53]. The salts were confined to a set of silica gels with the pore size ranging from 3 to 15 nm [21, 22]. This targeted investigation was aimed at elucidating the effect of the matrix pore size on the hydration of confined nitrates. The following reactions

$$Ca(NO_3)_2 + 2H_2O \rightleftarrows Ca(NO_3)_2 \cdot 2H_2O \quad (6.4)$$

$$LiNO_3 + 3H_2O \rightleftarrows LiNO_3 \cdot 3H_2O \quad (6.5)$$

were studied in the bulk and confined states. Reaction (6.4) is already considered in Secs. 5.2.3–5.2.5 to illustrate that the sorption properties of the composite Ca(NO$_3$)$_2$(45 wt.%)/KSK (SWS-8L) cannot be presented as the simple addition of the salt and the matrix properties (Fig. 5.5a [54]). It means that the salt's affinity for water vapor depends on the pore structure of the host matrix.

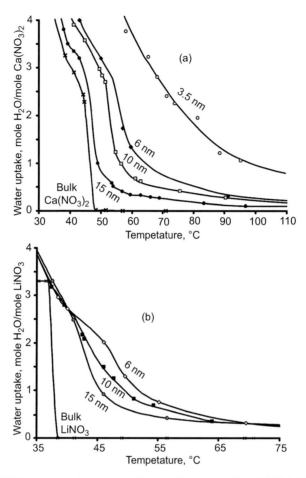

Figure 6.15 Isobars of water sorption on the composites Ca(NO$_3$)$_2$/silica (a) and LiNO$_3$/silica (b) with different pore sizes indicated near the curves (P_{H_2O} = 17 mbar) [57].

This is further confirmed by the water sorption isobars of the composites Ca(NO$_3$)$_2$/silica and LiNO$_3$/silica with different pore size (Fig. 6.15). The salt hydration isobars are shifted to higher temperatures in smaller pores; hence the salt's affinity for water vapor significantly increases, and the confined nitrates are more hydrophilic than the bulk ones. The same was reported for calcium chloride confined to middle and small pores (see Secs. 6.3 and 6.4). This effect can be used for adjustment of the salt's sorption properties to consumer demands. The increase in the hydration temperature of salt may be attributed to the contribution of the surface energy to the total Gibbs energy of the system (see Sec. 6.6).

We remind here the other two important findings reported for the involved composites (Chapter 5):

- Significant reduction in hydration/dehydration hysteresis (Fig. 5.6);
- Acceleration of hydration reaction (Fig. 5.7a), so that its rate is controlled by intra-particle water diffusion rather than by the vapor–salt chemical reaction.

6.6 "Water–Salt" Sorption Equilibrium inside the Matrix Pores

The results reported in this chapter give evidence that the confinement of a salt to a matrix's pores can change the salt's affinity for water vapor if the pore size is reduced enough. This effect is likely to be caused by the dispersion of salt into tiny particles rather than by the salt's interaction with the matrix as follows from the comparison of SWS-1A (Sec. 6.3.2) and SWS-1C (Sec. 6.3.3) composites. Perceptible changes are observed in the sorption properties of the salt hydrates in the middle pores (<20–30 nm): (a) formation of hydrates, containing a small number of water molecules, at a higher temperature/lower vapor pressure and (b) melting of hydrates, containing a large quantity of water, at a lower temperature. The changes in hydrates are considered in Sec. 6.6.1. The properties of the confined salt solution remain unchanged. In whole, the isotherms still resemble mono-variant, stepwise curves typical for bulk salts, however become smoother (e.g., Figs. 5.8a and 6.7b). In small pores (<5–7 nm), the equilibrium curves become purely bi-variant and lie below appropriate curves for the bulk solution. Thus, the confined solution has lower affinity for water vapor than the bulk one. The possible reasons for that are considered in Sec. 6.6.2.

The border between conditionally large, middle, and small pores is not strictly defined. It is clear that this conventional classification is not universal and depends on both the salt and the matrix. For instance, SWS-1A and SWS-1C belong to the middle pore systems, but behave in the same way as SWS-1S, which is a small pore sorbent (compare Figs. 6.8 and 6.9 with Fig. 6.11). Probably, it is worthy to consider another ranking based on the degree of deviation of the salt's sorption properties in the pores from those in the bulk. If the properties quantitatively differ from those in the bulk but qualitatively resemble them, the pores are considered "middle size." In "small" pores, the sorption properties of the confined salt become essentially different.

6.6.1 Salt Hydrates in Middle Pores

As shown above, hydrates of $CaCl_2$, $Ca(NO_3)_2$, and $LiNO_3$ in the middle pores perceptibly differ from the bulk hydrates: they are formed at higher temperature and melted at lower temperature. The latter phenomenon is well known in nanoscale and commonly attributed to the contribution of the surface energy to the total Gibbs free energy $\Delta_r G_o$ of the system [55]. The melting of hydrates in confined geometry is considered further in Chapter 11. Here we only remind that the melting point depression $\Delta T_m(d) = T_m(\infty) - T_m(d)$ for a small crystal of typical size d is determined as [56]

$$\Delta T_m(d) = 4\sigma T_m(\infty)/(d \Delta H \rho) = A/d \qquad (6.6)$$

where $T_m(\infty)$ is the normal (bulk) melting point, σ is the surface tension of the solid–liquid interface, ΔH is the bulk fusion enthalpy, and ρ is the solid density. For instance, the temperature of $LiNO_3$ melting in the silica pores follows Eq. (6.6): If plot ΔT_m versus the inverse pore size $1/r$ (Fig. 6.16a) [57], the slope A equals 215 K·nm. As the actual size d of the confined salt particles is unknown, it is reasonable to consider it as the first approximation, proportional to the pore size, $d \sim r$.

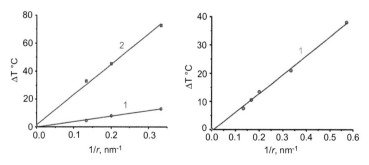

Figure 6.16 Increment in hydration (1) and melting (2) temperatures ΔT as a function of the inverse pore size $1/r$ for confined $LiNO_3$ (a) and $Ca(NO_3)_2$ (b) [57].

Surprisingly, the increment in hydration temperature T_h for $LiNO_3$ (a) and $Ca(NO_3)_2$ confined to silica pores is also satisfactorily described by Eq. (6.6), but with smaller slope $A = 40$ and 65 K·nm, respectively (Fig. 6.16). This indicates that increase in the T_h value can also be caused by the surface energy effects, because dividing the salt/hydrate into nano-pieces results in an enormous increase in the surface energy [58].

For estimating this effect, variation in the Gibbs free energy $\Delta G°$ of the hydration reaction (e.g., reaction 6.4) consists of the bulk and surface (interfacial) terms

$$\Delta_r G° = \Delta_b G° + \Delta_s G° \tag{6.7}$$

The bulk Gibbs energy can be written as

$$\Delta_b G° = \Delta G°_p - \Delta G°_r$$

$$= \Delta G° \text{ (hydrate)} - [\Delta G° \text{ (salt)} + n \cdot \Delta G° (H_2O)] \tag{6.8}$$

where $\Delta G°_p$ and $\Delta G°_r$ are the formation free energies of the bulk products and reagents, respectively. The equilibrium hydration temperature in bulk T_b is determined by the condition $\Delta_b G° = 0$ or $\Delta G°_p = \Delta G°_r$. It corresponds to the intersection of the lines $\Delta G°_p(T)$ and $\Delta G°_r(T)$ (point A in Fig. 6.17). The hydration occurs at $T < T_b$.

When confined to the matrix pores, the solid phases are dispersed, so that their surface free energies $\Delta G_{sur}(prod)$ and $\Delta G_{sur}(reag)$ must be taken into account. If they are equal, $\Delta G_{sur}(prod) = \Delta G_{sur}(reag)$, the hydration temperature (point B in Fig. 6.17) does not change (case b in Fig. 6.17). If $\Delta G_{sur}(prod) < \Delta G_{sur}(reag)$, the hydration temperature T_d in the dispersed state is lower than in bulk (case c). Accordingly, $T_d > T_b$, if $\Delta G_{sur}(prod) > \Delta G_{sur}(reag)$ (case a). The latter case is likely to correspond to $Ca(NO_3)_2$/silica and $LiNO_3$/silica composites. Hence, for reactions (6.4) and (6.5), the surface energy of the anhydrous salt is larger than that of the salt hydrate. This is true with high confidence if the hydrate melts during the hydration process as it happens with $LiNO_3 \cdot 3H_2O$ and $Ca(NO_3)_2 \cdot 2H_2O$ inside the silica pores smaller than 15 nm. Indeed, the XRD technique reveals that the peaks of $Ca(NO_3)_2$ disappear without the appropriate formation of $Ca(NO_3)_2 \cdot 2H_2O$ peaks [57]. Similarly, $CaCl_2 \cdot 4H_2O$ melts in the silica KSK pores during the $CaCl_2 \cdot 2H_2O$ hydration via reaction (6.3).

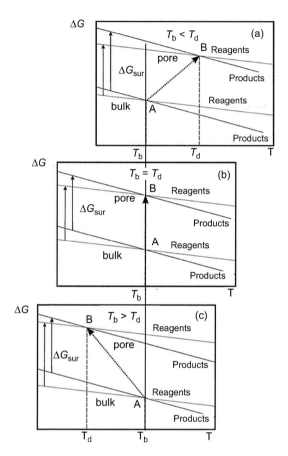

Figure 6.17 Schematics of the effect of surface energy ΔG_{sur} on the hydration temperature: (a) $\Delta G_{sur}(\text{prod}) > \Delta G_{sur}(\text{reag})$; (b) $\Delta G_{sur}(\text{prod}) = \Delta G_{sur}(\text{reag})$; (c) $\Delta G_{sur}(\text{prod}) < \Delta G_{sur}(\text{reag})$ [57].

The $\Delta G(T)$ graph is calculated for reaction (6.4) from the thermodynamic data for the bulk salt and hydrate (Fig. 6.18) [57]. The intersection of the lines $\Delta G°_p(T)$ and $\Delta G°_r(T)$ (point A in Fig. 6.18) gives the hydration temperature in bulk $T_b(\text{calc}) = 325$ K $= 51.6°C$, which is close to the experimentally obtained $T_b(\text{exp}) = 321$ K $= 48°C$ (Fig. 6.15a). The contributions of the surface energy $\Delta G = 4\sigma V/d$, where V is the molar volume, cannot be calculated for both the anhydrous salt and dihydrate, because the surface tension σ and the particle size d are unknown. The author [57] arbitrarily set the salt and hydrate surface energies equal to 3.2 and

1.5 kJ/mol, respectively, and demonstrated a reasonable increase in the hydration temperature T_d of the confined salt up to 56.5°C (Fig. 6.18).

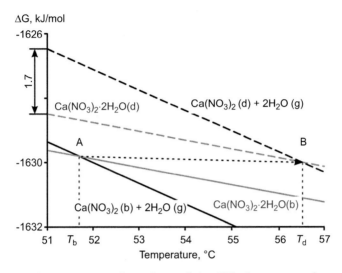

Figure 6.18 Temperature dependence of the Gibbs free energy of reaction (6.4) in bulk and the imaginary dispersed state (see text for explanation) [57]. P_{H_2O} = 17 mbar.

Figure 6.16 shows that the hydration temperature increases in smaller pores as $\Delta T_d(r) = A/r = (65 \text{ K·nm})/r$, and a small pore size variation Δr leads to $\Delta(\Delta T_d) = (-A/r^2) \cdot \Delta r$. Thus, the same pore size variation Δr results in much larger ΔT_d increment in small pores. Since the silica gels used as hosts for the studied nitrates have a wide PSD, the hydration occurs at a certain temperature range, which broadens out in smaller pores (Fig. 6.15).

6.6.2 Salt Solutions in Small Pores

It was experimentally found that in small pores (<5–7 nm), the equilibrium curves become purely bi-variant and lie below appropriate curves for the bulk solution. Thus, the confined solution has lower affinity for water vapor than the bulk one. Confined solutions may be affected by the strong potential of the pore wall, ion adsorption, wetting phenomenon, or just by restricted pore geometry [59].

Thermodynamic properties of electrolyte solutions are known to be determined mainly by electrostatic interactions between positive and negative ions of a dissociated salt. Debye and Huckel suggested an electrostatic model describing thermodynamic properties of dilute solutions of strong electrolyte [60]. They concluded that any ion in solution is surrounded by an envelope of other ions with the characteristic radius R [60, 61]

$$R = A \, [(D \, T)/(\varepsilon^2 \, I)]^{1/2} \qquad (6.9)$$

where A is a coefficient, D is the ion diffusion coefficient, T is the solution temperature, I is its ionic strength, and ε is the solvent dielectric constant. Although the solutions are not dilute under our conditions, Eq. (6.9) can be used for brief estimation of the ion atmosphere diameter $2R$, which appears to be 0.2–0.6 nm. This value is much lower than the average pore diameter D_{av} of the mesoporous silica KSK ($D_{av}/2R = 15$ nm/(0.2–0.6 nm) $= 25$–$75 >> 1$). As a result, the mesopore walls do not influence the ion spatial distribution in the solution and do not change its chemical potential and the water vapor pressure. For smaller KSM pores, the ratio $D_{av}/2R = 6$–17 still looks small. Nevertheless, the wall effect is expected to be significant if one takes into account that the near-wall layer of 0.4 nm thickness contains 23, 40, or 54% of the solution volume for 3.5 nm pores of the slit, cylindrical or spherical shape, respectively. Thus, a large portion of ions is in the wall vicinity and their "true" (bulk) spatial distribution may be strongly disturbed by near-wall geometrical restrictions.

Indeed, any ion located near an uncharged pore surface has an energy excess with respect to a bulk ion, since it has no neighboring ions in the half-space behind the wall. As a result, the ions avoid locations in the wall vicinity. It makes the ion escape from the micropore space. This effect is strongly enhanced for charged surfaces [62]. It is generally described on the basis of the ideal Donnan theory [63]. A more detailed model of the electrolyte exclusion is given by the Poisson–Boltzmann equation [64] and the Monte Carlo simulation [64, 65]. All these approaches lead to non-uniform ion distributions along the micropore radius. This results in the depression of the electrolyte concentration inside micropores and in deviations of the confined solution properties from the bulk ones. The observed increase in the partial vapor pressure over the solution confined to the KSM silica gel can also be driven by the solution inhomogeneity in the small pores. Since our experiments have been performed at high electrolyte concentrations where the solution properties differ from those predicted by the Debye–Huckel and Donnan models, direct Monte Carlo simulations of the effect discussed could be highly desirable. Unfortunately, such a simulation seems highly conjectural because an average distance between ions in highly concentrated electrolyte solutions is of atomic scale. The solvent can no longer be considered a uniform dielectric medium with a certain ε value, and its molecular nature has to be taken into account (molecule size, hydrogen binding, etc.).

References

1. Aristov, Yu. I., Tokarev, M. M., Cacciola, G., and Restuccia, G. (1996) Selective water sorbents for multiple applications: 1. $CaCl_2$ confined in mesopores of the silica gel: Sorption properties, *React. Kinet. Cat. Lett.*, **59**, pp. 325–334.

2. Aristov, Yu. I., Tokarev, M. M., Di Marco, G., Cacciola, G., Rectuccia, G., and Parmon, V. N. (1997) Properties of the system "calcium chloride–water" confined in pores of the silica gel: Equilibria "gas-condensed state" and "melting-solidification," *Rus. J. Phys. Chem.*, **71**, pp. 253–258.

3. Okada, K., Nakanome, M., Kameshima, Y., Isobe, T., and Nakajima, A. (2010) Water vapor adsorption of $CaCl_2$-impregnated activated carbon, *Mater. Res. Bull.*, **45**, pp. 1549–1553.

4. Cortés, F. B., Chejne, F., Carrasco-Marín, F., Pérez-Cadenas, A. F., and Moreno-Castilla, C. (2012) Water sorption on silica- and zeolite-supported hygroscopic salts for cooling system applications, *Energy Convers. Manag.*, **53**, pp. 219–223.

5. Liu, C. Y., Morofuji, K., Tamura, K., and Aika, K. (2004) Water sorption of $CaCl_2$-containing materials as heat storage media, *Chem. Lett.*, **33**, pp. 292–293.

6. Daou, K., Wang, R. Z., and Xia, Z. Z. (2006) Development of a new synthesized adsorbent for refrigeration and air conditioning applications, *Appl. Therm. Eng.*, **26**, pp. 56–65.

7. Aristov, Yu. I., Tokarev, M. M., Cacciola, G., and Restuccia, G. (1996) Selective water sorbents for multiple applications: 2. $CaCl_2$ confined in micropores of the silica gel: Sorption properties, *React. Kinet. Cat. Lett.*, **59**, pp. 335–342.

8. Tokarev, M. M., Gordeeva, L. G., Romannikov, V. N., Glaznev, I. S., and Aristov, Yu. I. (2002) New composite sorbent "$CaCl_2$ in mesopores of MCM-41" for sorption cooling/heating, *Int. J. Therm. Sci.*, **41**, pp. 470–474.

9. Mrowiec-Bialon, J., Jarzebskii, A. B., Lachowski, A., Malinovski, J., and Aristov, Yu. I. (1997) Effective inorganic hybrid adsorbents of water vapor by the sol-gel method, *Chem. Mater.*, **9**, pp. 2486–2490.

10. Tokarev, M. M. (2003) PhD thesis, Boreskov Institute of Catalysis, Novosibirsk.

11. Liu, H., Nagano, K., Sugiyama, D., Togawa, J., and Nakamura, M. (2013) Honeycomb filters made from mesoporous composite material for an open sorption thermal energy storage system to store low-temperature industrial waste heat, *Int. J. Heat Mass Transfer*, **65**, pp. 471–480.

12. Ponomarenko, I. V., Glaznev, I. S., Gubar, A. V., Aristov, Yu. I., and Kirik, S. D. (2010) Synthesis and water sorption properties of a new composite "$CaCl_2$ confined to SBA-15 pores," *Micropor. Mesopor. Mater.*, **129**, pp. 243–250.

13. Glaznev, I. S., Ponomarenko, I. V., Kirik, S. D., and Aristov, Yu. I. (2011) Composite $CaCl_2$/SBA-15 for adsorptive transformation of low temperature heat: Pore size effect, *Int. J. Refrigeration*, **34**, pp. 1244–1250.

14. Jänchen, J., Ackermann, D., Stach, H., and Brösicke, W. (2004) Studies of the water adsorption on zeolites and modified mesoporous materials for seasonal storage of solar heat, *Sol Energy*, **76**, pp. 339–344.

15. Gordeeva, L. G., Grekova, A. D., and Aristov, Yu. I. (2016) Composites "Li/Ca halogenides inside multi-wall carbon nano-tubes" for adsorptive heat storage, *Sol. Energy Mater. Sol. Cells*, **155**, pp. 176–183.

16. Zhang, H., Yuan, Y., Yang, F., Zhang, N., and Cao, X. (2015) Inorganic composite adsorbent $CaCl_2$/MWNT for water vapour adsorption, *RSC Adv.*, **5**, pp. 38630–38639.

17. Tokarev, M. M., private communication.

18. Jänchen, J., Ackermann, D., and Weiler, E. (2005) Calorimetric investigation on zeolites, $AlPO_4$'s and $CaCl_2$ impregnated attapulgite for thermochemical storage of heat, *Thermochim Acta*, **434**, pp. 37–41.

19. Aristov, Yu. I., Restuccia, G., Tokarev, M. M., Buerger, H.-D., and Freni, A. (2000) Selective water sorbents for multiple applications. 11. $CaCl_2$ confined to expanded vermiculite, *React. Kinet. Cat. Lett.*, **71**, pp. 377–384.

20. Casey, S. P., Elvins, J., Riffat, S. B., and Robinson, A. (2014) Salt impregnated desiccant matrices for 'open' thermochemical energy storage: Selection, synthesis and characterisation of candidate materials, *Energy Buildings*, **84**, pp. 412–425.

21. Simonova, I. A. and Aristov, Yu. I. (2005) Sorption properties of calcium nitrate dispersed in silica gel: The effect of pore size, *Rus. J. Phys. Chem.*, **79**, pp. 1307–1311.

22. Simonova, I. A. and Aristov, Yu. I. (2007) Dehydration of salt hydrates for storage of low temperature heat: Selection of promising reactions and nanotailoring of innovative materials, *Alternative Energy Ecology*, **10**, pp. 62–69.

23. Isobe, H. (1929) Dehydrating substance, US Patent N 1,740,351.

24. Chi, C. W. and Wasan, D. T. (1970) Fixed bed adsorption drying, *AIChE J.*, **16**, pp. 23–31.

25. Heiti, R. V. and Thodos, G. (1986) Energy release in the dehumidification of air using a bed of $CaCl_2$-impregnated celite, *Ind. Eng. Chem. Fundam.*, **25**, pp. 768–771.

26. Vrana, L. M. (2014) Calcium chloride. In: *Kirk-Othmer Encyclopedia of Chemical Technology*, John Wiley & Sons, Inc (Ed.), pp. 1–13.

27. Sinke, G. C., Mossner, E. H., and Curnutt, J. L. (1985) Enthalpies of solution and solubilities of calcium chloride and its lower hydrates, *J. Chem. Thermodyn.*, **17**, pp. 893–899.

28. Erich Pietsch, E. H. (Ed.) (1957) *Gmelins Handbuch der Anorganischen Chemie, Calcium Teil B–Lieferung 2* (Verlag Chemie GmbH) (in German).

29. Kirgintsev, A. N., Trushnikova, L. N., and Lavrentieva, V. G. (1972) *Solubility of Inorganic Substances in Water: Handbook* (Khimia, Leningrad, USSR) (in Russian).

30. Meissingset, K. K. and Gronvold, F. (1986) Thermodynamic properties and phase transitions of salt hydrates between 270 and 400 K IV. $CaCl_2 \cdot 6H_2O$, $CaCl_2 \cdot 4H_2O$, $CaCl_2 \cdot 2H_2O$, and $FeCl_3 \cdot 6H_2O$, *J. Chem. Therm.*, **18**, pp. 159–173.

31. Ananthaswamy, J. and Atkinson, G. (1985) Thermodynamics of concentrated electrolyte mixtures. 5. A review of the thermodynamic properties of aqueous calcium chloride in the temperature range 273.15–373.15 K, *J. Chem. Eng. Data*, **30**, pp. 120–128.

32. Conde, M. R. (2004) Properties of aqueous solutions of lithium and calcium chlorides: Formulations for use in air conditioning equipment design, *Int. J. Therm. Sci.*, **43**, pp. 367–382.

33. Yamaguchi, T., Hayashi, S., and Ohtaki, H. (1989) X-ray diffraction study of calcium (ii) chloride hydrate melts: $CaCl_2$-RH_2O (R = 4.0, 5.6, 6.0, 8.6), *Inorg. Chem.,* **28**, pp. 2434–2439.

34. Fedorov, N. F., Ivahnyuk, G. K., and Babkin, O. E. (1990) Modified sorbents based on porous carbons, *Russ. J. Appl. Chem.,* **6**, pp. 1275–1279.

35. Levitskiy, E. A., Parmon, V. N., Moroz, E. M., Bogdanov, S. V., Kovalenko, O. N., and Bogdanchikova, N. E. Heat accumulating material and method of its synthesis, Patent RF 4839454.

36. Ulyanova, M. A., Gurova, A. S., and Shreder, V. E. (2007) Water-resistant silica gels for life support systems, *Russ. Chem. Phys.,* **26**, pp. 71–76.

37. Anthony, J. W., Bideaux, R. A., Bladh, K. W., and Nichols, M. C. (1997) *Handbook of Mineralogy,* vol. 3 (Mineral Data Publishing).

38. Lyakhov, N. Z. and Boldyrev, V. V. (1972) Decomposition of salt hydrate, *Russ. Chem. Rev.,* **41**, pp. 919–947.

39. Delmon, B. (1969) *Introduction a la Cinetique Heterogene* (Technip, Paris).

40. Gurvich, B. M., Karimov, R. R., and Mezheritskii, S. M. (1986) Heat conductivity of $CaCl_2$ aqueous solutions, *Rus. J. Appl. Chem.,* **59**, pp. 2692–2696.

41. Iijima, S. (1991) Helical microtubules of graphitic carbon, *Nature,* **354**, pp. 56–58 .

42. Yan, T., Li, T. X., Li, H., and Wang, R. Z. (2014) Experimental study of the ammonia adsorption characteristics on the composite sorbent of $CaCl_2$ and multi-walled carbon nanotubes, *Int. J. Refrig.,* **46**, pp. 165–172.

43. Zhao, D., Feng, J., Huo, Q., Melosh, N., Fredrickson, G. H., Chmelka, B. F., and Stucky, G. D. (1998) Triblock copolymer syntheses of mesoporous silica with periodic 50 to 300 Angstrom pores, *Science,* **279**, pp. 548–552.

44. Aristov, Yu. I. (2013) Challenging offers of material science for adsorption heat transformation: A review, *Appl. Therm. Eng.,* **50**, pp. 1610–1618.

45. Schueth, F., Sing, K. S. W., and Weitkamp, J. (Eds.) (2002) *Handbook of Porous Solids* (Wiley-VCH, Weinheim, Germany).

46. Greg, S. J. and Sing, K. S. W. (1982) *Adsorption, Surface area and Porosity* (Academic Press, London).

47. Jang, R. T. (2003) *Adsorbents: Fundamentals and Applications* (John Wiley & Sons, New Jersey).

48. Iler, R. (1978) *The Chemistry of Silica* (Wiley, New York).

49. Schueth, F., Sing, K. S., and Weitkamp, J. (Eds.) (2002) *Handbook of Porous Solids*, vol. 1–5 (Wiley-VCH).

50. Wilson, S. T., Lok, B. M., Messina, C. A., Cannan, T. R., and Flanigen, E. M. (1982) Aluminophosphate molecular sieves: A new class of microporous crystalline inorganic solids, *J. Am. Chem. Soc.*, **104**, pp. 1146–1147.

51. Humphrey, J. L. and Keller, G. E. (1997) *Separation Process Technology* (McGraw-Hill, New York).

52. Okada, K., Yamamoto, N., Kameshima, Y., and Yasumori, A. (2003) Porous properties of activated carbons from waste newspaper prepared by chemical and physical activation, *J. Colloid Interf. Sci.*, **262**, pp. 179–193.

53. Erich Pietsch, E. H. (Ed.) (1957) *Gmelins Handbuch der Anorganischen Chemie, Calcium Tail B–Lieferung 2* (Verlag Chemie GmbH), S. 342–372.

54. Simonova, I. A., Freni, A., Restuccia, G., and Aristov, Yu. I. (2009) Water sorption on the composite "silica modified by calcium nitrate": Sorption equilibrium, *Micropor. Mesopor. Mater.*, **122**, pp. 223–228.

55. Defay, R., Prigogine, I., Bellemans, A., and Everett, D. H. (1966) *Surface Tension and Adsorption* (Wiley, New York).

56. Thomson (Lord Kelvin) (1871) *Philos. Magazine*, **42**, p. 448.

57. Simonova, I. A. (2008) PhD thesis, Boreskov Institute of Catalysis, Novosibirsk.

58. Gibbs, J. W. (1928) *Collected Works*, New York.

59. Bellissent-Funel, M.-C. (2003) Status of experiments probing the dynamics of water in confinement, *Eur. Phys. J. E*, **12**, pp. 83–92.

60. Debye, P. and Huckel, E. (1923) Zur theorie der electrolyte, *Phys. Z.*, **24**, 185.

61. Chattoraj, D. K. and Birdi, K. S. (1984) *Adsorption and the Gibbs Surface Excess* (Plenum, New York), p. 99.

62. Marcinkowsky, A. E., Kraus, K. A., Phillips, H. O., Johnson, J. S., and Shor, A. J. (1966) Hyperfiltration studies. IV. Salt rejection by dynamically formed hydrous oxide membranes, *J. Am. Chem. Soc.*, **88**, pp. 5744–5746.

63. Donnan, F. G. (1911) The theory of membrane equilibrium in presence of a non-dialyzable electrolyte, *Z. Electrochem.*, **17**, pp. 572.

64. Vlachy, V. and Haymet, A. D. J. (1990) Ion distributions in a cylindrical capillary, *J. Electroan. Chem.,* **283**, pp. 77–85.

65. Vlachy, V. and Haymet, A. D. J. (1989) Electrolytes in charged micropores, *J. Am. Chem. Soc.,* **111**, pp. 477–481.

Chapter 7

Composite Sorbents of Methanol

In the two previous chapters, we have comprehensively considered the sorption of water vapor on composites "salt in porous matrix" (CSPMs) and revealed the main features of these materials. Here we focus on the other adsorbate—methanol (CH_3OH)—which can also form complexes with a salt (S):

$$S + nCH_3OH = S \cdot nCH_3OH \tag{7.1}$$

which may result in a large amount of methanol absorbed [1–4]. For example, in the course of the reaction $LiCl + 3CH_3OH = LiCl \cdot 3CH_3OH$, 1 g of LiCl absorbs 2.26 g of CH_3OH, which far exceeds the typical values of methanol adsorption by common adsorbents.

Similar to salt hydrates, reaction (7.1) in the bulk is limited by (a) hindered reorganization of the crystalline structure "salt to solvate;" (b) slow diffusion of methanol vapor through a layer of the formed solvate; and (c) swelling of salt during the formation of solvate. For a neat solution to these problems, a new family of efficient methanol sorbents were synthesized by inserting various inorganic salts into porous matrices.

In this chapter, we survey the current state of the art of the CSPMs developed for methanol sorption. We, first, represent their sorption properties and discuss tools available for nanotailoring the affinity for methanol vapor. We consider how to choose the active

---

Nanocomposite Sorbents for Multiple Applications
Yu. I. Aristov
Copyright © 2020 Jenny Stanford Publishing Pte. Ltd.
ISBN 978-981-4267-50-2 (Hardcover), 978-981-4303-15-6 (eBook)
www.jennystanford.com

salt, host matrix, and synthesis conditions, and how these factors contribute to the sorption properties of the final composites. The target-oriented design of methanol sorbents for various practical applications, e.g., adsorptive cooling and shifting the equilibrium of catalytic methanol synthesis, is considered in Chapters 16 and 20.

7.1 Common Methanol Adsorbents

Methanol is a volatile organic compound harmful to human health, and its removal from gases is the main area where methanol adsorbents are widely utilized [5]. Adsorptive cooling is a modern environmentally benign and energy-saving technology [6, 7], and methanol is one of the most efficient working fluids. Therefore, novel sorbents of methanol adapted to this application are welcome. Catalytic methanol synthesis is one of the most important processes in the chemical industry, where the conversion degree is limited by unfavorable equilibrium. Shifting the synthesis equilibrium to the target product yield is a promising way to increase the conversion, intensify the process, and reduce the methanol cost [8, 9]. It can be performed by means of methanol removal from the reaction-product mixture due to its adsorption.

All these applications ask for appropriate methanol adsorbents adapted to a particular process and its conditions. For instance, methanol removal from gaseous mixtures requires a material that adsorbs methanol vapor at extremely low partial pressure ($P/P_0 < 10^{-5}$). Requirements for an adsorbent optimal for heat pumps and chillers depend on the end use (ice making, air conditioning, heating, etc.), the climatic conditions of the area where the unit is used, etc. The desirable P/P_0 range is 0.1–0.3. A suitable adsorbent for intensifying methanol synthesis should adsorb methanol at typical conditions of the industrial catalytic process (see Sec. 3.2.4 and Chapter 20). Hence, various applications ask for quite different adsorbent properties that may significantly vary. Despite this, only a few types of adsorbents are currently used in the mentioned processes: activated carbons [10, 11], zeolites [12], and alumina-based materials [13]. The methanol uptake of these adsorbents is typically 0.2–0.5 g/g. Several materials with advanced methanol adsorption properties are being developed [14], such as mesoporous mesophase silicates, polymers, pillared clays, functionalized silica gels, metal-substituted aluminophosphates, metal–organic frameworks, etc. Adsorption equilibrium of these new materials with methanol was measured. An exceptionally large methanol adsorption capacity up to 1.4 g/g was found for activated carbon MaxSorb III [15].

Yet, new efficient methanol sorbents are highly welcome especially those having stepwise sorption isotherms, which are desirable for many applications (see Chapter 3).

7.2 Intent Design of CSPMs for Methanol Sorption

In the course of reaction (7.1), a large amount of methanol can be bound with the formation of a solid solvate $S \cdot nCH_3OH$, where n may reach 6 [1–4]. Due to this, solvate decomposition—the reaction reverse to (7.1)—was suggested for chemical storage of heat with its posterior use for heating/cooling aims. Specifically, lithium and calcium chlorides as well as lithium bromide [16] were considered to be promising.

7.2.1 Effect of Active Salt

First of all, these three salts were inserted in the pores of various silica gels, and sorption properties of these composites were studied [17–21]. Silica gels KSK, Davisil Grade 646, Grace SP2, Grace Davison, and I254 were used as host matrices (Table 7.1). Moreover, a brief screening of other inorganic salts, which are able to form complexes with methanol vapor, was performed as displayed in Table 7.2 [17].

Table 7.1 Main characteristics of the silica gels used

Silica gel	V_p (cm^3/g)	S_{sp} (m^2/g)	d_{av} (nm)
KSK	1.0	350	15
Davisil Grade 646	1.15	300	15
Grace SP2	1.5	326	15
Grace Davison	1.3	360	14
Grace I254	0.9	500	6

The composites were synthesized by dry impregnation of the silica gels with an aqueous solution of the salt followed by thermal

Composite Sorbents of Methanol

drying at 473 K. The salt content C in the composites was determined by weighting dry samples before and after impregnaton.

Table 7.2 Methanol partial pressure P_{MeOH}, temperature T_r, and the value η_r of dissociation of methanolates of various salts, as well as the amount of methanol bound with the salt n (mol of methanol/mol of the salt) and w (g methanol/g salt)

Salt	n (mol/mol)	T_r (°C)	P_{MeOH} (mbar)	η_r	w (g/g)
LiCl	3	46	96	0.21	2.26
	1	54	96	0.15	0.76
$CaCl_2$	4	20	92	0.72	1.15
CaI_2	6	20	62	0.49	0.65
	2	20	26	0.21	0.22
$Ca(NO_3)_2$	2	20	20	0.16	0.39
$MgCl_2$	6	15	64	0.66	2.02
$Mg(NO_3)_2$	6	20	7	0.058	1.30
$CuCl_2$	2	20	89	0.70	0.48
$CdCl_2$	3	20	124	0.97	0.52
	2	20	38	0.30	0.35
	1.5	20	25	0.20	0.20
$CdBr_2$	3	20	124	0.97	0.35
	2	20	34	0.27	0.23
	1.5	20	20	0.15	0.18
$NiBr_2$	6	20	22	0.18	0.88
NaI	3	20	51	0.40	0.64
$SrBr_2$	1.5	20	33	0.26	0.19
	0.5	20	14	0.11	0.65
$CoCl_2$	3	20	39	0.31	0.74
	2	20	20	0.15	0.49
$CoBr_2$	6	20	41	0.32	0.88
	3	20	22	0.18	0.44
	2	20	13	0.10	0.29

Source: Ref. [21].

The dispersion of the salt inside the pores to nanoscale size allows a combination of good sorption dynamics of porous adsorbents with a large methanol capacity of inorganic salts, as well as overcoming the drawbacks listed at the very beginning of this chapter. Indeed, the encapsulation of the salts inside the pores of silica gel results in a huge increase in the methanol sorption capacity of the composite up to 0.7–0.9 g/g as compared with 0.1–0.2 g/g for the neat silica (Fig. 7.1) [17]. This sorption far exceeds that for most of the common methanol sorbents. And what is more, the isotherms are S-shaped due to the methanol–salt reaction. The step position depends on the salt's nature and may vary in the P/P_0 range of 0.01–0.05 (for LiBr and CaBr$_2$) to 0.2–0.4 (for LiCl and CaCl$_2$). This reveals that the salt is the main sorbing component of the composites, which is similar to the CSPMs for sorbing water vapor (Chapters 5 and 6).

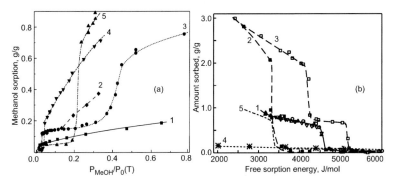

Figure 7.1 Temperature-invariant curves of methanol sorption on (a) neat KSK silica gel (1) and composites based on CaBr$_2$ (2), CaCl$_2$ (3), LiBr (4), and LiCl (5) confined to pores of this silica gel; (b) LiCl (31 wt.%)/SiO$_2$ composite (1), bulk LiCl (2 and 3), pure silica Grace SP2 (4), and curve simulated as a linear addition of sorption by bulk LiCl and the silica (5). Figure (b) reprinted from Ref. [19], Copyright 2008, with permission from Elsevier. Sorption: solid symbols; desorption: open symbols.

As the salt is the primary sorbing component of the composites involved, its chemical nature dramatically affects the composite's sorption properties. A great number of salts react with methanol vapor forming crystalline solvates (methanolates) due to reaction (7.1). The temperature T_r at which reaction (7.1) occurs at a fixed methanol pressure P_{MeOH} is determined by the integral van't Hoff Eq. (5.7). The thermodynamic parameters $\Delta H°$ and $\Delta S°$ of reaction (7.1) may substantially change for various salts resulting in a wide variation in the equilibrium temperature T_r at fixed P_{MeOH}, or the equilibrium relative pressure of methanol vapor $\eta_r = P_{MeOH}/P_0(T_r)$.

Table 7.2 shows that the relative pressure, at which the methanol solvates dissociate, bridges the whole gap between η_r = 0.06 and 0.97 for various salts. This provides a fair chance to select a salt with a proper temperature of the methanolate formation/decomposition. According to the nanotailoring concept (Chapter 1), this salt should be inserted into the pores of the appropriate host matrix.

The composites based on LiCl, LiBr, and $CaCl_2$ are expected to be most promising as they offer the largest methanol mass exchanged.

7.2.2 Effect of Host Matrix

The role of the matrix is of high importance. In the first place, the matrix performs a dispersing function and prevents the natural tendency of the salt particles to aggregate as well as accommodates their swelling when they react with the vapor. Second, the matrix pores provide transport of vapor molecules to nanoparticles of the confined salt. Finally, the dispersion of the salt inside the pores may lead to change in its properties as demonstrated right below. It would give an additional tool for the intent design of CSPMs specialized for methanol sorption.

Figure 7.1b presents data on the methanol sorption equilibrium on silica gel Grace SP2, bulk LiCl and composite LiCl (31 wt%)/(Grace SP2) [19]. The sorption and desorption branches of the temperature-invariant curve for the bulk LiCl differ significantly, demonstrating a strong hysteresis. Moreover, the synthesis and decomposition routes are different ($0 \rightarrow 3$ and $3 \rightarrow 1 \rightarrow 0$, respectively). Indeed, the hysteresis leads to much higher temperature for the decomposition reaction, which may impede the practical use of the material.

Quite the contrary, no sorption–desorption hysteresis is found for the confined LiCl. The dispersion of the salt (or its solvate) to tiny particles inside the pores may result in the formation of defective and irregular particles of the new phase, which probably makes its formation easier. That collapses the hysteresis loop. Thus, the encapsulation of the salt inside the matrix's pores overcomes one of the most severe drawbacks of gas–solid reactions in bulk.

Another important matrix effect may be revealed by comparing the methanol sorption by this composite with a linear superposition of the methanol absorption by bulk LiCl and its adsorption on the host silica gel taken with appropriate weight coefficients. Figure 7.1b shows that sorption by the composite cannot be represented as the mentioned linear addition. Thus, the salt's sorption properties may dramatically change due to its confinement to the matrix pores.

The adsorption branch of the temperature-invariant curve for the embedded salt is similar to that of the bulk salt but shifted along the free sorption energy axis by 1150 J/mol toward higher values of the free energy ΔF (Fig. 7.1b). Thus, LiCl's affinity for methanol increases in the confined state. This effect promotes deeper removal of the methanol vapor, which can be favorable for purification of gases from methanol.

The rise in the sorption ability of the salt due to dispersion in pores was reported for water sorption on CSPMs as well (Chapters 5 and 6). The reason behind this rise may be related to an increase in the fraction of surface atoms in tiny salt particles and, hence, a rise in the contribution of the surface energy to the total Gibbs free energy of the composite, as discussed in Chapter 6.

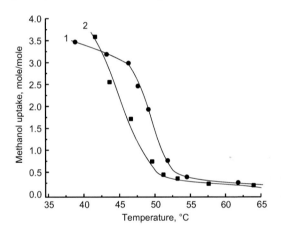

Figure 7.2 Isobars of methanol sorption on the composites LiCl/SiO$_2$ (6 nm) (1) and LiCl/SiO$_2$ (15 nm) (2). P_{MeOH} = 97 mbar. Plotted from data of Ref. [21].

A steep rise in methanol sorption from n = 0–0.5 to 3 mol/mol caused by the reaction

$$\text{LiCl} + 3\text{CH}_3\text{OH} = \text{LiCl·3CH}_3\text{OH} \tag{7.2}$$

is observed for the composites LiCl/silica with an average pore size d_{av} = 15 and 6 nm (Fig. 7.2). The methanolate LiCl·3CH$_3$OH appears to melt inside the silica pores and transform to an LiCl–CH$_3$OH solution [19]. It is seen that the temperature of formation of LiCl·3CH$_3$OH depends on the pore size. The salt embedded in smaller pores absorbs methanol at a higher temperature, implying a higher affinity for methanol. Similar effects were observed for water sorption on Ca(NO$_3$)$_2$/SiO$_2$ [22] and CaCl$_2$/SBA-15 [23] composites. Thus, the pore size of the host matrix can certainly affect the methanol sorption equilibrium of the composites, which can be used for intent modification of the sorption properties of CSPMs. Hereby, a proper choice of the host matrix is, indeed, vital for synthesizing composite sorbents of methanol of a CSPM type.

7.2.3 Effect of Synthesis Condition

Besides the aforementioned tools, certain synthesis conditions may significantly affect the sorbent properties as well. The procedure for preparing composite sorbents of methanol is the same as that for composite sorbents of water. It consists of dry impregnation of a preliminary dehydrated matrix with an aqueous salt solution followed by its thermal drying. As described in Chapter 4, this procedure results in the formation of two salt phases inside the pores, namely, the X-ray amorphous surface phase and the volume crystalline phase. The relative content of the phases depends on the nature of metal cation, the salt concentration, and the pH of the impregnating solution.

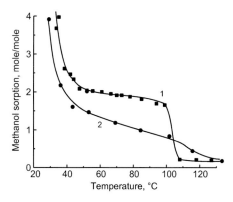

Figure 7.3 Isobars of methanol sorption on composites CaCl$_2$ (23%)/SiO$_2$ (15 nm) prepared at pH = 5.5 (1) and 8 (2), P_{MeOH} = 101 mbar [21].

In full agreement with this model, the alteration of the phase composition of CaCl$_2$/SiO$_2$ composites prepared at different pH of the solution dictates their methanol sorption properties (Fig. 7.3).

For the composite prepared in a neutral salt solution (pH = 5.5), the sorption isobar has a plateau corresponding to the formation of the stable solvate $CaCl_2 \cdot 2CH_3OH$, which undergoes a stepwise decomposition to $CaCl_2$ at $T = 100-110°C$. This behavior is typical for CSPMs containing a crystalline phase of the salt and, therefore, of the solvate. On the contrary, the isobar of methanol sorption by the composite prepared from an alkalescent solution (pH = 8) is a smooth curve typical for sorption by the X-ray amorphous salt. Interestingly, the latter composite ensures a higher sorption ability at $T > 105°C$ (Fig. 7.3).

7.2.4 Effect of Supplementary Salt

It is known that certain properties of a salt can be changed by a supplementary salt that forms a solid solution with the main salt. For instance, the addition of KCl and $CaBr_2 \cdot 6H_2O$ shifts the melting point of $CaCl_2 \cdot 6H_2O$ to that necessary for heat storage application [24]. The equilibrium of ammonia absorption by alkaline earth metal halides, $CaBr_2$ and $CaCl_2$, was modified by an additional salt in order to meet requirements of ammonia separation and storage during its synthesis [25]. This approach can be used to vary the methanol sorption properties of a single-salt CSPM. New composites based on binary salt systems (LiCl + LiBr) and ($CaCl_2$ + $CaBr_2$) confined to the silica gel pores with different salts fraction were prepared, and their sorption equilibrium with methanol was studied in Refs. [26–28]. Here we give just one demonstration that this method opens new opportunities in nanotailoring CSPMs. Other composite sorbents of water and ammonia based on binary salt systems are detailed below and in the literature [27, 28].

Each of the methanol sorption isotherms for the single-salt composites $CaCl_2/SiO_2$ and $CaBr_2/SiO_2$ has two steep segments (see curves 1 and 5 in Fig. 7.4), which correspond to the following solvation reactions:

$$CaCl_2 + 2CH_3OH = CaCl_2 \cdot 2CH_3OH \qquad (7.3)$$

$$CaCl_2 \cdot 2CH_3OH + 2CH_3OH = CaCl_2 \cdot 4CH_3OH \qquad (7.4)$$

$$CaBr_2 + 2CH_3OH = CaBr_2 \cdot 2CH_3OH \qquad (7.5)$$

$$CaBr_2 \cdot 2CH_3OH + 2CH_3OH = CaBr_2 \cdot 4CH_3OH \qquad (7.6)$$

CaBr$_2$ has a higher affinity for methanol vapor than CaCl$_2$. Indeed, the formation of the solvates CaBr$_2$·2CH$_3$OH and CaCl$_2$·2CH$_3$OH occurs at the methanol pressure 3–8 and 7–12 mbar, respectively. The formation pressure of CaBr$_2$·4CH$_3$OH (31–40 mbar) is much lower than that of CaCl$_2$·4CH$_3$OH (115–137 mbar) (Fig. 7.4). What is important, isotherms of the composites (CaCl$_2$ + CaBr$_2$)/SiO$_2$ lie between the curves for the single-salt composites, and the transition pressure gradually increases with the rise in the CaCl$_2$ content. This effect is due to the formation of a homogeneous solid solution of the salts [26, 27].

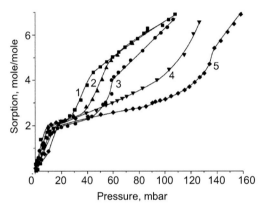

Figure 7.4 Isotherms of methanol sorption by composites (CaCl$_2$ + CaBr$_2$)/SiO$_2$ at T = 35°C and various molar ratio CaCl$_2$/CaBr$_2$ = 0:1 (1), 3:1 (2), 1:1 (3), 1:3 (4), and 1:0 (5). Reprinted from Ref. [27], Copyright 2013, with permission from Elsevier.

7.3 Methanol Sorption Properties of CSPMs

In this section, we first present the methanol sorption properties of several CSPMs and the effects of the active salt and the host matrix. Then we compare the basic feature of methanol sorption with those of water. Based on the data in Table 7.2, we selected potentially promising salts and prepared appropriate CSPMs whose properties are listed in Table 7.3. Silica gels KSK, Davisil Grade 646, Grace SP2, and Grace I254 were used as host matrices (Table 7.1) because they have large pore volume and relatively narrow pore size distribution.

All the composites were prepared by the dry impregnation of the silica gel with a saturated salt solution in order to increase the final salt content (Table 7.3). The majority of CSPMs, whose advanced properties are presented in this section, are based on halides of alkali- and alkali-earth metals that have enhanced affinity for methanol vapor as ascertained above (see Sec. 7.2.1).

Table 7.3 Salt content and pore volume of the composites for methanol sorption

Composite	Matrix	C (wt.%)	V_p (cm^3/g)
LiCl/SiO$_2$(a)	KSK	25.6	0.72
LiCl/SiO$_2$(b)	Grace SP2	30.6	1.05
LiBr/SiO$_2$	Grace SP2	29.2	1.05
MgCl$_2$/SiO$_2$	Davisil Grade 646	25.4	0.85
Ca(NO$_3$)$_2$/SiO$_2$	"	28.4	0.82
NiBr$_2$/SiO$_2$(a)	"	28.2	0.82
MnCl$_2$/SiO$_2$	"	28.2	0.82
CuCl$_2$/SiO$_2$	"	28.0	0. 83
CoCl$_2$/SiO$_2$	"	27.0	0.84
MgBr$_2$/SiO$_2$	"	20.8	0.91
BaCl$_2$/SiO$_2$	"	26.4	0.85
CaBr$_2$/SiO$_2$(a)	KSK	25.5	0.86
LiCl/SiO$_2$(c)	Grace I254	24.6	0.68
NiBr$_2$/SiO$_2$(b)	Grace SP2	44.6	0.83
CaBr$_2$/SiO$_2$(b)	"	45.9	0.84
CaCl$_2$/SiO$_2$(a)	Davisil Grade 646	25.0	0.85
CaCl$_2$/SiO$_2$(b)	Grace I 254	23.0	0.69

Source: Ref. [21].

7.3.1 Composites "CaCl$_2$–Silica Gel"

Isobars of methanol desorption for composite CaCl$_2$ (25 wt.%)/silica (15 nm) have a plateau corresponding to $n \approx 2$ (Fig. 7.5) [18], which indicates the formation of a solid crystalline solvate (methanolate) CaCl$_2$·2CH$_3$OH due to reaction (7.3).

This complex is stable over a wide temperature range. At a temperature higher than the right boundary of the plateau, the solvate undergoes decomposition toward the salt. The CaCl$_2$–CH$_3$OH system (at $n \leq 2$) consists of two components and three phases (CaCl$_2$, CaCl$_2$·2CH$_3$OH, CH$_3$OH) and, according to the Gibbs phase rule, is mono-variant. That reveals as a sharp rise in the sorption isobars (Fig. 7.5). As the temperature becomes lower than the left boundary of the plateau, the system exhibits a sharp increase in the methanol uptake up to $n \approx 8$–10.

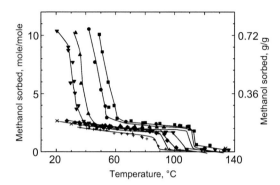

Figure 7.5 Isobars of methanol desorption for $CaCl_2$ (25 wt.%)/silica (15 nm) composite at P_{MeOH} = 21 (+), 41 (×), 53 (♦), 101 (▼), 153 (▲,), 226 (•), and 300 (■) mbar. Reprinted by permission from Springer Nature from Ref. [18], Copyright 2007.

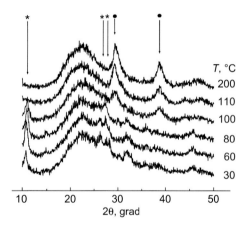

Figure 7.6 In situ X-ray diffraction patterns of $CaCl_2$ (25 wt.%)/silica (15 nm)–CH_3OH system at P_{MeOH} = 101 mbar, * –$CaCl_2·2CH_3OH$ [29], • –$CaCl_2$. Reprinted by permission from Springer Nature from Ref. [18], Copyright 2007.

The formation of crystalline phase $CaCl_2·2CH_3OH$ due to the methanol sorption on the $CaCl_2$ (25 wt.%)/silica (15 nm) composite is confirmed by XRD in situ measurements (Fig. 7.6). At temperature $T > 80°C$, the sample comprises a dry $CaCl_2$ located inside the pores. As the temperature decreases, the intensity of reflexes attributed to $CaCl_2$ diminishes and the reflexes of $CaCl_2·2CH_3OH$ appear [18, 29].

Hence, dry CaCl$_2$ transforms to di-methanolate CaCl$_2$·2CH$_3$OH, which agrees well with the adsorption data. At a temperature lower than the left boundary of the plateau (T = 30°C), the intensity of the CaCl$_2$·2CH$_3$OH reflexes decreases, but no other reflex appears. That is likely to indicate the formation of CaCl$_2$–CH$_3$OH solution due to further methanol sorption. The temperatures of the CaCl$_2$·2CH$_3$OH transitions toward the dry salt and CaCl$_2$–CH$_3$OH solution correlate well with those evaluated from the sorption measurements (Fig. 7.5). No formation of other crystalline solvates CaCl$_2$·nCH$_3$OH with n = 1, 3, 4, and 6 typical of the bulk system is detected by the XRD analysis of the CaCl$_2$ (25 wt.%)/silica (15 nm) composite.

When presenting the sorption data as a function of the methanol relative pressure η, the temperature-invariant curve of methanol sorption can be obtained (Fig. 7.7). This curve for the CaCl$_2$ (25 wt.%)/silica (15 nm) composite repeats the typical features of the isobars: a stepwise rise in sorption at $0.01 \leq \eta \leq 0.02$, the plateau at $0.02 \leq \eta \leq 0.4$, and a steep increase at $\eta > 0.45$. Again the plateau is attributed to the formation of stable crystalline methanolate CaCl$_2$·2CH$_3$OH. Further methanol sorption results in the formation of the methanolate CaCl$_2$·6CH$_3$OH, which melts forming CaCl$_2$–CH$_3$OH solution inside the pores. The composition of the solution changes continuously, which results in the gradual growth of methanol uptake.

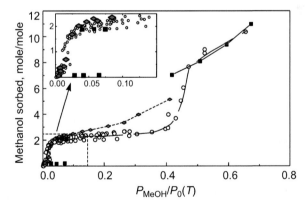

Figure 7.7 Temperature-invariant curves of methanol sorption on the CaCl$_2$ (25 wt.%)/silica (15 nm) (o) and CaCl$_2$ (23 wt.%)/silica (6 nm) (♦) composites as well as literature data for the bulk CaCl$_2$–MeOH system (■). Adapted by permission from Springer Nature from Ref. [18], Copyright 2007.

The temperature-invariant curve of methanol sorption on the composite with smaller pores, CaCl$_2$ (23 wt.%)/silica (6 nm), qualitatively differs from that for the CaCl$_2$ (25 wt.%)/silica (15 nm) (Fig. 7.7). A step-like rise in sorption at $0.01 \leq \eta \leq 0.02$ is followed by a gradual increase in the uptake at $\eta > 0.03$, which is likely to indicate the formation of a CaCl$_2$–methanol solution inside the pores. No plateau at $n = 2$ is observed. The inflection at $n \approx 2$ is probably caused by the fact that the crystalline methanolate CaCl$_2\cdot$2CH$_3$OH coexists in the silica pores with a salt–methanol solution.

Both these sorbents demonstrate higher affinity for methanol as compared to the bulk salt (see the insert in Fig. 7.7), which favors deeper methanol removal from gaseous mixtures. At $\eta > 0.45$, the sorption equilibrium of the CaCl$_2$ (25 wt.%)/SiO$_2$ (15 nm) composite coincides well with that for the bulk solution [21]. Confinement of the CaCl$_2$–methanol solution into the 15 nm silica pores does not affect its sorption properties, which agrees well with a CaCl$_2$–water solution in the same pores. At $0.15 < \eta < 0.43$, the composite CaCl$_2$ (23 wt.%)/silica (6 nm) is superior to the CaCl$_2$ (25 wt.%)/silica (15 nm) composite in the methanol sorption capacity (Fig. 7.7).

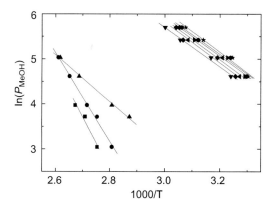

Figure 7.8 The isosters of methanol desorption from CaCl$_2$ (25 wt.%)/silica (15 nm) at n = 0 (■), 1 (●), 2 (▲), 3 (▼), 4 (♦), 5 (◄), 6 (►), 7 (●), and 8 (★). Reprinted by permission from Springer Nature from Ref. [18], Copyright 2007.

The isosters of methanol desorption from the CaCl$_2$ (25 wt.%)/silica (15 nm) composite plotted from the isobaric data are shown in Fig. 7.8 for n = 0–8 in the ln(P_{MeOH}) versus $1/T$ presentation. They are straight lines with the slope dependent on the methanol uptake. The isosteric heat of methanol sorption, ΔH_{is}, calculated from the slope is 38 ± 2 kJ/mol at $n > 2$, which is close to the heat of methanol evaporation $\Delta H_{ev} = 35.4$ kJ/mol. As the methanol uptake decreases down to $n < 2$, ΔH_{is} rises up to 81 ± 4 kJ/mol. The increase in the isosteric heat can be attributed to the formation of crystalline solvate $CaCl_2 \cdot 2CH_3OH$ with strong chemical bonds between $CaCl_2$ and methanol molecules. The sorption isosters of the $CaCl_2$ (23 wt.%)/silica (6 nm) composite follow the same trend. The ΔH_{is} value increases from 35 to 89 kJ/mol, when the sorption n decreases from 4 to 1 mol/mol (not presented).

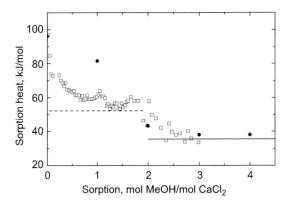

Figure 7.9 Dependence of the calorimetric (□) and isosteric (●) heats of methanol sorption on $CaCl_2$/silica (15 nm). Dashed line represents the enthalpy of reaction (7.3) in bulk; solid line indicates the heat of methanol evaporation in bulk. Reprinted by permission from Springer Nature from Ref. [18], Copyright 2007.

The data on the isosteric heat of methanol sorption are quite consistent with the direct measurement of the heat ΔH_{cal} of methanol sorption on the $CaCl_2$ (25 wt.%)/silica (15 nm) composite by a Tian–Calvet calorimeter (Fig. 7.9). As n increases from 0 to 1, the sorption heat decreases from 97 to 60 kJ/mol. The high value of the sorption heat at $n < 0.5$ is due to the fact that the first portions of methanol are likely to adsorb on active centers of the silica gel surface. Thus, the sorption heat depends on the chemical nature of these centers. At $0.5 < n < 2$, the sorption heat is essentially constant, $\Delta H_{cal} = 60 \pm 3$ kJ/mol, and is likely to be attributed to the formation of crystalline methanolate $CaCl_2 \cdot 2CH_3OH$. The value of ΔH_{cal} is close to the formation enthalpy of the bulk methanolate $CaCl_2 \cdot 2CH_3OH$:

ΔH = 51.7 kJ/mol [4]. Thus, the sorption heat measured at 0.5 < n < 2 is associated with the formation of $CaCl_2 \cdot 2CH_3OH$ in the silica pores. Further increase in n decreases the sorption heat to ΔH_{cal} = 38 ± 3 kJ/mol, which is close to the heat of methanol evaporation.

The specific methanol sorption capacity of the $CaCl_2$ (25 wt.%)/ SiO_2 (15 nm) composite may reach 0.7 g/g at η = 0.7. It is pertinent to note that most of the methanol sorbed (75–80%) can be desorbed by heating the sorbent by some 15–30°C (Fig. 7.5). The remaining methanol can be removed at a much higher temperature of 100–120°C.

7.3.2 Composites "Lithium Halides–Silica Gel"

The following composites with various salt content were prepared by the dry impregnation method: LiCl (31 wt.%)/silica, LiCl (21 wt.%)/silica, LiBr (29 wt.%)/silica, and LiBr (24 wt.%)/silica. The commercial silica gel Grace Davison (d_{av} = 14 nm, S_{sp} = 360 m²/g, V_p = 1.3 cm³/g) was used as the host matrix. Other details of the composite synthesis and study may be found elsewhere [19].

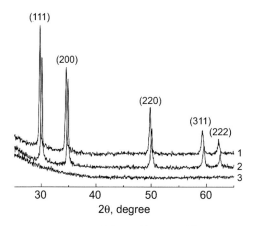

Figure 7.10 XRD in situ pattern of the LiCl (31 wt.%)/SiO_2 composite measured in a pure helium flow at 100°C (1), as well as in the helium flow saturated with methanol at T = 45 (2) and 30°C (3). Reprinted from Ref. [19], Copyright 2008, with permission from Elsevier.

The XRD patterns of the composites LiCl (31 wt.%)/silica and LiBr (29 wt.%)/silica recorded under helium flow at 100°C indicate that a crystalline phase of LiCl (Fig. 7.10) and LiBr (not presented) with a cubic structure forms inside the silica pores. The size of coherently scattering domains is 20–30 nm, which is close to the silica pore size. Bulk crystals of LiCl and LiBr are not detected in these composites. The XRD in situ patterns of the composite LiCl (31 wt.%)/silica recorded under methanol vapor at $T > 40°C$ (Fig. 7.10) are similar to those recorded for the dry sample. The broadening of the spacing parameter from 0.5149 nm to 0.5171 nm at 45 and 100°C, respectively, is attributed to the thermal expansion of the material. The composite saturated with methanol at $T < 45°C$ appears to become X-ray amorphous, which may indicate the formation of an LiCl–methanol solution inside the pores. Surprisingly, intermediate phases of crystalline solvates LiCl·CH$_3$OH or LiCl·3CH$_3$OH formed in the bulk LiCl–CH$_3$OH system are not detected during methanol sorption on the composite.

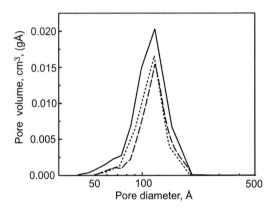

Figure 7.11 Pore size distribution of the starting silica gel (solid line), LiCl (21 wt.%)/SiO$_2$ (dotted line) and LiBr (29 wt.%)/SiO$_2$ (dashed line) composites. Reprinted from Ref. [19], Copyright 2008, with permission from Elsevier.

The data on the porous structure of the raw silica gel and the composite sorbents LiCl (21 wt.%)/SiO$_2$ and LiBr (29 wt.%)/SiO$_2$ are displayed in Fig. 7.11 and Table 7.4. Modification of the silica gel by the salt results in a decrease in the total pore volume and the surface area. The pore volume of the composites per 1 g of the raw silica gel is measured to be V_p = 1.13 and 1.11 cm^3/g for the LiCl- and LiBr-based composites, respectively. Taking into account the densities of LiCl and LiBr, r = 2.068 and 3.464 g/cm^3, the volume occupied by the salt in 1 g of the silica gel can be calculated as V_S = 0.13 and 0.12 cm^3/g. Subsequently, the empty pore volume of the composite per 1 g of the silica can be evaluated as V_p(comp) = V_p(SiO$_2$) − V_s = 1.16 and 1.17 cm^3/g for LiCl (21 wt.%)/SiO$_2$ and LiBr (29 wt.%)/SiO$_2$, respectively. The calculated values are very close to those experimentally obtained from the nitrogen adsorption data (Table 7.4). The surface areas of the composites are measured to be S_{sp}(comp) = 201 and 204 m^2/g for the LiCl- and LiBr-based composites, respectively, which are much smaller than S_{sp} (silica) = 361 m^2/g for the raw silica. To quantify possible blocking of the silica surface by the salt, the normalized surface area (NSA) was calculated as

$$\mathrm{NSA} = \frac{S_{sp}(\mathrm{comp})}{(1-y) \cdot S_{sp}(\mathrm{silica})} \tag{7.7}$$

where y is the weight fraction of the salt in the sample. The calculated values of NSA are 0.73 and 0.79 for the LiCl- and LiBr-based composites, respectively. The NSA < 1 is probably caused by the fact that the salt particles inside the silica pores can block a part of the narrow pores, which contributes more to the pore area than to the pore volume of the sample [30]. The increase in the average pore diameter of the composite (Fig. 7.11 and Table 7.4) from 14 nm for the raw silica gel to 16–17 nm for the composites confirms the assumption on the partial blocking of the narrow pores of silica gel by the incorporated salt.

Table 7.4 Pore structure parameters of the initial silica gel, composites LiCl (21 wt.%)/silica and LiBr (29 wt.%)/silica obtained from the nitrogen adsorption isotherms

| Sample | $V_p{}^a$ (cm^3/g) | | $S_{sp}{}^b$ (m^2/g) per gram of sample | NSA^c | $d_{av}{}^d$ (nm) |
	Per gram of the composite	Per gram of silica			
SiO$_2$	1.29	1.29	331	—	14
LiCl/SiO$_2$	0.88	1.13	209	0.73	17
LiBr/SiO$_2$	0.79	1.11	203	0.79	16

^aV_p — pore volume; $^bS_{sp}$ — BET surface area; ^{c}NSA — normalized surface area; $^dd_{av}$ — average pore diameter
Source: Reprinted from Ref. [19], Copyright 2008, with permission from Elsevier.

Thus, the data on XRD and nitrogen adsorption prove that the salt is likely encapsulated in the silica gel pores and partially blocks the narrow pores [19].

The isobars of methanol sorption on LiCl (31 wt.%)/SiO$_2$ and LiBr (29 wt.%)/SiO$_2$ composites (Fig. 7.12) demonstrate their strong affinity for methanol. They have a regular shape but shift toward higher temperatures with an increase in the methanol vapor pressure. For both composites, the maximum amount of methanol sorbed w_{max} = 0.75–0.80 g/g is 2–5 times larger than that for non-modified silica gel (0.15 g/g, Fig. 7.1b) as well as for conventional zeolites and activated carbons (see Sec. 7.1). Again, this great enhancement is due to the predominant contribution of the salt embedded in the silica gel matrix to methanol sorption.

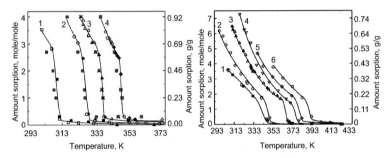

Figure 7.12 Isobars of methanol sorption (solid symbols) and desorption (open symbols) for (a) LiCl (31 wt.%)/silica at P_{MeOH} = 47 (1), 114 (2), 186 (3), and 300 (4) mbar. The crossed symbols denote equilibrium data taken from the experimental isosters for LiCl (21 wt.%)/silica; (b) LiBr (29 wt.%)/silica at P_{MeOH} = 30 (1), 43 (2), 100 (3) and 160 (4), 227 (5) and 347 (6) mbar. Reprinted from Ref. [19], Copyright 2008, with permission from Elsevier.

At a high temperature, composite LiCl (31 wt.%)/silica sorbs a small amount of methanol ($w \leq 0.05$ g/g, Fig. 7.12a), which can be attributed to the methanol adsorption on active centers of the silica gel surface. As the temperature decreases, the salt starts to absorb methanol, and the uptake reaches w = 0.7 g/g or $n \approx 3$ mol/mol within an extremely narrow temperature range of 3–5°C. This steep rise in the methanol uptake is caused by reaction (7.2). The formed complex LiCl·3CH$_3$OH melts inside the silica pores and transforms into an LiCl–CH$_3$OH solution [19]. Afterward, a gradual rise in methanol sorption is observed (Fig. 7.12a) due to methanol sorption by the solution. The sorption and desorption isobars nearly coincide, so that no sorption/desorption hysteresis is revealed.

The main features of methanol sorption by the LiBr (29 wt.%)/silica composite (Fig. 7.12b) are similar to those found for LiCl (31 wt.%)/silica. The sharp increase in methanol uptake from $n \approx 0$ to 1 mol/mol is followed by a gradual increase in the range $n = 1–7$ mol/mol, which is typical for an LiBr–CH$_3$OH solution [31, 32]. The formation of stable crystalline solvates LiBr·nCH$_3$OH ($n = 1, 3, 4$), known for the bulk system [31], is not detected. The sorption hysteresis is absent.

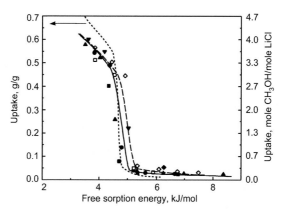

Figure 7.13 Temperature-independent curves of methanol sorption on (solid symbols,———) and desorption from (open symbols, – – –) LiCl (20%)/silica and LiCl (31%)/silica (Grace 8926.02) composites (------). Reprinted with permission from Ref. [20], Copyright 2009, American Chemical Society.

All sorption isobars of LiCl/silica composites appear to lie satisfactorily along the same curve when presented as the uptake n versus the free sorption energy $\Delta F = -RT\ln(P_{\text{MeOH}}/P_s)$, where P_s is the saturation methanol pressure at temperature T (Fig. 7.1b and 7.13). Consequently, the methanol sorption process on the composites tends to obey the Polanyi principle of temperature invariance [33].

Comparison of the sorption data of the composite and bulk LiCl (Fig. 7.1b) shows the following changes in sorption equilibrium [19]:
- The salt's affinity for methanol increases due to its confinement to the pores;
- No sorption–desorption hysteresis is found;

- The desorption branch of the bulk salt demonstrates a stepwise shape with a pronounced plateau at $\Delta F = 4250-5100$ J/mol, which corresponds to $n \approx 1$ mol/mol, indicating the formation of stable $LiCl \cdot CH_3OH$ solvate. This agrees with the data of Ref. [34], which describe the formation of $LiCl \cdot CH_3OH$ during the decomposition of $LiCl \cdot 3CH_3OH$. At $\Delta F > 5100$ J/mol, the bulk solvate $LiCl \cdot CH_3OH$ decomposes giving LiCl.

As already mentioned above (see Sec. 7.2.2), the methanol sorption by the composite cannot be represented as a simple addition of the sorption by the bulk salt and the host material (silica) taken with appropriate weight coefficients (Fig. 7.1b). The reason for this non-additivity could be the dispersion of LiCl to nano-sized particles inside the silica pores, resulting in quantitative and qualitative altering of its sorption properties (the size effect). With a decrease in particle size, the bulk properties are lost as the fraction of surface atoms becomes large.

The isosteres of methanol sorption on the LiCl (21 wt.%)/silica and LiBr (24 wt.%)/silica composites are straight lines in the Clapeyron–Clausius coordinates. The standard enthalpy ΔH° and entropy ΔS° of methanol sorption were evaluated using the equation $\ln P_{MeOH} = \Delta H^\circ/RT - \Delta S^\circ/R$ (Table 7.5). These values are nearly constant for the whole methanol uptake range for both composites. This indirectly confirms that crystalline methanolates do not form during methanol sorption, since in these complexes methanol molecules would be linked by stronger chemical bonds with the salt. The values $\Delta H_{is} = -41.7 \pm 2.5$ and -41.8 ± 2.5 kJ/mol, obtained for LiCl (21 wt.%)/silica and LiBr (24 wt.%)/silica, respectively, exceed the enthalpy of methanol condensation, -35.2 kJ/mol. This indicates relatively strong sorbate–sorbent interaction, which is probably caused by solvation of ions by methanol molecules in the solution and the subsequent formation of $Li^+(CH_3OH)_n$, $X^-(CH_3OH)_k$, $Li^+(Li^+X^-)_s(CH_3OH)_m$, and $X^-(Li^+X^-)_p(CH_3OH)_r$ clusters, where X^- is Cl^- or Br^- [35, 36]. The standard entropies of sorption $\Delta S^\circ = -105 \pm 10$ and -102 ± 10 J/(mol·K) are found for LiCl (21 wt.%)/silica and LiBr (24 wt.%)/silica, respectiely.

164 | *Composite Sorbents of Methanol*

Table 7.5 Isosteric enthalpy and standard entropy of methanol sorption on LiCl (21 wt.%)/silica and LiBr (24 wt.%)/silica

n (mol/mol)	w (g/g)	$-\Delta H_{is}$ (kJ/mol)	$-\Delta S^{\circ}$ (J/(mol K))
	LiCl (21 wt.%)/silica		
0.3	0.05	39.3	99.8
1.0	0.16	41.6	108.8
1.3	0.21	43.1	107.2
2.0	0.32	42.4	109.7
2.7	0.43	44.7	101.4
4.0	0.63	41.8	108.0
4.5	0.71	38.7	99.7
	LiBr (24 wt.%)/silica		
0.5	0.04	40.9	85.8
1.1	0.10	41.7	95.5
2.0	0.18	42.9	103.0
2.8	0.25	43.7	110.5
4.1	0.36	40.7	108.8
5.0	0.45	40.8	109.9

Source: Reprinted from Ref. [19], Copyright 2008, with permission from Elsevier.

Taking into account the methanol sorption capacity of the composites w_{max} = 0.8 g/g and the isosteric enthalpy of methanol sorption ΔH_{is} = −41.7 kJ/mol, their heat storage capacity can be evaluated as E_{max} = 1.0 kJ/g, which far exceeds that of conventional methanol adsorbents. The results obtained demonstrate that the LiCl/silica and LiBr/silica composites may be promising for heat transformation applications such as heating, cooling, and energy storage (see Chapters 16 and 17).

To resume, in this chapter we have presented new composites "salt in porous matrix" for methanol sorption. It has been proved that the phase composition and the sorption equilibrium of the composites with methanol vapor can be intently managed by an intelligent choice of a suitable salt, the porous structure of the host matrix, synthesis conditions, and the supplementary salt. All these tools can be used to adjust the real sorbent to the optimal one, which has been theoretically predicted before the synthesis.

This study of methanol sorption on new CSPMs based on lithium and calcium halides confined to various silica gels has revealed the following regularities:

- Modification by the salt greatly increases (by 2–5 times) the methanol sorption capacity (up to 0.8–1.0 g/g). Hence, these composites may be interesting for methanol removal from gas mixtures. The main sorbing agent is the salt, which forms with methanol solid solvates and then solution;
- Confinement of the salt to the matrix pores may dramatically change the salt sorption properties:
  a. Certain crystalline solvates detected in bulk are not observed in the confined state (e.g., $LiCl \cdot CH_3OH$ and $CaCl_2 \cdot nCH_3OH$ at $n = 1, 3,$ and 4);
  b. The formation of the salt solvates occurs at lower relative methanol pressure than in bulk so that the confined salt is a better agent for methanol removal from gas mixtures;
  c. Sorption–desorption hysteresis is essentially more narrow for the dispersed salt or even disappears.

All these regularities are qualitatively similar to those revealed for the composite sorbents of water presented in Chapters 5 and 6.

References

1. Erich Pietsch, E. H. (Ed.) (1957) *Gmelins Handbuch der Anorganischen Chemie, Calcium Eraganzungs Band* (Verlag Chemie GmbH), Sn. 28, Tl. B, Rf. 2, p. 523.

2. Gmelin Data: 2000–2005 Gesellschaft Deutscher Chemiker licensed to MDL Information Systems GmbH; 1988–1999: Gmelin Institut fuer Anorganische Chemie und Grenzgebiete der Wissenschaften.

3. Loid, E., Brown, C. B., Bonnel, D. G. R., and Jones, W. J. (1928). Equilibrium between alcohols and salts. Part II, *J. Chem. Soc.*, pp. 658–666.

4. Offenhartz, P. O'D., Brown, F. C., Mar, R., and Carling, R. W. (1980) A heat pump and thermal storage system for solar heating and cooling based on the reaction of calcium chloride and methanol vapour, *J. Sol. Energy Eng.*, **102**, pp. 59–65.

5. Dekany, I., Szanto, F., Armin, W., and Lagaly, G. (1986) Interactions of hydrophobic layer silicates with alcohol-benzene mixtures, *Ber. Bunsen-Ges. Phys. Chem.*, **90**, pp. 422–427.

6. Meunier, F. (2013) Adsorption heat powered heat pumps, *Appl. Therm. Eng.*, **61**, pp. 830–836.

7. Deng, J., Wang, R. Z., and Han, G. Y. (2011) A review of thermally activated cooling technologies for combined cooling, heating and power systems, *Prog. Energ. Combust.*, **37**, pp. 172–203.

8. Kuczynski, M., Browne, W. I., Fontein, H. I., and Westerterp, K. R. (1987) Methanol synthesis in a counter current gas–solid trickle flow reactor an experimental study, *Chem. Eng. Sci.*, **42**, pp. 1887–1898.

9. Kruglov, A. V. (1994) Methanol synthesis in a simulated counter-current moving-bed adsorptive catalytic reactor, *Chem. Eng. Sci.*, **49**, pp. 4699–4716.

10. Yang, R. (2003) *Adsorbents: Fundamentals and Applications* (Wiley), p. 410.

11. Carrott, P. J. M., Ribeiro Carrott, M. M. L., and Cansado, I. P. P. (2001) Reference data for the adsorption of methanol on carbon materials, *Carbon*, **39**, pp. 193–200.

12. Janchen, J., van Wolput, J. H. M. C., van Well, W. J. M., and Stach, H. (2001) Adsorption of water, methanol and acetonitrile in ZK-5 investigated by temperature programmed desorption, microcalorimetry and FTIR, *Thermochim. Acta*, **379**, pp. 213–225.

13. Kuczynski, M. and Westerterp, K. R. (1986) Retrofit methanol plants with the converter system, *Hydrocarb. Process.*, **65**, pp. 80–83.

14. Schueth, F., Sing, K. S. W., and Weitkamp, J. (Eds.) (2002) *Handbook of Porous Solids*, vol. 1–5 (Wiley-VCH, Weinheim, Germany).

15. Otowa, T., Tanibata, R., and Itoh, M. (1993) Production and adsorption characteristics of MAXSORB: High-surface-area active carbon, *Gas Separat. Purifi.*, **7**, pp. 241–245.

16. Lourdudoss, S. and Stymne, H. (1987) Energy storing absorption heat pump process, *Int. J. Energy Res.*, **11**, pp. 263–274.

17. Gordeeva, L., Freni, A., Restuccia, G., and Aristov, Yu. I. (2007) Methanol sorbents "salt inside mesoporous silica": The screening of salts for adsorptive air conditioning driven by low temperature heat, *Ind. Eng. Chem. Res.*, **46**, pp. 2747–2752.

18. Aristov, Yu. I., Gordeeva, L. G., Pankratiev, Yu. D., Plyasova, L. M., Bikova, I. V., Freni, A., and Restuccia, G. (2007) Sorption equilibrium of methanol on new composite sorbent "CaCl$_2$/Silica Gel", *Adsorption*, **13**, pp. 121–127.

19. Gordeeva, L. G., Freni, A., Kriger, T. A., Resticcia, G., and Aristov, Yu. I. (2008) Composite sorbents "lithium halides in silica gel pores": Methanol sorption equilibrium, *Micropor. Mesopor. Mat.*, **112**, pp. 254–261.

20. Gordeeva, L. G., Freni, A., Aristov, Yu. I., and Restuccia, G. (2009) Composite sorbent of methanol "lithium chloride in mesoporous silica gel" for adsorption cooling machines: Performance and stability evaluation, *Ind. Chem. Engn. Res.*, **48**, pp. 6197–6202.

21. Gordeeva, L. G. (2013) Nano-tailoring new composite sorbents "salt in porous matrix," Doctoral Thesis, Boreskov Institute of Catalysis.

22. Simonova, I. A. and Aristov, Yu. I. (2005) Sorption properties of calcium nitrate dispersed in silica gel: the effect of pore size, *Rus. J. Phys. Chem.*, **79**, pp. 1477–1481.

23. Ponomarenko, I. V., Glaznev, I. S., Gubar, A. V., Aristov, Yu. I., and Kirik, S. D. (2010) Synthesis and water sorption properties of a new composite "$CaCl_2$ confined to SBA-15 pores," *Micropor. Mesopor. Mater.*, **129**, pp. 243–250.

24. Feilchenfeld, H., Fuchs, J., Kahana, F., and Sarig, S. (1985) The melting point adjustment of calcium chloride hexahydrate by addition of potassium chloride or calcium bromide hexahydrate, *Solar Energy*, **34**, pp. 199–201.

25. Liu, C. Yi. and Aika, K. (2004) Ammonia absorption into alkaline earth metal halide mixtures as an ammonia storage material, *Ind. Eng. Chem. Res.*, **43**, pp. 7484–7491.

26. Gordeeva, L. G., Grekova, A. D., Krieger, T. A., and Aristov, Yu. I. (2009) Adsorption properties of composite materials (LiCl + LiBr)/silica, *Micropor. Mesopor. Mater.*, **126**, pp. 262–267.

27. Gordeeva, L. G., Grekova, A. D., Krieger, T. A., and Aristov, Yu. I. (2013) Composite sorbents "binary salts inside porous matrix" for adsorption heat transformation, *Appl. Therm. Eng.*, **50**, pp. 1633 1638.

28. Grekova, A. D., Veselovskaya, J. V., Tokarev, M. M., Krieger, T. A., Shmakov, A. N., and Gordeeva, L. G. (2014) Ammonia sorption on the composites ($BaCl_2$ + $BaBr_2$) inside vermiculite pores, *Colloids Surf. A*, **448**, pp. 169–174.

29. Gillier-Pandraud, H. and Philoche-Levisalles, M. (1979) Structure cristalline du compose, *C. R. Acad. Sci.*, **273**, pp. 949–951.

30. Vradman, L., Landau, M. L., Kantorovich, D., Koltypin, Y., and Gedanken, A. (2005) Evaluation of metal oxide phase assembling mode inside the nanotubular pores of mesostructured silica, *Micropor. Mesopor. Mater.*, **79**, pp. 307–318.

31. Skabichevski, P. (1969) Osmotic coefficients of LiCl and LiBr solutions in methanol, *Russ. J. Phis. Chem.*, **XLIII**, p. 2556.

32. Conde, M. R. (2004) Properties of aqueous solutions of lithium and calcium chlorides: Formulations for use in air conditioning equipment design, *Int. J. Therm. Sci.*, **43**, pp. 367–382.

33. Polanyi, M. (1932) Theory of adsorption of gases and vapours, *Trans. Faraday Society.*, **28**, p. 316.

34. Gmelin Data: 2000–2005 Gesellschaft Deutscher Chemiker licensed to MDL Information Systems GmbH; 1988–1999: Gmelin Institut fuer Anorganische Chemie und Grenzgebiete der Max-Planck-Gesellschaft zur Foerderung der Wissenschaften. Vol. Li: SVol.; pp. 395–440.

35. Megyes, T., Radnay, T., and Wakisaka, A. (2002) Complementary relation between ion–counterion and ion–solvent interaction in lithium halide–methanol solutions, *J. Phys. Chem. A*, **106**, p. 8059.

36. Mochizuki, S. and Wakisaka, A. (2002) Solvation for ions and counterions: Complementary relation between ion–counterion and ion–solvent interaction, *J. Phys. Chem. A*, **106**, pp. 5095–5100.

Chapter 8

Composite Sorbents of Ammonia

Ammonia (NH_3) is produced on a large scale by the Haber–Bosch process [1], which is one of the most important inventions of the 20th century. Ammonia is mainly used for fertilizing agricultural crops, production of plastics, fibers, explosives, intermediates for dyes and pharmaceuticals. NH_3 vapor has a harmful potential for human beings due to its high water solubility and basicity. Therefore, the adsorptive abatement of ammonia is an important practical process [2]. Ammonia adsorption was also suggested for adsorptive cooling, heat pumping [3], and hydrogen storage [4]. Therefore, new effective ammonia adsorbents are highly welcome.

This chapter addresses new composites "salt in porous matrix" (CSPMs) for ammonia sorption that utilize a reversible chemical reaction

$$S + nNH_3 = S{\cdot}nNH_3 \qquad (8.1)$$

where the complex $S{\cdot}nNH_3$ retains a large mass of ammonia because n may reach 12 [5–7]. One of the main problems in using inorganic salts as chemosorbents of ammonia is salt swelling [8] due to reaction (8.1). This results in mechanical destruction of the material, which reduces bed porosity and slows down the rate of ammonia sorption.

Nanocomposite Sorbents for Multiple Applications
Yu. I. Aristov
Copyright © 2020 Jenny Stanford Publishing Pte. Ltd.
ISBN 978-981-4267-50-2 (Hardcover), 978-981-4303-15-6 (eBook)
www.jennystanford.com

To overcome this problem, it was suggested to separate the salt particles by mixing with a soft material that can accommodate the swelling, e.g., expanded graphite [9], or intercalating a salt with a carbonaceous matrix, such as activated carbons and carbon fibers [10–12]. Filling the matrix pores with the salt provides space separation of salt microcrystals, which prevents agglomeration and improves mass transfer. The confinement of the salt in the pores of the host matrix is an important step in the intelligent design of composites with sorption properties required for specific applications [13].

First, we comprehensively consider synthesis and sorption properties of CSPMs based on a single salt ($CaCl_2$, $SrCl_2$, and $BaCl_2$) confined to alumina, activated carbons, and vermiculite as host matrices [14]. Variation in the salt's sorption properties due to its confinement to the pores is studied and compared with CSPMs for sorption of water and methanol presented in the previous chapters. Then we briefly present new CSPMs based on two confined salts.

8.1 CSPMs for Ammonia Sorption

Common ammonia adsorbents are zeolites, alumina, and activated carbons [15–17]. Usually, ammonia is not well retained by physical adsorption on these adsorbents; therefore, new effective NH_3 sorbents are welcome. In this section, we briefly discuss the effects of the active salt and the host matrix on the ammonia sorption properties of the new CSPMs.

8.1.1 Effect of Active Salt

Equilibrium pressure P and temperature T for reaction (8.1) are linked by the integral van't Hoff equation:

$$\ln P = \frac{\Delta H^\circ}{RT} - \frac{\Delta S^\circ}{R} \tag{8.2}$$

where ΔH° and ΔS° are changes in the standard enthalpy and entropy during the chemical reaction, and R is the universal gas constant. These values depend on the salt's nature. Here we mainly consider the CSPMs based on $CaCl_2$ [18–20], $SrCl_2$ [14], and $BaCl_2$ [14, 21–23], because these salts can bind a large quantity of NH_3 and appropriate composites appear to be promising for adsorptive cooling, ammonia scrubbing, and other important applications.

The considered bulk salts react with ammonia forming crystalline ammines (see solid lines in Figs. 8.1–8.3). Calcium chloride forms four stable ammines by the following reactions [5] (Fig. 8.1):

$$CaCl_2 + NH_3 \leftrightarrows CaCl_2 \cdot NH_3 \qquad (8.3)$$

$$CaCl_2 \cdot NH_3 + NH_3 \leftrightarrows CaCl_2 \cdot 2NH_3 \qquad (8.4)$$

$$CaCl_2 \cdot 2NH_3 + 2NH_3 \leftrightarrows CaCl_2 \cdot 4NH_3 \qquad (8.5)$$

$$CaCl_2 \cdot 4NH_3 + 4NH_3 \leftrightarrows CaCl_2 \cdot 8NH_3 \qquad (8.6)$$

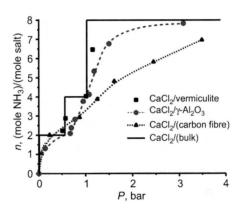

Figure 8.1 Isotherms of ammonia sorption at 30°C for bulk CaCl$_2$ and the composites "CaCl$_2$/matrix."

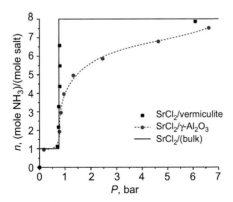

Figure 8.2 Isotherms of ammonia sorption at 30°C for bulk SrCl$_2$ and the composites "SrCl$_2$/matrix."

Strontium chloride forms only two stable ammines [6, 14] (Fig. 8.2):

$$SrCl_2 + NH_3 \leftrightarrows SrCl_2 \cdot NH_3 \qquad (8.7)$$

$$SrCl_2 \cdot NH_3 + 7NH_3 \leftrightarrows SrCl_2 \cdot 8NH_3 \qquad (8.8)$$

Barium chloride reacts with ammonia and forms only one stable ammine [14, 21] (Fig. 8.3):

$$BaCl_2 + 8NH_3 \leftrightarrows BaCl_2 \cdot 8NH_3 \qquad (8.9)$$

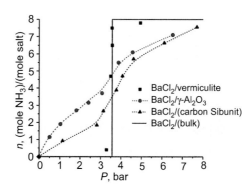

Figure 8.3 Isotherms of ammonia sorption at 30°C for bulk $BaCl_2$ and the composites "$BaCl_2$/matrix." Reprinted from Ref. [21], Copyright 2010, with permission from Elsevier.

For the bulk salts, the affinity for ammonia changes in the row $CaCl_2 \approx SrCl_2 < BaCl_2$. Ammonia sorption isotherms for the composites strongly depend on the nature of the matrix.

8.1.2 Effect of Host Matrix

New CSPMs were prepared by using these salts as active substances and alumina, activated carbons, and vermiculite as host matrices (their composition is listed in Table 8.1) [14]. The synthesis was performed by dry impregnation of the matrix with a salt solution of the desired concentration. The new CSPMs were studied by various physicochemical methods.

8.1.2.1 CaCl₂-based composites

X-ray diffraction (XRD) analysis shows that crystallinity of the confined salt depends on both the salt content and the porous structure of the matrix. Composite Ca-V-63 based on expanded vermiculite is characterized by intensive peaks on the XRD pattern matching the theoretical XRD pattern for bulk calcium chloride (Fig. 8.4). Hence, the salt in this composite is well crystallized.

Table 8.1 Description of the composites studied

Composite	Active salt	Host matrix	Salt content (wt.%)
Ca-A-10	CaCl$_2$	γ-Al$_2$O$_3$	9.7
Ca-A-20		γ-Al$_2$O$_3$	19.6
Ca-A-30			29.9
Ca-V-63		Vermiculite	63.4
Ca-ACF-31		Carbon fiber	31.0
Sr-A-15	SrCl$_2$	γ-Al$_2$O$_3$	14.5
Sr-V-50		Vermiculite	49.6
Ba-A-16	BaCl$_2$	γ-Al$_2$O$_3$	15.7
Ba-V-45		Vermiculite	49.9
Ba-Sib-19		Carbon Sibunit	18.6

Source: Ref. [14].

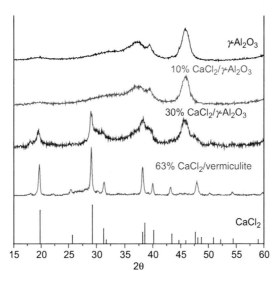

Figure 8.4 X-ray diffraction patterns of various CSPMs based on calcium chloride [14].

The X-ray pattern for Ca-A-10 repeats in whole the pattern for γ-Al$_2$O$_3$ and does not show any reflections corresponding to the crystalline calcium chloride in the pores (Fig. 8.4); hence, the salt is in the X-ray amorphous (XRA) state. The X-ray patterns for composites Ca-A-20 and Ca-A-30 with the salt content higher than 10% show reflections corresponding to an orthorhombic phase of CaCl$_2$ (hydrophilite, PNNM space group), but their integral intensities are low (Fig. 8.4).

Noteworthy that the integral intensities of (111) reflection of the CaCl$_2$ orthorhombic phase in Ca-A-20, Ca-A-30, and Ca-V-63 are not proportional to the salt content in these composites (Tables 8.1, 8.2 and Fig. 8.5). The percentages of the crystalline and XRA phases of CaCl$_2$ in Ca-A-20 and Ca-A-30 composites were estimated under the assumption that all the salt inside the expanded vermiculite (Ca-V-63) is in a crystalline state. It turned out that only about 20% of CaCl$_2$ is crystalline in Ca-A-20 and about 70% in Ca-A-30. Stabilization of the confined salt in the XRA state was previously observed for LiBr, Na$_2$SO$_4$, CuSO$_4$, and MgSO$_4$ confined to the pores of γ-Al$_2$O$_3$ [24, 25]. It is possible that a bi-variant type of the sorption equilibrium observed for many alumina-based composites is caused by the formation of XRA phase of the confined salt.

Table 8.2 Phase structure of the CaCl$_2$-based composites

Composite	Matrix	Crystal lattice	Reflection (111)	
			Intensity (a.u.)	d_{scr} (nm)
Ca-A-10	γ-Al$_2$O$_3$	XRA	—	—
Ca-A-20		Orthorhombic (PNNM)	8	23
Ca-A-30			44	22
Ca-V-63	Expanded vermiculite		135	29

The coherent scattering region size (d_{scr}) is estimated for (111) reflection of crystalline calcium chloride in the Ca-A-20, Ca-A-30, and Ca-V-63 composites (Table 8.2). Although, in average, vermiculite has much larger pores than alumina, the d_{scr} values for CaCl$_2$ crystals confined in both matrices are almost equal. This may indicate that the size of the salt crystals is essentially determined by the precipitation conditions rather than exclusively by the matrix's pore size.

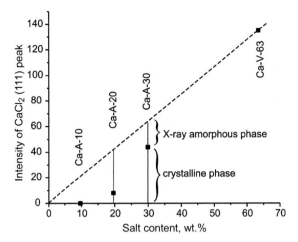

Figure 8.5 Intensity of (111) peak versus salt content for various composites.

8.1.2.2 BaCl₂-based composites

All BaCl₂-based composites contain two crystalline phases, namely, orthorhombic and hexagonal (Fig. 8.6). The orthorhombic phase (space group Pnma, a = 7.878 Å, b = 4.714 Å, c = 9.415 Å), also known as α-modification, is a stable phase of the bulk salt up to T = 925°C. At this temperature, the orthorhombic phase transforms to a high-temperature cubic BaCl₂ (β-modification) [26]. The hexagonal phase (space group $P\bar{6}2m$, a = 8.133 Å, b = 4.675 Å, c = 9.339 Å) is metastable. It was registered in the bulk state by XRD in the temperature range of 52 to 370°C during the heating of crystalline BaCl₂·2H₂O [27]. It was also shown that the BaCl₂ hexagonal phase is stable up to 275°C in fluorozirconate glass containing nano- and microcrystals of the salt [28]. The metastable hexagonal phase is probably stabilized in the pores due to the dispersion of salt.

For many solids, it was experimentally confirmed that a decrease in the size of a crystal may be accompanied by stabilization of a non-equilibrium crystalline structure. For instance, stabilization of a high-temperature β-modification was observed during the growth of mercury di-iodide in the pores of nanostructured matrices at T = 77–295 K, which is far below the temperature of the $\alpha \rightarrow \beta$ phase transition in the bulk [29]. The formation of nanocrystals with a structure different from that of bulk crystals was observed also for copper halides [30, 31], gallium arsenide [32], cadmium sulfide, and selenide [33].

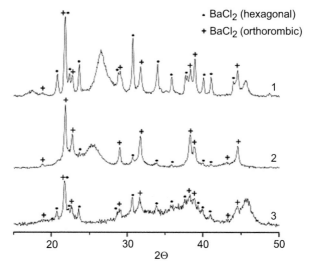

Figure 8.6 X-ray patterns for Ba-V-45 (1), Ba-Sib-19 (2), and Ba-A-16 (3) composites (λ = 1.54 Å).

The ratio between the integral intensities of (211) reflection for the orthorhombic phase and (201) reflection for the hexagonal phase was used for quantitative assessment of the relative percentage of these phases (Table 8.3). It turned out that this ratio depends on the porous media where $BaCl_2$ is placed in. The intensities of the reflections for alumina-based composite Ba-A-16 are close to one another. The orthorhombic phase predominates in the expanded vermiculite (Ba-V-45), while the hexagonal phase prevails in the composite based on carbon Sibunit (Ba-Sib-19). TEM images for Ba-Sib-19 composite (Fig. 8.7) show acicular-shaped crystals with the characteristic interplanar spacings d_{hkl} = 0.475, 0.385, 0.305, and 0.206 nm, which correspond to the hexagonal $BaCl_2$ phase. It confirms the stabilization of this non-equilibrium phase inside the Sibunit pores.

Thus, a structural study of the new composites showed that the salt is located inside the pores of a host matrix, and the degree of its dispersion depends on the matrix's porous structure and salt content. The salt located inside the vermiculite macropores is characterized by narrow reflexes similar to those of the bulk salt. In the alumina mesopores, the salt is XRA at the low salt content (C ≤ 10 wt.%) and has wide reflexes at the larger salt content (≥30 wt.%). This difference in the salt state results in strong changes in the composite sorption properties (Figs. 8.1–8.3).

Table 8.3 Phase structure of the BaCl$_2$-containing composites

Composite	Matrix	Orthorhombic reflection (211)		Hexagonal reflection (201)	
		Intensity	d_{scr} (nm)	Intensity	d_{scr} (nm)
Ba-A-16	γ-Al$_2$O$_3$	48	36	46	27
Ba-V-45	Vermiculite	200	35	98	26
Ba-Sib-19	Sibunit	46	43	150	23

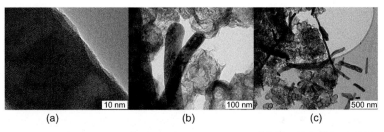

Figure 8.7 TEM images for Ba-Sib-19 composite: (a) × 1000, (b) × 5000, (c) × 50,000.

8.2 Ammonia Sorption Properties of New CSPMs

8.2.1 Composites "CaCl$_2$ inside Porous Matrices"

Isosters of ammonia sorption on Ca-V-63 composite (not presented) are straight lines ln P versus $1/T$ [14] with the slope close to those measured for the appropriate complexes CaCl$_2$·nNH$_3$ (n = 1, 2, 4, 8) in the bulk (reactions (8.3)–(8.6)). This indicates that the sorption properties of the salt confined to the vermiculite macropores are similar to the bulk salt.

Experimental isosters of NH$_3$ sorption by the salt confined to the mesoporous matrices of alumina [19] (Fig. 8.8a) and ACF [20] (Fig. 8.8b) represent a set of straight lines ln $P = A + B/T$, which correspond to the continuously changing uptake w (or n). The intersection $A = -\Delta S/R$ and the slope $B = \Delta H/R$ depend on the average uptake and allow the sorption enthalpy ΔH and entropy ΔS to be calculated (Tables 8.4 and 8.5). It is surprising that for Ca-ACF-31, the ΔH value is only slightly larger than the heat of ammonia evaporation (23.3 kJ/mol). In our opinion, the main reason behind the underestimation of the sorption enthalpy is the effect of the "dead" volume that leads to a decrease in uptake with an increase in temperature [20]. If the effect is taken into consideration, the true isosters of ammonia sorption can be plotted (see Fig. 3b in Ref. [20]) and the true sorption enthalpy and entropy can be calculated (Table 8.5). This effect is negligible for Ca-A-15; however, it is significant for Ca-ACF-31 with a low density, so the apparent ΔH and ΔS values must be corrected.

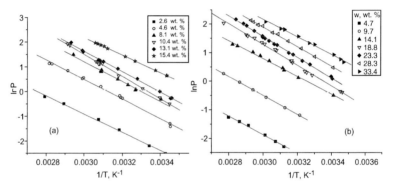

Figure 8.8 Experimental isosters of ammonia sorption on Ca-A-15 (a) and Ca-ACF-31 (b) at different NH_3 uptakes (see the uptakes in the legends).

Isotherms of NH_3 sorption on the $CaCl_2$-based composites are plotted from the experimental isosters (Fig. 8.1). It is found that the sorption properties of the composite Ca-V-63 [21] are close to those for the bulk salt. $CaCl_2$ confined to the alumina and ACF pores is characterized by the sorption isotherms that are much smoother than that for the bulk salt (Fig. 8.1).

To clarify the mechanism of ammonia sorption by the studied composites, X-ray in situ experiments are performed. It is found that reaction of the salt confined in the vermiculite pores (Ca-V-63) with ammonia at $P = 1$ bar and $T = 30°C$ (Fig. 8.9) leads to continuous lowering of the intensity of reflections corresponding to the salt and rising the intensity of reflections corresponding to a $CaCl_2 \cdot 2NH_3$ crystalline phase [14]. The reflections corresponding to a crystalline phase of $CaCl_2$ disappeared completely after 12 min of ammonia sorption. No evidence of the formation of $CaCl_2 \cdot NH_3$ is found. This phase was not revealed by the adsorption experiments as well [19]. This is not surprising, taking into account that $CaCl_2$ monoamine at 30°C is stable within the ammonia pressure range as narrow as 10^{-5} to 2×10^{-4} bar, while for $CaCl_2$ diamine, this range is much wider ($2 \times 10^{-4} \div 0.56$ bar) according to the thermodynamics of reactions (8.3) and (8.4) [5, 6]. Thus, at $P = 1$ bar, direct formation of $CaCl_2 \cdot 2NH_3$ occurs via reaction (8.4) as reported in Ref. [34].

Table 8.4 Enthalpy ΔH and entropy ΔS of NH_3 sorption on Ca-A-15 at different uptakes

Ammonia uptake		ΔH (kJ/mol)	ΔS (J/mol·K)
w (wt.%)	n (mol/mol)		
2.6	1.15	−30.5	−84.0
4.1	1.82	−34.5	−110.7
4.6	2.07	−33.0	−103.0
4.7	2.10	−35.7	−107.7
5.2	2.34	−33.5	−108.0
7.9	3.56	−31.7	−105.0
8.1	3.65	−35.8	−117.8
10.4	4.68	−37.0	−127.4
13.1	5.90	−34.1	−116.13
14.0	6.30	−29.7	−105.7
15.0	6.75	−29.0	−106.4

Source: Reprinted from Ref. [19], Copyright 2006, with permission from Oxford University Press.

Table 8.5 Apparent and corrected values of enthalpy and entropy of NH_3 sorption on Ca-ACF-31 at different uptakes

Uptake (wt.%)	ΔH_{app} (kJ/mol)	ΔS_{app} (J/mol·K)	ΔH (kJ/mol)	ΔS (J/mol·K)
5.7	−27.0	−65	−27.4	−67
11.5	−27.1	−77	−29.3	−84
16.1	−25.9	−84	−30.2	−99
20.9	−29.6	−100	−32.0	−107
27.4	−30.2	−104	−32.4	−110
31.8	−28.4	−101	−32.9	−114
36.9	−25.6	−95	−30.4	−109

Source: Reprinted from Ref. [20], Copyright 2010, with permission from Elsevier.

After 16 min of ammonia sorption, the formation of a new phase is observed. It can be tentatively attributed to CaCl$_2$·4NH$_3$, which is stable at P$_{NH3}$ = 0.56 ÷ 1.03 bar range at 30°C (Fig. 8.9). Further increase in pressure to 1.6 bar leads to the disappearance of the CaCl$_2$·4NH$_3$ phase and formation of CaCl$_2$·8NH$_3$ crystalline amine [14], which is in line with the results of adsorption measurements.

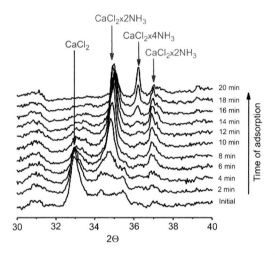

Figure 8.9 Temporal evolution of in situ X-ray pattern of Ca-V-63 during NH$_3$ sorption at 1 bar and 30°C. λ = 1.731 Å.

A similar general behavior is observed for the alumina-based composite Ca-A-20: the ammonia sorption at 1.6 bar and 30°C leads to the same sequential formation of the salt crystalline amines with n = 2, 4, and 8 (Fig. 8.10). It is remarkable that the confined calcium chloride, which has mainly been XRA (Fig. 8.5), forms the crystalline phases of amines in the course of chemical reactions (8.4)–(8.6).

After lowering the NH$_3$ pressure down to 1 bar and heating up to 60°C, CaCl$_2$·8NH$_3$ decomposes giving CaCl$_2$·4NH$_3$ (Fig. 8.11). Further heating up to 100°C results in the formation of CaCl$_2$·2NH$_3$.

Thus, the results obtained proved that the formation of CaCl$_2$ crystalline amines during ammonia sorption occurs both inside the macropores of expanded vermiculite and the mesopores of γ-Al$_2$O$_3$. On the one hand, this is in accordance with a step-like isotherm of ammonia sorption for Ca-V-63 composite. On the other hand, it could not explain the nature of the change in sorption equilibrium type observed for Ca-A-20. This effect can be possibly attributed to the presence of the XRA phase of the salt inside the alumina mesopores, which can absorb ammonia without forming stoichiometric crystalline amines.

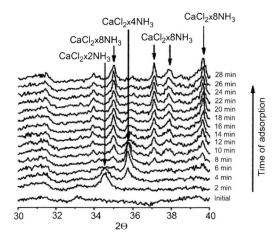

Figure 8.10 Temporal evolution of in situ X-ray pattern for Ca-A-20 during the ammonia sorption at 1.6 bar and 30°C. λ = 1.731 Å.

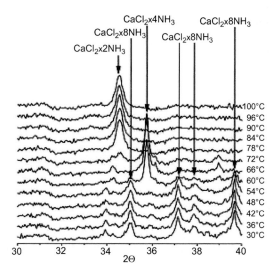

Figure 8.11 Temperature evolution of in situ X-ray pattern for Ca-A-20 during the ammonia desorption at 1 bar at the heating rate 3°C/min. λ = 1.731 Å.

8.2.2 Composite Sorbents "SrCl$_2$ inside Porous Matrices"

Isosters of ammonia sorption on Sr-V-50 composite are straight lines in the ln P versus $1/T$ presentation [14], which almost coincide regardless of the NH$_3$ uptake (Fig. 8.12a). Their slopes are close to that reported for the complex SrCl$_2$·8NH$_3$ in the bulk (reaction (8.8)): ΔH = −43.0 kJ/mol, ΔS = −138.8 J/mol·K [35]. Hence, the salt confined to the vermiculite macropores absorbs NH$_3$ in a way similar to the bulk salt.

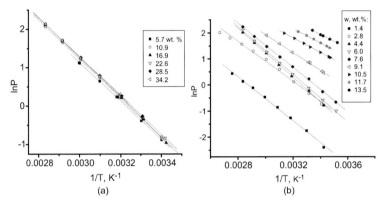

Figure 8.12 Experimental isosters of ammonia sorption on Sr-V-50 (a) and Sr-A-15 [20] (b) at different NH$_3$ uptakes (see the uptakes in the legends).

Table 8.6 Enthalpy ΔH and entropy ΔS of NH$_3$ sorption on Sr-A-15 at different ammonia uptakes

	Ammonia uptake	ΔH	ΔS
w (wt.%)	n (mol NH$_3$/mol SrCl$_2$)	(kJ/mol)	(J/mol·K)
1.4	0.9	−35.1	−100.9
2.8	1.8	−31.6	−102.1
4.4	2.8	−39.4	−128.4
6.0	3.9	−38.1	−125.4
7.6	4.9	−35.4	−119.0
9.1	5.8	−28.8	−102.6
10.5	6.8	−23.5	−90.2
11.7	7.6	−22.8	−90.9

Source: Ref. [14].

Experimental isosters of NH_3 sorption by $SrCl_2$ confined to the mesopores of alumina [14] (Fig. 8.12b) represent the straight lines $\ln P = A + B/T$ with the intersection A and slope B dependent on the NH_3 uptake. Appropriate values of the sorption enthalpy ΔH and entropy ΔS are collected in Table 8.6. Both these values are smaller than those reported for the bulk salt and, at $w > 4.4$ wt.%, gradually decrease with the rise in uptake (Fig. 8.13). This decrease is essential and reaches 20 kJ/mol and 50 J/mol·K for the enthalpy and entropy, respectively.

This variation in the thermodynamic parameters for the confined $SrCl_2$ gives a clear indication of the significant change in the salt sorption properties due to its confinement to the alumina pores.

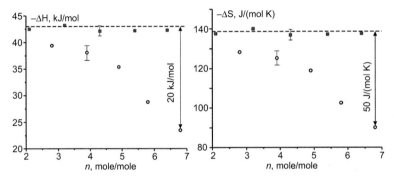

Figure 8.13 Isosteric heat (a) and entropy (b) of NH_3 sorption on Sr-V-50 (■) and Sr-A-15 (○) as a function of the uptake. Dashed lines indicate the literature data for bulk $SrCl_2$.

8.2.3 Composite Sorbents "BaCl₂ inside Porous Matrices"

Isosters of ammonia sorption on Ba-V-63 composite are straight lines in the $\ln P$ versus $1/T$ presentation [21], which coincide regardless of the NH_3 uptake (Fig. 8.14a). Their slopes are close to that reported for the complex $BaCl_2 \cdot 8NH_3$ in the bulk (reaction (8.9)): $\Delta H = -36.7$ kJ/mol, $\Delta S = -131.8$ J/mol·K [36]. Hence, the salt confined to the vermiculite macropores absorbs NH_3 in a way similar to the bulk salt.

Isotherms of the NH_3 sorption on Ba-V-45 were measured at the temperature range of 37–54.5°C by a TG method using a Rubotherm magnetic suspension balance [22]. It was found that the modification of the host matrix by the salt dramatically increases the ammonia uptake due to the formation of BaCl$_2$·8NH$_3$ complex due to reaction (8.9). Another important advantage of this composite is that the vermiculite can accommodate the swelling of salt caused by the reaction between ammonia and the salt. Hysteresis between the synthesis and decomposition reactions was observed and investigated.

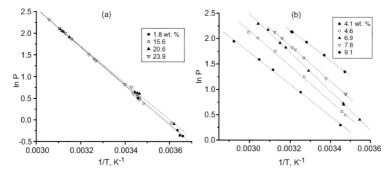

Figure 8.14 Experimental isosters of ammonia sorption on Ba-V-45 (a) and Ba-A-16 [20] (b) at different NH$_3$ uptakes (see the uptakes in the legends). Reprinted from Ref. [21], Copyright 2010, with permission from Elsevier.

Table 8.7 Enthalpy and entropy of ammonia sorption on composite Ba-A-16

Ammonia uptake		ΔH	ΔS
w (wt.%)	n (mol/mol)	(kJ/mol)	(J/mol·K)
4.1	3.2	−26.1	−92.8
4.6	3.6	−28.6	−103.9
6.9	5.4	−31.6	−115.6
7.8	6.1	−28.9	−108.0
9.1	7.1	−24.7	−97.0

Source: Reprinted from Ref. [21], Copyright 2010, with permission from Elsevier.

Experimental isosters of NH$_3$ sorption on Ba-A-16 (Fig. 8.14b) and Ba-Sib-19 (see Fig. 4 in Ref. [21]) are straight lines ln $P = A + B/T$ with the intersection A and slope B that are functions of the NH$_3$ uptake. The values of sorption enthalpy ΔH and entropy ΔS for Ba-A-16 are collected in Table 8.7. These values are significantly smaller than those for the bulk salt, which is typical for all composites based on mesoporous matrices.

Plotting the isotherms of ammonia sorption allows sorption properties of the confined salt to be compared with those of bulk $BaCl_2$ (Fig. 8.3). At $T = 30°C$, the theoretical sorption isotherm for bulk $BaCl_2$ consists of two plateaus at $n = 0$ and $n = 8$ (mol NH_3)/(mol $BaCl_2$) with a stepwise transition between them at $P = 3.6$ bar, which is due to reaction (8.9). The experimental isotherm for Ba-V-45 composite coincides with the theoretical one. Once again, the salt confined to the macropores of the expanded vermiculite has the same properties as the bulk one. The similarity of the properties of the bulk and confined $BaCl_2$ is probably due to the larger size of the salt crystallites inside the large pores of the vermiculite.

The sorption equilibrium for Ba-A-16 and Ba-S-19 is bi-variant, which means the NH_3 uptake increases continuously as the ammonia pressure rises. It is not clear whether this equilibrium is truly bi-variant and the complex $BaCl_2 \cdot nNH_3$ has a variable composition with $0 < n \leq 8$. Another reason for the smooth rise in the uptake with the ammonia pressure may be an inhomogeneity of the salt nanoparticles due to, for instance, their distribution on the particle size. In this case, each salt particle undergoes a mono-variant transformation to the complex $BaCl_2 \cdot 8NH_3$ at a certain pressure $P(r)$, which depends on the particle size r. As a result, the transformation occurs over some pressure interval. The mentioned salt inhomogeneity might be also due to the interaction of the guest salt with the host matrix. Probably, for Ba-A-16 and Ba-S-19, both the salt size effect and the guest–host interaction are important.

8.2.4 Summary of New Ammonia Sorbents "Salt–Matrix"

The new CSPMs based on a single salt ($CaCl_2$, $SrCl_2$, and $BaCl_2$) confined to various porous matrices (expanded vermiculite, alumina, Sibunit, ACF) were synthesized and studied as methanol sorbents. It is found that the confinement of the salt to the matrix pores results in an essential enhancement of the composite's sorption capacity. The latter depends on the nature of the salt and the matrix and is mainly defined by a solid–gas reaction between the salt and ammonia, while the sorption of ammonia by pure matrices is low (Fig. 8.15). As the salt is an active sorbing component, the sorption capacity of the new composites is a linear function of the salt content in the composite

(Fig. 8.16). The composite Ca-V-63, containing 63 wt.% CaCl$_2$ in the vermiculite macropores, ensures the NH$_3$ uptake as large as 0.7 g/g, which is of high practical interest, e.g., for adsorptive heat transformation [22, 23]. Moreover, placing the salt inside the porous matrix provides space separation of the salt microcrystals, which prevents agglomeration and improves mass transfer.

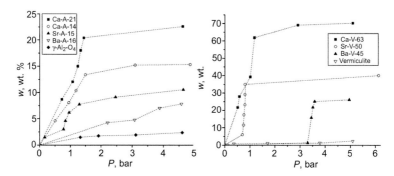

Figure 8.15 Isotherms of ammonia sorption on the salt/alumina composites and pure alumina (a); the salt/vermiculite composites and pure vermiculite (b); T = 30°C [14].

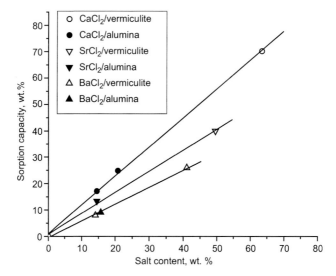

Figure 8.16 The maximal NH$_3$ sorption capacity versus the salt content in the composite. T = 30°C and P = 4 bar [14].

The study of NH_3 sorption equilibrium for the new CSPMs shows that the sorption properties of salts dispersed inside the macropores of the expanded vermiculite are close to those for the bulk salts. The salts inside the vermiculite macropores presumably absorb ammonia-forming crystalline amines according to chemical reactions (8.3)–(8.9). The confinement of the salts to mesoporous matrices (alumina, Sibunit, ACF) results in the bi-variant sorption equilibrium between the salt and ammonia and in lowering the enthalpy and entropy of sorption. The bi-variant type of sorption equilibrium with water and methanol was also observed for composites based on mesoporous matrices, such as silica gels, MCM-41, porous carbons, alumina, etc. (see Chapters 5–7). It was suggested that the *smooth* rise in the equilibrium ammonia (water, methanol) uptake with an increase in pressure may be caused by a size effect due to the dispersion of the salt nanoparticles inside the mesopores of the matrix or/and by a guest–host interaction between the salt and the matrix, but the essence of this phenomenon is still not definite.

8.3 Effect of Supplementary Salt

It has already been mentioned earlier (Chapter 7) that the sorption properties of a salt can be changed by a supplementary salt that forms a solid solution with the main salt. For instance, the equilibrium of ammonia absorption by alkaline earth metal halides, $CaBr_2$ and $CaCl_2$, in the bulk was modified by an additional salt [37]. This approach was used to vary the ammonia sorption properties of the single-salt composite $BaCl_2$/vermiculite considered in Sec. 8.2.3. New composites based on a binary salt system ($BaCl_2 + BaBr_2$) confined to the marcopores of the expanded vermiculite and the mesopores of silica gel were synthesized and studied [38–41].

A commercial silica gel Davisil gr. 646 (the average pore diameter $d_{av} = 15$ nm (BET), specific surface area $S_{sp} = 293$ m^2/g (BET), pore volume $V_p = 1.18$ cm^3/g) and an expanded vermiculite ($d_{av} = 6.5$ µm, $S_{sp} = 2.0$ m^2/g, $V_p = 2.7$ cm^3/g) were used as host matrices. The composites were synthesized by dry impregnation of the matrices with an aqueous solution of the two salts followed by thermal drying at 433 K. A number of new composites were prepared with different molar ratio n_{BaCl2}/n_{BaBr2}.

8.3.1 Composite Sorbents (BaCl$_2$ + BaBr$_2$)/Silica

The phase composition of both single-salt composites BaCl$_2$/SiO$_2$ and BaBr$_2$/SiO$_2$ is characterized by a mixture of the two modifications of the salts, namely, orthorhombic and hexagonal. With an increase in bromine content, the reflexes on the XRD patterns of the composites (BaCl$_2$ + BaBr$_2$)/SiO$_2$ transform gradually from those of BaCl$_2$ to the reflexes of BaBr$_2$ (Fig. 8.17). This shows the formation of the continuous row of homogeneous solid solutions in the whole composition range.

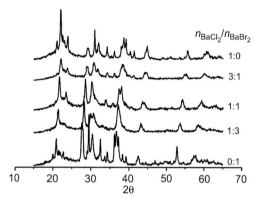

Figure 8.17 XRD patterns of the composites (BaCl$_2$ + BaBr$_2$)/SiO$_2$. Reprinted from Ref. [38], Copyright 2013, with permission from Elsevier.

Two steps are observed on the isotherms of ammonia sorption on both the single-salt composites (Fig. 8.18a). First, at P_{NH3} = 0–0.1 bar, the uptake w rises from 0 to 0.03–0.05 g/g due to the ammonia adsorption on active centers of the silica gel. The second rise in sorption from 0.05 to 0.20–0.23 g/g is attributed to the formation of complexes due to reaction (8.9) and

$$\text{BaBr}_2 + 8\text{NH}_3 = \text{BaBr}_2 \cdot 8\text{NH}_3 \qquad (8.10)$$

BaBr$_2$ has a higher affinity for ammonia, and the formation of BaBr$_2$·8NH$_3$ occurs at lower ammonia pressure (0.7–1.6 bar) than of BaCl$_2$·8NH$_3$ (7–8 bar). Isotherms of ammonia sorption on the composites (BaCl$_2$ + BaBr$_2$)/SiO$_2$ have a similar shape, and two steps are observed (Fig. 8.18a). As the chlorine content increases, the pressure corresponding to the formation of NH$_3$ complexes of the salts and their solid solution continuously moves from 0.7–1.6

to 5–6 bar. The average pressure corresponding to the formation of complexes SS$_{BaClnBr(2-n)}$·8NH$_3$ is plotted versus the relative content $n_{BaCl2}/(n_{BaCl2} + n_{BaBr2})$ in Fig. 8.18b. Such a graph allows determination of the solvation pressure for the composite with a fixed solid solution composition. On the other hand, it is very convenient for predicting the composition of the CSPM that sorbs ammonia at the predetermined conditions (P_{NH3} and T). This provides a reliable base for the target-oriented synthesis of new ammonia sorbents with desired properties. It was, for instance, demonstrated in Ref. [38] for three adsorptive cooling cycles specified for air conditioning, ice making, and deep freezing.

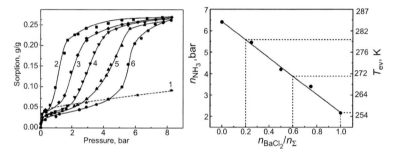

Figure 8.18 (a) Isotherms of NH$_3$ sorption on silica gel (1) and the composites (BaCl$_2$ + BaBr$_2$)/SiO$_2$ (2–6). $p = n_{BaCl_2}/n_{BaBr_2}$ = 0:1 (2), 1:3 (3), 1:1 (4), 3:1 (5), and 1:0 (6). (b) Ammonia pressure corresponding to the NH$_3$ uptake $w \approx 0.23$ g/g versus the BaBr$_2$ content in the various composites (BaCl$_2$ + BaBr$_2$)/SiO$_2$. T = 312 K. Reprinted from Ref. [38], Copyright 2013, with permission from Elsevier.

8.3.2 Composite Sorbents (BaCl$_2$ + BaBr$_2$)/Vermiculite

The phase composition and ammonia sorption equilibrium were studied for the new binary salts composites BaClBrV-1/1.2 and BaClBrV-3/2 in comparison with the single-salt composites BaCl$_2$V and BaBr$_2$V, whose composition is displayed in Table 8.8.

The data on XRD in situ (not presented) reveal that during the sorption run intensities of the reflexes attributed to solution BaCl$_x$Br$_{(2-x)}$ diminish continuously until they completely vanish [39]. Simultaneously, new reflexes appear, which can be assigned to the crystalline phase of ammonia complex BaCl$_x$Br$_{(2-x)}$·8NH$_3$ formed due to ammonia sorption. Continuous variation in the interplanar distances of this complex was detected during both ammonia sorption and desorption and was ascribed to the distortion of the crystalline lattice due to reactions (8.9) and (8.10).

Table 8.8 The composition of the sorbents ($BaCl_2$ + $BaBr_2$)/vermiculite

Composite	mol $BaCl_2$/mol $BaBr_2$	C_{BaCl_2} (wt.%)	C_{BaBr_2}, wt. %	C_{verm}, wt. %
$BaCl_2V$	1/0	41	0	59
BaClBrV-3/2	3/2	24	24	52
BaClBrV-1/1.2	1/1.2	19	32	49
$BaBr_2V$	0/1	0	59	41

Note: C_{BaCl_2}, C_{BaBr_2}, and C_{verm} are the weight contents of $BaCl_2$, $BaBr_2$, and vermiculite, respectively.
Source: Reprinted from Ref. [39], Copyright 2014, with permission from Elsevier.

The isotherms of ammonia sorption on the composites BaClBrV-1/1.2 and BaClBrV-3/2 are S-shaped curves (Fig. 8.19). At low ammonia pressure, the ammonia uptake w does not exceed 0.03 g/g, which is probably associated with adsorption on the active centers of the matrix (vermiculite) surface. With an increase in ammonia pressure, a rise in sorption up to $w \approx 0.27$ g/g or $n \approx 8$ mol/mol is observed for both the composites. This confirms the formation of complexes $BaCl_xBr_{(2-x)} \cdot 8NH_3$ due to the chemical reaction of solid solutions with ammonia.

The shapes of the ammonia sorption isotherms on BaClBrV-1/1.2 and BaClBrV-3/2 are quite different from those for $BaCl_2V$ composite and the bulk salts $BaCl_2$ and $BaBr_2$ (Fig. 8.19). The steps on sorption isotherms for BaClBrV-1/1.2 and BaClBrV-3/2 become smoother, and the formation of the complex $BaCl_xBr_{(2-x)} \cdot 8NH_3$ proceeds in a range of ammonia pressure of 1.5 to 3.5 bar, which is intermediate between the pure salts and allows the intentional modification of the composite sorption properties to be performed. According to Grekova et al. [39], the possible reason for the smoothening of isotherms is the different affinity of $BaCl_2$ and $BaBr_2$ toward NH_3. Because of this, formation of the complex $BaCl_xBr_{(2-x)} \cdot 8NH_3$ occurs non-stoichiometrically. First, the complex $BaCl_xBr_{(2-x)} \cdot 8NH_3$ enriched with bromine forms during ammonia sorption on the BaClBrV-1/1.2 and BaClBrV-3/2 composites. At increasing ammonia pressure, the chlorine content in the complex $BaCl_xBr_{(2-x)} \cdot 8NH_3$ gradually increases. That leads to the smoothening of the ammonia sorption isotherms on these composites.

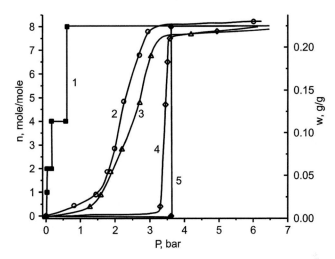

Figure 8.19 The isotherms of NH$_3$ sorption on bulk BaBr$_2$ (1) [42], composites BaClBrV-1/1.2 (2), BaClBrV-3/2(3), BaCl$_2$V (4), and bulk BaCl$_2$ (5). T = 30°C. Reprinted from Ref. [39], Copyright 2014, with permission from Elsevier.

Ammonia sorption isosteres, calculated from the equilibrium isotherms on ammonia sorption on the composites BaClBrV-1/1.2 and BaClBrV-3/2, are straight lines in the Clausius–Clapeyron coordinates (not presented), which allows the standard enthalpy and entropy of the NH$_3$ sorption to be evaluated. In the uptake range 0.034–0.238 g/g (or n = 1–7 mol/mol), the sorption enthalpy $\Delta H°$ = −35 ± 3 kJ/mol does not depend on the NH$_3$ uptake. For both the composites, it is close to the formation enthalpy of bulk BaCl$_2$·8NH$_3$ (−36.7 kJ/mol [42]). The standard entropy of ammonia sorption changes from −115 to −133 ± 2 J/mol·K with an increase in sorption from n = 1 to 7 mol/mol.

8.3.3 Discussion

Thus, the results obtained reveal some common features of sorption equilibrium of the composites based on binary salt systems confined to the porous matrices. The formation of solid solution SS in such binary systems leads to distortion of the crystalline lattice of the salts and to changes in the spacing parameter. That shifts the equilibrium temperature (pressure) of the complex formation SS + nV = SS·nV regarding the solvation of pure salt S + nV = S·nV. The dissolution of BaCl$_2$ in the lattice of BaBr$_2$ reduces the spacing parameter that probably hinders the incorporation of the sorbed NH_3 molecules in the lattice. Therefore, the transition of SS_{Br} to its solvates occurs at a lower temperature (higher pressure) than the $BaBr_2$ solvation. Dissolution of $BaBr_2$ in the lattice of $BaCl_2$ results in the opposite effect: the crystalline lattice expands, which promotes the incorporation of the sorbate molecules in the lattice and raises the solvation temperature (reduces the pressure). The shift in the temperature of the salt solvation due to dissolution of the additional salt can be used for the intent design of composite sorbents for particular adsorption cycles [39–41].

References

1. Appl, M. (2006) Ammonia, in *Ullmann's Encyclopedia of Industrial Chemistry* (Wiley-VCH, Weinheim).

2. Yanagi, H., Khong, M., and Ng, K. C. (2005) Performance testing of ammonia scrubbing system, *Proc. 2005 ASME Summer Heat Transfer Conference*, San Francisco, USA, July 17–22, 2005.

3. Spinner, B. (1993) Ammonia-based thermochemical transformers, *Heat Recov. Syst. CHP*, **13**(4), pp. 301–307.

4. Christensen, C. H., Sørensen, R. Z., Johannessen, T., Quaade, U. J., Honkala, K., Elmøe, T. D., Køhler, R., Nørskov, J. K. (2005) Metal ammine complexes for hydrogen storage, *J. Mater. Chem.*, **15**, pp. 4106–4108.

5. Carling, R. W. (1981) Dissociation pressures and enthalpies of reaction in $MgCl_2 \cdot nH_2O$ and $CaCl_2 \cdot nNH_3$, *J. Chem. Thermodyn.*, **13**, pp. 503–512.

6. Touzan, Ph. (1999) Thermodynamic values of ammonia-salts reactions for chemical sorption heat pumps, *Proc. Int. Sorption Heat Pump Conf.*, Munich, Germany, March 24–26, 1999, pp. 225–238.

7. Westman, S., Werner, P.-E., Schuler, T., and Radlow, W. (1981) X-Ray investigation of ammines of alkaline earth metal halides. I. The structures of $CaCl_2(NH_3)_8$, $CaCl_2(NH_3)_2$ and the decomposition product CaClOH, *Acta Chemica Scandinavica A*, **35**, pp. 467–472.

8. Fujioka, K., Kato, S., Fujiki, S., and Hirata, Y. (1996) Variations of molar volume and heat capacity of reactive solids of $CaCl_2$ used for chemical heat pumps, *J. Chem. Eng. Jpn.*, **29**(5), pp. 858–864.

9. Mauran, S., Lebrun, M., Prades, P. Moreau, M., Spinner, B., and Drapier, C. (1994) Active composite and its use as reaction medium, US Patent No 5,283,219.

10. Vasiliev, L. L., Kanonchik, L. E., Antujh, A. A., and Kulakov, A. G. (1996) NaX, Carbon fibre and $CaCl_2$ ammonia reactors for heat pumps and refrigerators, *Adsorption*, **2**, pp. 311–316.

11. Ye, H., Yuan, Z., Li, S., and Zhang, L. (2014) Activated carbon fiber cloth and $CaCl_2$ composite sorbents for a water vapor sorption cooling system, *Appl. Therm. Eng.*, **62**, pp. 690–696.

12. Aidoun, Z. and Ternan, M. (2002) Salt impregnated carbon fibres as the reactive medium in a chemical heat pump: The NH_3–$CoCl_2$ system, *Appl. Therm. Eng.*, **22**, pp. 1163–1173.

13. Aristov, Yu. I., Gordeeva, L. G., and Tokarev, M. M. (2008) *Composites "Salt in Porous Matrix": Synthesis, Properties, Applications* (SO RAN Publishing, Novosibirsk) (in Russian).

14. Veselovskaya, J. V. (2011) Sorption properties of new composite NH_3 sorbents based on dispersed chlorides of alkali-earth metals, Ph. D thesis, Boreskov Institute of Catalysis.

15. Helminen, J., Helenius, J., and Paatero, E. J. (2001) Adsorption equilibria of ammonia gas on inorganic and organic sorbents at 298.15 K, *J. Chem. Eng. Data*, **46**, pp. 391–399.

16. Valyon, J., Onyestak, G., and Rees, L. V. C. (1998) Study of the dynamics of NH_3 adsorption in ZSM-5 zeolites and the acidity of the sorption sites using the frequency-response technique, *J. Phys. Chem. B*, **102**, pp. 8994–9001.

17. Le Leuch, L. M. and Bandosz, T. J. (2007) The role of water and surface acidity on the reactive adsorption of ammonia on modified activated carbon, *Carbon*, **45**, pp. 568–578.

18. Sharonov, V. E. and Aristov, Yu. I. (2005) Ammonia adsorption by $MgCl_2$, $CaCl_2$ and $BaCl_2$ confined to porous alumina: The fixed bed adsorber, *React. Kinet. Catal. Lett.*, **85**, pp. 183–188.

19. Sharonov, V. E., Veselovskaya, J. V., and Aristov, Yu. I. (2006) Ammonia sorption on composites "$CaCl_2$ in inorganic host matrix": Isosteric chart and its performance, *Int. J. Low Carbon Tech.*, **1**, pp. 191–200.

20. Tokarev, M. M., Veselovskaya, J. V., Yanagi, H., and Aristov, Yu. I. (2010) Novel ammonia sorbents "porous matrix modified by active salt" for adsorptive heat transformation: 2. Calcium chloride in ACF felt, *Appl. Therm. Eng.*, **30**, pp. 845–849.

21. Veselovskaya, J. V., Tokarev, M. M., and Aristov, Yu. I. (2010) Novel ammonia sorbents "porous matrix modified by active salt" for adsorptive heat transformation: 1. Barium chloride in various matrices, *Appl. Therm. Eng.*, **30**, pp. 584–589.

22. Zhong, Y., Critoph, R. E., Thorpe, R. N., Tamainot-Telto, Z., and Aristov, Yu. I. (2007) Isothermal sorption characteristics of the $BaCl_2$–NH_3 pair in a vermiculite host matrix, *Appl. Therm. Eng.*, **27**, pp. 2455–2462.

23. Veselovskaya, J. V., Critoph, R. E., Thorpe, R. N., Metcalf, S., Tokarev, M. M., and Aristov, Yu. I. (2010) Novel ammonia sorbents "porous matrix modified by active salt" for adsorptive heat transformation: 3. Testing of "$BaCl_2$/vermiculite" composite in the lab-scale adsorption chiller, *Appl. Therm. Eng.*, **30**, pp. 1188–1192.

24. Gordeeva, L. G., Restuccia, G., Tokarev, M. M., Cacciola, G., and Aristov, Yu. I. (2000) Adsorption properties of the lithium bromide–water system in pores of extended graphite, Sibunit, and alumina, *Rus. J. Phys. Chem.*, **74**, pp. 2016–2020.

25. Gordeeva, L. G., Glaznev, I. S., and Aristov, Yu. I. (2003) Sorption of water by sodium, copper, and magnesium sulfates dispersed into mesopores of silica gel and alumina, *Rus. J. Phys. Chem.*, **77**, pp. 1715–1720.

26. Brackett, E. B., Brackett, T. E., and Sass, R. L. (1963) The crystal structure of barium chloride, barium bromide and barium iodide, *J. Phys. Chem.*, **67**, pp. 2132–2135.

27. Haase, A. and Brauer, G. (1978) Hydratstufen und Kristallstrukturen von Bariumchlorid, *Z. Anorg. Allg. Chem.*, **441**, pp. 181–195.

28. Edgar, A., Williams, G. V. M., Secu, M., Schweizer, S., and Spaeth, J.-M. (2004) Optical properties of a high-efficiency glass ceramic X-ray storage phosphor, *Radiat. Meas.*, **38**, pp. 413–416.

29. Akopyan, I. Kh., Labzovskaya, M. E., Novikov, B. V., and Smirnov, V. M. (2007) Metastable modifications in mercury diiodide nanocrystals, *Phys. Solid State*, **49**, pp. 1375–1381.

30. Akopyan, I. Kh., Gaisin, V. A., Loginov, D. K., Novikov, B. V., Tsagan-Manzhiev, A., Vasil'ev, M. I., and Golubkov, V. V. (2005) Stabilized high-temperature hexagonal phase in copper halide nanocrystals, *Phys. Solid State*, **47**, pp. 1372–1375.

31. Akopyan, I. Kh., Golubkov, V. V., Dyatlova, O. A., Novikov, B. V., and Tsagan-Mandzhiev, A. N. (2008) Structure of copper halide nanocrystals in photochromic glasses, *Phys. Solid State*, **50**, pp. 1352–1356.

32. Zanolli, Z., Fuchs, F., Furthmuller, J., von Barth, U., and Bechstedt, F. (2007) Model GW band structure of InAs and GaAs in the wurtzite phase, *Phys. Rev. B*, **75**, 245121 .

33. Akopyan, I. Kh., Ivanova, T. I., Labzovskaya, M. E., Novikov, B. V., and Erdni-Goryaev, A. (2010) Manifestation of metastable cubic modifications in finely dispersed A_2B_6 compounds, *Tech. Phys. Lett.*, **36**, pp. 240–243.

34. Mellor, J. W. (1960) *A Comprehensive Treatise on Inorganic and Theoretical Chemistry, Vol. III: Cu, Ag, Ca, Sr, Ba* (Longmans, London), p. 716.

35. Hüttig, G. F. (1922) Über die ammoniakate der strontiumhalogenide, *Z. Anorg. Allgem. Chem.*, **124**, pp. 322–332.

36. Destoky, C., Bougard, J., and Jadot, R. (1991) Research of solid-gas reacting media and intercalation compounds used in suitable structures of reactors to improve the performances of chemical heat pumps, First periodic report JOUE-0038C.

37. Liu, C. Yi. and Aika, K. (2004) Ammonia absorption into alkaline earth metal halide mixtures as an ammonia storage material, *Ind. Eng. Chem. Res.*, **43**, pp. 7484–7491.

38. Gordeeva, L. G., Grekova, A. D., Krieger, T. A., and Aristov, Yu. I. (2013) Composites "binary salts in porous matrix" for adsorption heat transformation, *Appl. Therm. Eng.*, **50**, pp. 1633–1638.

39. Grekova, A. D., Veselovskaya, J. V., Tokarev, M. M., Krieger, T. A., Shmakov, A. N., and Gordeeva, L. G. (2014) Ammonia sorption on the composites ($BaCl_2$ + $BaBr_2$) inside vermiculite pores," *Colloids Surf. A*, **448**, pp. 169–174.

40. Grekova, A. D., Veselovskaya, J. V., Tokarev, M. M., and Gordeeva, L. G. (2012) Novel ammonia sorbents "porous matrix modified by active salt" for adsorptive heat transformation: 5. Designing the composite adsorbent for ice makers, *Appl. Therm. Eng.*, **37**, pp. 80–86.

41. Veselovskaya, J. V., Tokarev, M. M., Grekova, A. D., and Gordeeva, L. G. (2012) Novel ammonia sorbents "porous matrix modified by active salt" for adsorptive heat transformation: 6. The ways of adsorption dynamics enhancement, *Appl. Therm. Eng.*, **37**, pp. 87–94.

42. Hodorowicz, S. A., Hodorowicz, E. K., and Eick, H. A. (1983) Preparation and characterization of the system $BaBr_xCl_{2-x}$: The structure of BaBrCl, *J. Solid. State Chem.*, **48**, pp. 351–356.

Chapter 9

Composite Sorbents of Carbon Dioxide

Having realized the gravity of the problems arising from CO_2 emissions and global warming, the world community has taken initiatives to alleviate or reverse the situation (Montreal and Kyoto protocols). A promising technology that reduces the emission of CO_2 involves its capture from exhaust gases, followed by sequestration in the ocean or in worked out mines [1]. In this connection, carbon dioxide sorbents with high capacity and selectivity under high temperature and humidity are in the focus of investigations all over the world.

The main technologies for CO_2 separation from exhausted gases include adsorption with activated carbons [2] and zeolites [3], absorption with amines and potassium carbonate solutions [4], cryogenic method [5], separation with membranes [6], and chemisorption [7]. In this chapter, we present the results of synthesis and testing of composite sorbents "K_2CO_3 confined to a porous matrix" [8–17], which can reversibly capture carbon dioxide following the reaction

$$K_2CO_3 + CO_2 + H_2O \leftrightarrow 2KHCO_3 \tag{9.1}$$

This reaction allows achieving high CO_2 capacity and selectivity in the presence of water. The salt's confinement to the pores results in its high dispersion, which helps in reducing the kinetic problems of "gas–solid" reaction (9.1) typical in bulk. Here we consider only

---

Nanocomposite Sorbents for Multiple Applications
Yu. I. Aristov
Copyright © 2020 Jenny Stanford Publishing Pte. Ltd.
ISBN 978-981-4267-50-2 (Hardcover), 978-981-4303-15-6 (eBook)
www.jennystanford.com

Composite Sorbents of Carbon Dioxide

the first steps in developing this approach, whereas the current state of the art can be found elsewhere [18].

9.1 Effect of the Nature of Matrix

Table 9.1 presents the characteristics of various matrices used for the synthesis of composites. The samples were prepared by filling the pores of the porous matrices by an aqueous solution of 40 wt.% K_2CO_3. Then the samples were dried at 100°C by pumping out water vapor until their weight remained constant. The potassium carbonate content in the sorbents ranged from 22.0 to 47.5 wt.% depending on the matrix pore volume.

Table 9.1 Texture parameters of pure porous materials used as host matrices, and the potassium carbonate content in the composite materials obtained (designated by the symbols in the last column)

Host matrix	S_{BET} (m^2/g)	Pore volume (cm^3/g)	Grains shape and size*	Salt content (mass %)	Symbol in the text
Al_2O_3-1	270	0.7	Spheres, 0.2 mm	28.4	A1
Al_2O_3-2	280	1.5	Irregular, 1–2 mm	45.9	A2
Carbon-1	450	0.6	Spheres, 2–3 mm	25.3	C1
Carbon-2	>1150	0.5	Cylinders, 1 mm	22.0	C2
Silica gel KSK	350	1.0	Irregular, 1–2 mm	36.0	S1
Vermiculite	2.3	1.6	Irregular, 1–2 mm	47.5	V1

*the largest size of the sorbent grains
Source: Reprinted by permission from Springer Nature from Ref. [10], Copyright 2000.

The sorption of carbon dioxide was studied at 40°C in a fixed-bed cylindrical absorber of 180 mm length and 6 mm diameter (with overall volume of 5 cm^3). An air flux preliminary saturated with water vapor up to 4.4 vol.% was mixed with carbon dioxide and put into the absorber at a flow rate of 150 cm^3/min. The inlet CO_2 concentration was 2.0 vol.%. The effluent from the absorber was taken intermittently (usually 0.3 ml each at 120 s). The CO_2 concentration was measured by a chromatograph with the accuracy of about 2% in the range from 20 to 20,000 ppm. The entrapped carbon dioxide was removed, and thus the sorbents were recovered by heating them in air at 200–350°C for 2 h.

9.1.1 Breakthrough Curves of Carbon Dioxide under Moist Conditions

As illustrated in Fig. 9.1, the shape of the CO_2 breakthrough curves under a fixed-bed operation is quite different for the studied composites. For aluminas A1 and A2 (Table 9.1), almost complete CO_2 absorption (the residual CO_2 concentration C_{out} is less than 15–20 ppm) is observed during retention time t_{rt} = 40–50 min; then the outlet CO_2 concentration C_{out} rapidly grows up to the inlet CO_2 concentration [10]. The dynamic capacity of these sorbents can reach 0.06–0.094 g/g at C_{out} = 50 ppm, and 0.07–0.12 g/g at C_{out} = 10,000 ppm. These account for 64–82% of the limiting capacity value calculated according to the stoichiometry of reaction (9.1) (Table 9.2). Indeed, the CO_2 sorption is attributed to its interaction with the impregnated salt since the pure matrices are almost inactive in the carbon dioxide sorption (t_{rt} < 2 min for alumina, Fig. 9.1).

For K_2CO_3-on-carbons (Fig. 9.1), the entrapping carbon dioxide was less effective and a breakthrough was observed shortly after the supply of the feed stream; then the outlet CO_2 concentration increased monotonically. For the C1 and C2 sorbents, the dynamic capacity is smaller than for A1 and A2 (Table 9.2).

The pertinent curves of the absorption of CO_2 by K_2CO_3/silica gel and K_2CO_3/vermiculite composites (S1 and V1, Fig. 9.1) show the immediate breakthrough of carbon dioxide. For sample V1, a plateau at the outlet concentration of 0.8–1.0 vol.% is observed. The dynamic capacities of S1 and V1 are very low and do not exceed 0.006 g/g (Table 9.2). The low activity of K_2CO_3/silica was explained in Ref. [8] by the reaction of potassium carbonate with the matrix, which results in the formation of an inactive potassium silicate phase. A lower activity of the vermiculite-based sorbent originates, possibly, from the large pore size of this sorbent and, consequently, a low dispersion of the impregnated salt.

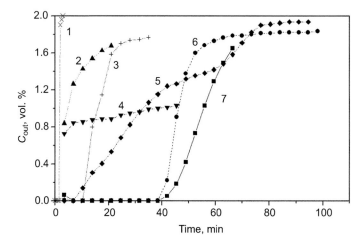

Figure 9.1 The carbon dioxide breakthrough curves of the fixed-bed operation of the composite sorbents at 40°C over fresh samples of pure Al_2O_3 (1), S1 (2), C2 (3), V1 (4), C1 (5), A1 (6), A (7). Reprinted by permission from Springer Nature from Ref. [10], Copyright 2000.

Table 9.2 Outlet CO_2 concentration before the breakthrough, C_{out}, and dynamic sorption capacity, A_{dyn}, for the composite sorbents tested

Sorbent	C_{out} (ppm)	A_{dyn} (g/g) (% from maximal*) at 50 ppm	at 10,000 ppm
A1	15	0.063 (73)	0.067 (78)
A2	15	0.094 (64)	0.121 (82)
C1	—	0.008 (10)	0.044 (55)
C2	40	0.025 (35)	0.034 (48)
S1	—	0 (0)	< 0.001 (< 1)
S1	—	0 (0)	0.006 (4)

*calculated according to the stoichiometry of reaction (9.1)
Source: Reprinted by permission from Springer Nature from Ref. [10], Copyright 2000.

Thus, among the sorbents tested, only the A1 and A2 composites based on porous alumina could be of practical importance because of the high dynamic capacity, low outlet concentration, and sharp sorption front.

9.1.2 Composite Sorbent K₂CO₃/Alumina

The X-ray diffraction study of the initial K₂CO₃/alumina composite reveals two phases, namely, a basic phase of K₂CO₃ and a small quantity of hydroaluminum carbonate KAl(CO₃)₂·1.5H₂O, the latter being formed during the synthesis. The first phase transforms into KHCO₃ in the course of reaction (9.1). The second one does not react with CO₂ for 3–10 min., however, completely decomposes in the atmosphere of humid CO₂ for 30–70 h, giving KHCO₃.

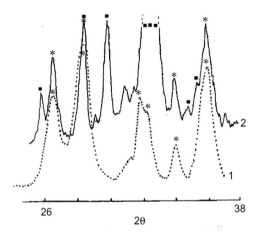

Figure 9.2 X-ray patterns of composite K₂CO₃/alumina (2) and KAl(CO₃)₂·1.5H₂O (1). * denotes reflexes of KAl(CO₃)₂·1.5H₂O, ■ - K₂CO₃·1.5H₂O. Reprinted by permission from Springer Nature from Ref. [13], Copyright 2003.

The heat released during reaction (9.1) is measured via a Tian–Calvet calorimeter. Successive and dosated portions of CO₂ and H₂O are injected into the calorimeter measuring cell loaded with the composite. The sorption of these gases occurs jointly with the molar ratio of 1:1, which corresponds to the stoichiometry of reaction (9.1). The sorption heat depends on the reaction degree and reaches 220 kJ/mol for the first gas portions (Fig. 9.3a). Then it reduces to 140 kJ/mol, which is close to the standard enthalpy of reaction (9.1) in bulk. This heat variation is probably caused by inhomogeneity of the salt particles, which can be due to their polydispersity [14]. This is indirectly confirmed by the fact that c.a. 20% of the salt remains unreacted.

The inhomogeneity of the salt's properties leads to a gradual reduction in the CO_2 sorption capacity a at increasing temperature (Fig. 9.3b), so that the equilibrium of reaction (9.1) for the confined salt is bi-variant. For this reaction in the bulk, the classical chemical thermodynamics predicts a mono-variant equilibrium with a stepwise transformation at $T = 120°C$. It is interesting that c.a. 30% of the confined salt can sorb CO_2 at $T > 120°C$, thus having a stronger affinity for CO_2 than the bulk salt.

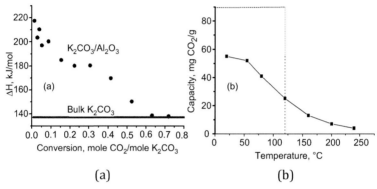

Figure 9.3 (a) Heat of CO_2 and H_2O sorption versus conversion in reaction (9.1) [14]. (b) Dynamic capacity of K_2CO_3/alumina versus temperature [13]. Dotted line indicates the theoretical capacity calculated from the stoichiometry of reaction (9.1) at $P(CO_2) = P(H_2O) = 0.02$ bar. Figure (b) reprinted by permission from Springer Nature from Ref. [13], Copyright 2003.

Another manifestation of the inhomogeneity of salt inside the pores is the finding that the composite dynamic capacity a (mg CO_2/g) depends on the input CO_2 concentration C_0 (mol/l) (not presented) as

$$a(C_0) = \frac{83 \cdot 2355 C_0}{1 + 2355 C_0} \tag{9.2}$$

The maximal dynamic capacity can be evaluated by approximation of this equation to $C_0 \to \infty$ and equals 83 mg/g, which is 90–95% of the total salt capacity calculated according to the stoichiometry of reaction (9.1).

In addition to the matrix's chemical nature and thermodynamic properties of the confined salt, the composite sorption capacity also depends on the water content. The maximal dynamic capacity is reached if the composite contains 1.5 mol of H_2O per 1 mol of K_2CO_3, which corresponds to crystalline potassium carbonate sesquihydrate $K_2CO_3 \cdot 1.5H_2O$ [19]. This allows the following reaction mechanism to be suggested [13, 14]:

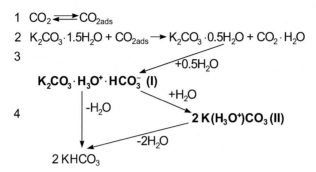

Scheme 9.1

The CO_2 adsorbed on the composite surface (stage 1) interacts with the water molecules of $K_2CO_3 \cdot 1.5H_2O$ giving, first, the hydrate $CO_2 \cdot H_2O$ (stage 2) and then the intermediate compound I. This compound can transform to $KHCO_3$ either directly or through the intermediate compound II. This mechanism well agrees with the data obtained by infrared spectroscopy [13] and isotope exchange [14] (not presented). Transformation of the anhydrous salt into hydrate $K_2CO_3 \cdot 1.5H_2O$ is accompanied by strong reconstruction of the crystalline lattice and is likely to be a limiting stage of the CO_2 sorption. Hence, for fast CO_2 sorption, potassium carbonate should form crystalline hydrate $K_2CO_3 \cdot 1.5H_2O$; therefore, the composite sorbent has to be regenerated in the presence of water vapor at the pressure and temperature corresponding to the hydrate stability field. For instance, at $P(H_2O) = 1$ bar, the optimal regeneration temperature is 170°C [14].

Another obstacle that the X-ray diffraction analysis revealed is the formation of hydroaluminum carbonate $KAl(CO_3)_2 \cdot 1.5H_2O$ at each regeneration run due to the reaction of the active salt with the host matrix. This carbonate is gradually accumulated and, being inactive in the target reaction (9.1), makes the composite less efficient in sorbing CO_2. To overcome this obstacle, the initial host material (alumina) was suggested to be treated with an alkaline solution [14].

9.2 Brief Comparison with Common CO_2 Adsorbents

The dynamic capacity of the new CO_2 sorbent K_2CO_3/alumina was compared with that for common CO_2 adsorbents, namely, commercial zeolites 4A и 13X. These zeolites were regenerated at 300°C, which is significantly higher than the regeneration temperature of the composite (170°C). The composite capacity is comparable with or superior to that of zeolites, especially at $T > 60°C$ (Fig. 9.4). The special procedure for regenerating the composite discussed above allows almost pure CO_2 to be obtained after the condensation of water vapor. For instance, the new sorbent was used for the one-stage extraction of CO_2 from the ambient air and its concentrating up to 90% purity [14].

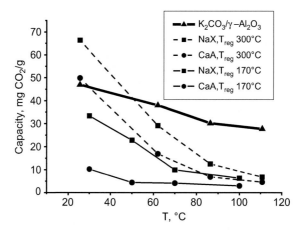

Figure 9.4 Temperature dependence of the CO_2 dynamic capacity of common zeolites CaA and NaX, and the new composite K_2CO_3/alumina [14].

Quick sorption of CO_2 by the composite results in a high degree of CO_2 removal from the ambient air. For example, the stationary concentration at the outlet of the flow adsorber until the breakthrough remains 20–50 ppm at the contact time t_c = 1–3 s, and just 1–2 ppm at t_c = 10–20 s. These residual CO_2 concentrations completely meet the requirements imposed on the purity of air to feed alkaline fuel cells.

9.3 Conclusion

New composite sorbents of carbon dioxide were prepared by impregnation of potassium carbonate in various porous matrices. The dynamic capacity of the synthesized sorbents measured in a flow absorber at 40°C can reach 0.08–0.12 g CO_2 per 1 g of the sorbent. It is much larger than the CO_2 sorption by the pure matrix; hence, the salt is the main sorbing component due to reaction (9.1). The recorded sorption corresponds to 64–82% of the limiting salt capacity calculated according to the stoichiometry of reaction (9.1). Nevertheless, the dynamic CO_2 sorption capacity depends dramatically on the chemical nature and porous structure of the host matrix and decreases in the sequence *alumina > activated carbon > vermiculite > silica gel.* This indicates that the host matrix should not be assumed as an inert support and can undergo chemical transformations caused by its interaction with the impregnated salt.

The composite based on porous alumina (K_2CO_3/alumina) could be of practical importance because of its high dynamic capacity, low outlet concentration of CO_2, and sharp sorption front. The dynamic capacity of this sorbent depends on temperature and the inlet CO_2 concentration, which is probably caused by the inhomogeneity of the salt particles inside the pores. Based on X-ray diffraction, infrared spectroscopy, and isotope exchange techniques, the tentative mechanism of CO_2 sorption was suggested. This scheme manifests an important role of the crystalline sesquihydrate $K_2CO_3 \cdot 1.5H_2O$ to ensure fast CO_2 sorption. A proper procedure was suggested for the regeneration of the sorbent.

The dynamic capacity of the new composite K_2CO_3/alumina was found to be comparable with or superior to that of common CO_2 adsorbents (commercial zeolites 4A and 13X), especially at $T > 60°C$. The new composite sorbent of CO_2 could be promising for extracting CO_2 from air and other gases to concentrate the CO_2 or to obtain CO_2-free air for fuel cell technology.

References

1. IPCC (Intergovernmental Panel on Climate Change) (2007) Climate change 2007: Synthesis report. In: Pachauri, K. P. and Reisinger, A. (Eds.), *Contribution of Working Groups I, II and III to the Fourth*

Assessment Report of the Intergovernmental Panel on Climate Change (IPCC, Geneva, Switzerland).

2. Sircar, S., Golden, T. C., and Rao, M. B. (1996) Activated carbon for gas separation and storage, *Carbon*, **34**(1), pp. 1–12.

3. Bűlow, M. (2002) Complex sorption kinetics of carbon dioxide in NaX-zeolite crystals, *Adsorption*, **8**, pp. 9–14.

4. Leci, C. L. (1996) Development requirements for absorption processes for effective CO_2 capture from power plants, *Energy Convers. Manage.*, **38**, pp. S45–S50.

5. Yao, J. and Goel, R. K. (1984) *Cryogenic Carbon Dioxide Fractionation*, An energy workshop, Huston, Texas, September 1984.

6. Srivastava, M. L., Shukla, N. K., Singh, S. K., and Jaiswal, M. R. (1996) Studies on DL-α-tocopherol liquid membranes, *J. Membr. Sci.*, **117**, pp. 39–44.

7. Yong, Z., Mata, V. G., and Rodrigues, A. E. (2001) Adsorption of carbon dioxide on chemically modified high surface area carbon-based adsorbents at high temperature, *Adsorption*, **7**, pp. 41–50.

8. Hayashi, H., Taniuchi, J., Furuyashiki, N., Sugiyama, S., Hirano, S., Shigemoto, N., and Nonaka, T. (1998) Efficient recovery of carbon dioxide from flue gases of coal-fired power plants by cyclic fixed-bed operations over K_2CO_3-on-carbon, *Ind. Eng. Chem. Res.*, **37**, pp. 185–191.

9. Hirano, S., Shigemoto, N., Yamada, S., and Hayashi, H. (1995) Cyclic fixed-bed operations over K_2CO_3-on-carbon for the recovery of carbon dioxide under moist conditions, *Bull. Chem. Soc. Jpn.*, **68**, pp. 1030–1035.

10. Okunev, A. G., Sharonov, V. E., Aristov, Yu. I., and Parmon, V. N. (2000) Sorption of carbon dioxide from wet gases by K_2CO_3 in porous matrix: Influence of the matrix nature, *React. Kinet. Catal. Lett.*, **71**, pp. 355–362.

11. Sharonov, V. E., Tyshchishchin, E. A., Moroz, E. M., Okunev, A. G., and Aristov, Yu. I. (2001) Sorption of CO_2 from humid gases on potassium carbonate supported by porous matrix, *Russ. J. Appl. Chem.*, **74**, pp. 400–405.

12. Sharonov, V. E., Okunev, A. G., and Aristov, Yu. I. (2004) Kinetics of carbon dioxide sorption by a composite material "K_2CO_3 in Al_2O_3," *React. Kinet. Cat. Letts.*, **82**, pp. 363–370.

13. Okunev, A. G., Sharonov, V. E., Gubar, A. V., Danilova, I. G., Paukshtis, E. A., Moroz, E. M., Kriger, T. A., Malakhov, V. V., and Aristov, Yu. I. (2003)

Sorption of carbon dioxide by the composite sorbent "potassium carbonate in porous matrix," *Rus. Chem. Bull. Int. Ed.*, **52**, pp. 359–363.

14. Sharonov, V. E. (2004) Regenerable absorbents of carbon dioxide based on carbonates of alkali-earth metals, Ph.D thesis, Boreskov Institute of Catalysis, Novosibirsk.

15. Zhao, C., Chen, X., and Zhao, C. (2009) CO_2 absorption using dry potassium-based sorbents with different supports, *Energy Fuels*, **23**(9), pp. 4683–4687.

16. Veselovskaya, J. V., Derevschikov, V. S., Kardash, T. Y., Stonkus, O. A., Trubitsina, T. A., and Okunev, A. G. (2013) Direct CO_2 capture from ambient air using K_2CO_3/Al_2O_3 composite sorbent, *Int. J. Greenhouse Gas Control*, **17**, pp. 332–340.

17. Derevschikov, V. S., Veseloskaya, Z. V., Kardash, T. Y., Trubitsyn, D. A., and Okunev, A. G. (2014) Direct CO_2 capture from ambient air using K_2CO_3/Y_2O_3 composite sorbent, *Fuel*, **127**, pp. 212–218.

18. Zhao, C., Chen, X., Anthony, E. J., Jiang, X., Wu, Y., and Dong, W. (2013) Capturing CO_2 in flue gas from fossil fuel-fired power plants using dry regenerable alkali metal-based sorbent, *Prog. Energy Combustion Sci.*, **39**, pp. 515–534.

19. Duan, Y. H., Luebke, D. R., Pennline, H. W., Li, B. Y., Janik, M. J., and Halley, J. W. (2012) Ab initio thermodynamic study of the CO_2 capture properties of potassium carbonate sesquihydrate, $K_2CO_31.5H_2O$, *J. Phys. Chem. C*, **116**, pp. 14461–14470.

Chapter 10

Thermal Properties of CSPMs

In the previous chapters, we have considered the adsorption properties of composites "salt in porous matrix" (CSPMs). Here we survey their thermal properties, namely, heat capacity and heat conductivity. Both these values are fundamental characteristics of CSPMs and create a base for calculating their thermodynamic and dynamic performances in various applications, some of which are considered in this book. Tabulating these properties is strictly necessary for theoretical analysis of the applications, making estimations, engineering calculations, and mathematical modeling.

In this chapter, we consider these important characteristics for various CSPMs, first of all for the composite "CaCl$_2$ in silica gel" (SWS-1L), which was successfully tested for several practical applications such as gas drying, heat transformation and storage, regeneration of heat and moisture in ventilation systems, etc.

10.1 Heat Capacity

The heat capacity C_p is a fundamental property of any individual substance. Its measurement in a wide temperature range from very low temperatures allows determination of other important thermodynamic parameters, e.g., calorimetric entropy and enthalpy. In addition, experimental curves $C_p(T)$ give very useful information

Nanocomposite Sorbents for Multiple Applications
Yu. I. Aristov
Copyright © 2020 Jenny Stanford Publishing Pte. Ltd.
ISBN 978-981-4267-50-2 (Hardcover), 978-981-4303-15-6 (eBook)
www.jennystanford.com

about phase transitions and characteristic temperatures at which different degrees of freedom are liberated.

This section addresses experimental measurements of the heat capacity of the composite "CaCl$_2$ in the mesoporous silica gel." We, first, reported data on the low-temperature heat capacity of SWS-1L at anhydrous and hydrated states. These measurements were performed in a temperature range of 6 to 300 K with a vacuum adiabatic calorimeter [1]. The heat capacities C_p (298.15 K) in both states are compared with those calculated as a linear addition of the bulk salt and matrix to understand whether the confined salt and its dihydrate have the same heat capacity as in bulk or any changes occur due to the small size of the confined phase or its interaction with the silica surface. The increase in heat capacity [2, 3] as well as the decrease in the Debye temperature [4, 5] and the melting point [6] of nanoparticles as compared with the bulk values have been reported for many metals and alloys.

In the second part of this section, we present the heat capacity of the composite $C_p(w, T)$ at various adsorbate uptakes w and temperatures T. It was measured calorimetrically for the composites based in CaCl$_2$ [7] and Ca(NO$_3$)$_2$ [8]. An analytical equation was suggested for approximating the function $C_p(w, T)$ at 40°C < T < 110°C.

10.1.1 Low-Temperature Capacity of CaCl$_2$/Silica Gel (SWS-1L)

A synthesis procedure of SWS-1L was described in Ref. [9]. The salt content in the dry composite was 35.7 ± 0.1 mass %. The hydrated sample was prepared by equilibrating the dry sample with water vapor at the relative pressure P/P_o = 0.06. The equilibrium water content was 10.57 wt.%, which corresponded to n = 2.04 ± 0.01 mol H$_2$O per 1 mol CaCl$_2$.

The heat capacity C_p was measured with a low-temperature vacuum adiabatic calorimeter [10] in a wide temperature range from 6 to 303 K. The accuracy of the heat capacity measurements was estimated to be about 2% at 15 K, 1% at 25 K, and 0.4% at T = 40–300 K. Experimental details can be found elsewhere [1].

Primary experimental data on the heat capacity of anhydrous (**I**) and hydrated (**II**) composites are displayed in Fig. 10.1 as a function of temperature. The value of heat capacity, 0.73 J/K g, measured by the DSC method in Ref. [7] at T = 314 K for sample **I** well agrees with the present data. For none of the two samples, there is any sign of any transition or thermal anomaly.

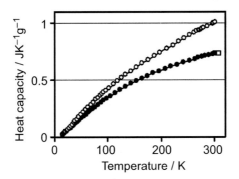

Figure 10.1 Heat capacity of the anhydrous (black circles) and hydrated (open circles). Open square corresponds to the heat capacity of the CaCl$_2$–SiO$_2$ system at T = 314 K taken from Ref. [7]. Figure adapted by permission from Springer Nature from Ref. [1], Copyright 2010.

To calculate the thermodynamic parameters, the obtained experimental data are extrapolated to the origin following the Debye law. The values of calorimetric entropy and enthalpy increments, $S^0_{298} - S^0_0$ and $\Delta H^0_{298} - \Delta H^0_0$, are derived by integrating the complete curve $C_p(T)$ and tabulated in Ref. [1]. For convenience of practical calculations, the experimental curves are also approximated by the following analytical expression:

$$C_p(T) = A + BT + CT^2 + DT^3 + F\sqrt{T} + G\ln(T) \qquad (10.1)$$

Appropriate fitting coefficients are presented in Table 10.1.

Table 10.1 Coefficients of Eq. (10.1), which ensure the best fit to the experimental C_p curves

Sample	A	B·10³	C·10⁶	D·10¹⁰	F	G
I	0.02019	6.160	−15.4629	163.532	0	−0.0332
II	0.08872	−4.301	2.43605	−2.03538	0.2183	−0.3134

Source: Reprinted by permission from Springer Nature from Ref. [1], Copyright 2010.

Useful comparisons can be made at the standard temperature 298.15 K. The heat capacity of confined salt C_p(CaCl$_2$) can be calculated as $[C_p(I) - \alpha_1 C_p(SiO_2)]/\alpha_2$. Here $C_p(I) = 0.7277 \pm 0.0030$

J/g K is the experimental heat capacity of the anhydrous composite, $C_p(SiO_2) = (0.7850 \pm 0.0030)$ J/g K is the heat capacity of silica KSK, α_1 and α_2 are the weight fractions of the silica ($\alpha_1 = 0.643$) and the salt ($\alpha_2 = 0.357$). This gives $C_p(CaCl_2) = 0.624 \pm 0.005$ J/g K, which is smaller than the heat capacity of the bulk salt (0.6537 J/g K) taken from the data of Kelley and Moore [11] measured in the temperature range of 52.6–295.1 K (Fig. 10.2). Although the heat capacity of dispersed substance is commonly larger than that of the bulk material, it can either decrease or increase for nanoparticles embedded in a matrix [5, 12]. In our case, its reduction may result from the strong chemical interaction between the confined salt and the host silica.

A similar estimation is made for the hydrated composite "$CaCl_2 \cdot 2.04H_2O$–silica gel" [1]. Using the data on the heat capacity of $CaCl_2$ dihydrate and tetrahydrate [13], one can calculate the heat capacity of the bulk sample with the composition of sample **II** $C_p(bulk) = 0.948$ J/g K. This value is significantly smaller that the experimental one at 298.15 K, $C_p^0(298\ K) = 1.001$ J/g K; hence, the heat capacity of the dispersed dihydrate is larger than that of the bulk one. The former heat capacity can be estimated as 1.29 J/g K, which is larger than the heat capacity (1.17 J/g K) of the bulk dihydrate by 0.12 J/g K [13]. This difference can be attributed to the enhancement of the water mobility in the dispersed hydrate as compared with the bulk one.

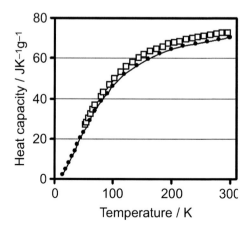

Figure 10.2 Heat capacity of calcium chloride: □ – data for the bulk salt taken from [11], ● – calculated values for the confined salt. Reprinted by permission from Springer Nature from Ref. [1], Copyright 2010.

This was confirmed by the analysis of the solid-state ^{2}H NMR line shape [14]. It was shown that the temperature of the entire transformation of the solid to liquid-like NMR signal (T_{NMR} = 453 K) is in good correspondence with the melting point for the bulk dihydrate (T_m = 449 K [13]). On the contrary, for the dispersed dihydrate, this transformation occurs in a much broader temperature range, and the solid-like signal vanishes at T_{NMR} = 323 K, which is much lower than the melting temperature of the dispersed dihydrate (T_m = 397 K). Hence, water molecules are more mobile in the dispersed dihydrate, and this mobility is not connected with the melting of hydrate.

To single out the heat capacity $c_{p,m}^0(T)$ associated only with water in the hydrated composite, we calculate this value as the difference between the curves $C_{p,m}^0(T)$ of hydrated and anhydrous composites. The effective heat capacity of confined water is a gradually increasing function of temperature (Fig. 10.3). It displays no phase transitions in the studied range of temperature and depends mainly on the molecular mobility of water. This mobility in the bulk hydrate $CaCl_2 \cdot 2H_2O$ is analyzed in Ref. [15]: the translational modes of water are observed below 50 meV; the wagging, twisting, and rocking modes are found between 50 and 100 meV, while the deformation vibrations are active above ca. 200 meV. The QENS and INS spectra of the bulk dihydrate are typical of a solid with well-defined crystallographic positions for the water molecules [16], whereas the spectra for the dispersed hydrate display broad bands characteristic of more disordered substance [15].

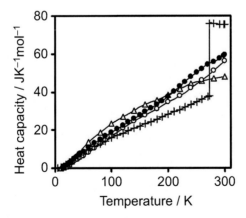

Figure 10.3 Effective heat capacity of water in the dispersed $CaCl_2 \cdot 2H_2O$ (•), analcime (Δ), paranatrolite (o), and the heat capacity of bulk H_2O (+). Reprinted by permission from Springer Nature from Ref. [1], Copyright 2010.

The effective heat capacity of water in the hydrated composite was found to be close to that for mineral zeolites containing molecular water—analcime $NaAlSi_2O_6 \cdot H_2O$ [17] and paranatrolite $Na_2Al_2Si_3O_{10} \cdot 3H_2O$ [18]—and sufficiently differs from that for bulk water and ice (Fig. 10.3). The effective heat capacity of water in the hydrated composite is larger than the heat capacity of bulk ice and is significantly smaller than that of liquid water. Hence, the water molecules in the salt dihydrate are more mobile than in the bulk ice.

Comparisons of the molecular mobility of water in the confined and bulk dihydrates of calcium chlorides are made in Chapter 12. The calorimetric data of Ref. [1] are used for calculating the change in entropy in the course of the calcium chloride dihydrate formation in the confined state: $CaCl_2 + 2H_2O = CaCl_2 \cdot 2H_2O$. This entropy change appears to be (6 ± 1) J/mol·K lesser as compared with this reaction in the bulk state. This difference can be responsible for the twofold decrease in the equilibrium pressure over the confined dihydrate, which was experimentally observed.

10.1.2 Heat Capacity of SWS-1L as Function of Water Uptake and Temperature

For engineering calculations and mathematical modeling of various adsorption units and processes employing a CSPM, the dependence of heat capacity $C_p(w, T)$ at various adsorbate uptakes w and temperatures T has to be known and tabulated [19].

Commonly, this value is calculated as a linear addition of the specific heat C_p(dry adsorbent) of a dry adsorbent and the specific heat C_p(adsorbate) of the adsorbate in the adsorbed state with appropriate weight fractions: $C_p(w) = (1 - w)C_p$(dry adsorbent) $+ wC_p$(adsorbate). The value of C_p(adsorbate) is not known and is arbitrarily considered to be equal to its specific heat either in gas or liquid phase. Theoretical calculation of the C_p(adsorbate) value for the "$CaCl_2$/silica gel + water" system showed that it is closer, but not equal, to the C_p value in the liquid state [20]. However, the current state of theoretical chemistry does not allow sufficient accuracy of such calculations, and direct experimental measurements give at present the only reliable way to obtain the necessary data.

Here we mainly consider experimental data on the heat capacity of the composites based in $CaCl_2$ (SWS-1L) [7] in the silica gel pores.

These data are approximated by analytical equations that describe the function $C_p(w, T)$ in the temperature range of 40–110°C. Similar measurements were performed for the composite $Ca(NO)_3$/(silica gel), SWS-8L. These data were reported in Ref. [8]. The heat capacity of wetted composites was measured using a DSC Mettler TA 4000. To avoid evaporation of the sorbed water and to maintain a fixed uptake w, a measuring pan with the sample was tightly closed. All details of these measurements can be found elsewhere [7].

A set of experimental dependencies $C_p(T)$ were measured over the temperature range of 40–110°C at a fixed water content $w = m_{H_2O}/m_{ads}$, where m_{H_2O} is the mass of sorbed water, m_{ads} is the mass of dry sorbent (between 0 and 0.414 g/g). Under these conditions, the heat capacity at w = const. is a linear function of temperature (Fig. 10.4)

$$C_p(w, T) = a(w) + b(w) \cdot T$$

where the coefficients $a(w)$ and $b(w)$ depend on the water content w and can be approximated by the polynomial expressions

$$a(w) = a_0 + a_1 w + a_2 w^2$$
$$b(w) = b_0 + b_1 w + b_2 w^2$$

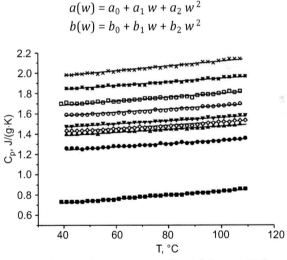

Figure 10.4 Specific heat of the composite "CaCl$_2$/silica gel KSK" as a function of temperature measured at various water uptakes w (from bottom to top): 0, 0.196, 0.226, 0.256, 0.270, 0.285, 0.331, 0.379, and 0.414 g/g. Reprinted from Ref. [19], Copyright 2012, with permission from Elsevier.

Thus, the overall equation is

$$C_p(w,T) = (a_0 + a_1 w + a_2 w^2) + (b_0 + b_1 w + b_2 w^2)T \qquad (10.2)$$

Table 10.2 Coefficients of Eq. (10.2), which ensure the best fit to the experimental C_p curve of SWS-1L

a_0 (J/g K)	a_1 (J/g K^2)	a_2 (J/g K)	$b_0 \cdot 10^3$ (J/g K^2)	$b_1 \cdot 10^3$ (J/g K^2)	$b_2 \cdot 10^3$ (J/g K^2)
0.6441	2.73789	−0.55671	1.92126	−5.67143	16.0362

Source: Reprinted from Ref. [19], Copyright 2012, with permission from Elsevier.

The coefficients a_i and b_i are obtained by fitting the experimental values with Eq. (10.2). These values are displayed in Table 10.2 [19]. The accuracy of measurement and approximation is 0.02 J/g K, which is quite sufficient for the majority of engineering calculations.

10.2 Thermal Conductivity

The thermal conductivity λ is another fundamental characteristics of CSPMs, which is important for various practical applications. Here we, first, report this value for new composite sorbents "salt in porous matrix" as a function of the water uptake w [21]. These measurements were performed under atmospheric pressure with the composites consolidated in dense bricks. Three hygroscopic salts (CaCl$_2$, MgCl$_2$, and LiBr) were confined to the mesopores of commercial silica gel (SWS-1L, SWS-2L, and SWS-3L, respectively). For CaCl$_2$, a porous aluminum oxide was used as host matrix, too. The λ value was measured at various water contents w, and this dependence was analyzed in the frame of the Luikov–Bjurström model.

CSPMs that are used for adsorption heat transformation operate at low pressure ($P < 100$ mbar). Because of this, the thermal conductivity of the composites CaCl$_2$/SiO$_2$ and LiBr/SiO$_2$ was measured as a function $\lambda(T, P_{H2O}, w)$ of the three parameters: the vapor pressure P_{H2O}, temperature T, and water uptake w. The experimental conditions were chosen according to the boundaries of a typical adsorption cooling cycle (10 mbar < P_{H2O} < 70 mbar, 40°C < T < 130°C) [22]. Finally, the influence of the thermal conductivity on the specific power of the sorption chiller is shortly discussed.

10.2.1 Thermal Conductivity of CSPMs at Atmospheric Pressure

The composites were molded into rectangular bricks of dimension $70{\times}30{\times}10$ mm^3 under a pressure (P) of 150–200 bar. The bricks were then dried at $T = 150°C$ until their weight became constant. After drying, the samples were kept in contact with water vapor until the desired equilibrium water content w was reached. A standard "hot wire" method [23] was used to measure the thermal conductivity. All the experimental details can be found elsewhere [21, 24].

Experimental dependencies $\lambda(w)$ of the thermal conductivity of SWS-1L, SWS-2L, and SWS-3L upon the water content are qualitatively similar (Fig. 10.5). The main features of these curves are (a) a smooth raise or constancy of the λ value at the water uptake lower than the "threshold" uptake w^* and (b) its sharp increase within the narrow w range near the "threshold" uptake w^* (Fig. 10.5). The minimal thermal conductivity $\lambda = 0.13$–0.16 W/m K was measured for the dry samples ($w = 0$). At low water uptake ($w \leq 0.20$), the dispersed salt transformed into solid crystalline hydrates "salt·nH$_2$O" (Chapter 5). For instance, the hydrates of calcium chloride contain 0.33, 1, or 2 molecules of water per salt molecule and the dihydrate can coexist with the salt solution inside the pores [9]. For MgCl$_2$, hydrates with $n = 2, 4, 6$, and 7, and for LiBr with $n = 1$ and 2 are known in the bulk state [25]. On the one hand, the specific heat conductivity of the hydrates is lesser than that of the anhydrous salt [26]; on the other hand, the sample density $\rho = \rho_0 (1 + w)$ increases with the rise in w. As a result, a smooth increase in λ value was observed at $w < 0.20$.

At a higher uptake, the salt solution appeared inside the silica pores, whose thermal conductivity is, at least, two times lower than that of the mentioned crystalline hydrates [25, 27]. Therefore, it should decrease the composite thermal conductivity; however, this fall can be completely or partially compensated by the growth in λ due to the composite density. The balance of these opposite tendencies resulted in about the same λ value in the range $0.2 > w > (0.4$–$05)$ (Fig. 10.5).

Further portions of the sorbed water caused a steep increase in λ. The increase was observed within a narrow interval of $\Delta w \approx 0.1$; however, the "threshold" value w^* differed for the studied composites: c.a. 0.45, 0.39, and 0.51 for SWS-1L, SWS-2L, and

SWS-3L, respectively. Further increase in the uptake led to a monotonous growth in the composite thermal conductivity up to 0.4–0.5 W/m K, which is 3–4 times that of the dry sorbents (Fig. 10.5).

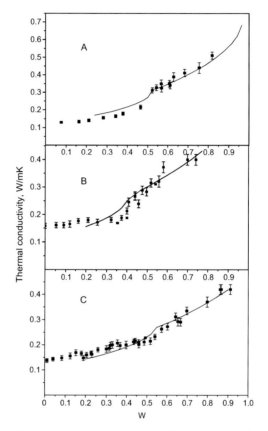

Figure 10.5 Thermal conductivity of the SWS samples as a function of the sorbed water content. Symbols—experimental data; lines—calculations with the Luikov–Bjurström model [21, 24]. A—SWS-1L; B—SWS-2L; C—SWS-3L. Adapted from Ref. [24], Copyright 2013, with permission from Elsevier.

For a multiphase system, the heat transfer depends on the conductivity and volume fraction of each single component [28]. Figure 10.6 displays the composite thermal conductivity as a function of v, which is the fraction of silica pore volume occupied by the salt solution. It appears that regardless of the salt confined, the steep increase in the composite thermal conductivity occurs at the same fraction of the pore volume occupied by the salt solution, $v^* = 0.60–0.64$ (Fig. 10.6). This probably indicates a general mechanism of the threshold rise of the thermal conductivity of silica-based composites.

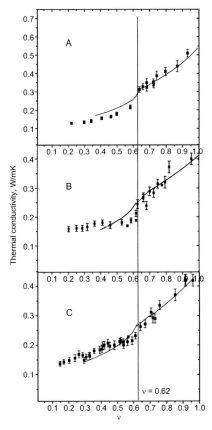

Figure 10.6 Thermal conductivity of the composites as a function of the fraction of silica pore volume occupied by the salt solution. Symbols—experimental data; lines—calculations with the Luikov–Bjurström model. A—SWS-1L; B— SWS-2L; C—SWS-3L. Reprinted from Ref. [24], Copyright 2013, with permission from Elsevier.

Such threshold effects in porous wetted media can, in particular, be explained by the percolation theory [29]. It assumes that there is a "percolation threshold," i.e., a volume fraction v_c of the conducting component in the multicomponent system so that at $v > v_c$, an infinite cluster of this component is formed. As a result, the effective conductivity of this multicomponent system drastically increases. For the systems under study, such a cluster may be formed by the salt solution distributed inside the pore space. Indeed, at $v \approx 0.62$, the mesopores of silica gel are completely filled with the salt solution. The solution starts to cover the external surface of the primary silica particles, and a continuous film of the solution begins to form. Evidently, the presence of the film facilitates heat contact and heat transfer between the adjacent silica particles. The number of such "connected" granules rapidly increases, and at the "threshold" value v^*, the water sorption is likely to result in the formation of a continuous solution network, thus enhancing the heat conductivity of the whole system.

This effect can be quantitatively described in the frame of the Luikov–Bjurström model [30, 31]. Our calculations (not presented) show that the model used describes the main features of the experimental dependences $\lambda(w)$ (solid lines in Figs. 10.5 and 10.6). For example, the model predicts a steep increase in the composite thermal conductivity at $v \approx 0.62$.

10.2.2 Thermal Conductivity of CaCl₂/Alumina Composite

No such "threshold" behavior was found for $CaCl_2$/alumina composite (SWS-1A) for which the thermal conductivity gradually (without steep segments) increases from 0.12 to 0.40 W/m K with the water uptake increasing up to 0.65 (Fig. 10.7). Since the confined salt is the same as for SWS-1L, the reason behind the different behavior could be the difference in the pore structure and/or chemical nature of the two host matrices (silica and alumina). The average pore size of alumina is twice smaller than that for KSK silica. This, according to the data in Chapter 6, can lead to a bivariant type of water sorption equilibrium for SWS-1A instead of mono-variant for SWS-1L. The bivariant type of equilibrium means that the liquid salt solution is formed inside alumina pores during water sorption. If both the external and internal surfaces of primary alumina particles are well wetted by the solution, the latter can form a continuous film distributed both inside and outside the primary particles. The film's thickness gradually increases with water sorption, thus causing a monotonous rise in the composite thermal conductivity. As the solution simultaneously occupies both meso- and macropores of alumina, no "threshold" effect takes place.

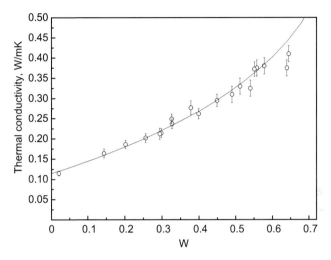

Figure 10.7 Dependence of the SWS-1A thermal conductivity on the water uptake and the fraction of alumina pore volume occupied by the salt solution. Line—calculations with the Luikov–Bjurström model. Reprinted from Ref. [24], Copyright 2013, with permission from Elsevier.

Thus the nature of the porous host matrix can affect the $\lambda(w)$ function for CSPMs due to at least two effects: the influence of the matrix pore structure on the type of water sorption equilibrium and the different surface ability of various matrices to be wetted by the confined salt solution.

10.2.3 Thermal Conductivity under Real Conditions of AHT Cycle

CSPMs are considered to be promising candidates for adsorption heat transformers (AHTs) (see Ref. [32] and Chapters 16, 17). These units operate at a low vapor pressure of 10–100 mbar and T = 40–130°C. Evidently, for proper dynamic analysis of AHT cycles, the λ value has to be known just under the mentioned conditions. The λ data reported above at ambient pressure and temperature can be considered a reference point. A decrease in the adsorbent's thermal conductivity may be predicted when the pressure goes down, while it should increase with a rise in temperature. The scale of these changes will be found out further below.

The thermal conductivity of SWSs was measured by a standard "hot wire" method [23] inside a special vacuum chamber in order to fix the required working conditions (P_{H2O}, T). All experimental details can be found in Ref. [22].

The thermal conductivity of the dry composites at 10 mbar is measured to be 0.12–0.125 W/m K. The λ value only slightly depends on the air pressure and increases by some 5–20% at 100 mbar. It looks reasonable as the contribution of the molecular diffusion mechanism to the heat transfer through the gas phase inside the pores increases with a rise in the pressure. Indeed, this mechanism becomes predominant in pores with diameter larger than the mean free path in air (c.a. 40 μm at P = 10 mbar) and provides higher heat transfer respect to the Knudsen mechanism. Hence, under typical operating conditions of an AHT unit, the gas transport inside the pores only slightly influences the thermal conductivity of the system because the majority of heat is transferred through the solid skeleton of the sorbent and, thus, the SWS thermal conductivity should strongly depend on the amount of sorbed water.

It is hardly possible to separately consider the contributions of water uptake, pressure, and temperature to the overall value of SWS thermal conductivity. Therefore, we linked the measured λ values directly with the appropriate points of the AHT thermodynamic cycle (Fig. 10.8). General information about AHT cycles is available in Chapters 16 and 17.

The weak isoster describes the isosteric cooling, which follows the water desorption. It is characterized by a minimal water uptake of 5.3 wt.% for SWS-1L and 14.0 wt.% for SWS-2L. It turned out that along this isoster (Fig. 10.8), the thermal conductivity of SWS-1L is close to that of the dry composite (λ = 0.12–0.16 W/m K) measured in Ref. [21]. It means that at low uptake, the sorbed water almost does not affect the composite thermal conductivity.

Significantly larger conductivity (up to 0.2–0.3 W/m K, Fig. 10.8) was recorded along the rich isoster that follows the water adsorption stage. At high water content (w = 25–30 wt.%), the liquid phase of aqueous salt solution gives a considerable contribution to the total λ value of SWSs due to the formation of the solution layer on the surface of the solid skeleton, as discussed in Sec. 10.2.1.

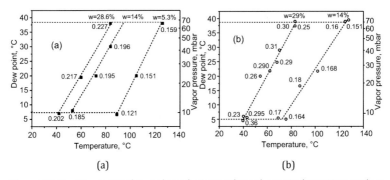

Figure 10.8 Composite thermal conductivity along the typical sorption cooling cycle: (a) SWS-1L and (b) SWS-2L. Reprinted from Ref. [22], Copyright 2002, with permission from Elsevier.

Besides, at a fixed uptake, the tendency for the thermal conductivity to be higher with an increase in temperature and pressure was observed (Fig. 10.8). The λ values of SWS-2L are, in average, higher than that of SWS-1L because the former sorbes a larger mass of water along the AHT cycle considered. It is interesting to note that the measured λ values are consistent with those obtained for the composites SWS-1L and SWS-2L under the ambient pressure and temperature in Ref. [21].

The λ values obtained can be used as input data for simulating the dynamic behavior of adsorption chiller based on the SWS-1L–water pair. The influence of λ value on the cycle cooling power was analyzed in the frame of a simple mathematical model presented in Ref. [22]. It was demonstrated that for adsorbent beds with enhanced heat transfer properties, it is very important to use accurate λ values, which account for effects of the uptake, pressure, and temperature. If, instead, one uses an average (and constant) λ value, as it is commonly done, it could give an error as high as 30% in the calculation of the specific cooling power of the cycle.

References

1. Aristov, Yu. I., Kovalevskaya, Yu. A., Tokarev, M. M., and Paukov, I. E. (2011) Low temperature heat capacity of the system "silica gel–calcium chloride–water," *J. Ther. Anal. Calorimetry*, **103**, pp. 773–778.
2. Bai, H. Y., Luo, Y. L., Jin, D., and Sun, J. R. (1996) Particle size and interfacial effect on the specific heat of nanocrystalline Fe, *J. Appl. Phys.*, **79**, pp. 361.

3. Herr, U., Geigl, M., and Samwer, K. (1998) Debye temperature of nanocrystalline $Zr_{1-x}Al_x$ solid solutions with different grain sizes, *Phil. Magazine A*, **77**, pp. 641–652.

4. Childress, J. R., Chien, C. L., Zhou, M. J., and Sheng, P. (1991) Lattice softening in nanometer-size iron particles, *Phys. Rev. B*, **44**, 11689.

5. Koops, G. E. J., Pattyn, H., Vantomme, A., Nauwelaerts, S., and Venegas, R. (2004) Extreme lowering of the Debye temperature of Sn nanoclusters embedded in thermally grown SiO_2 by low-lying vibrational surface modes, *Phys. Rev. B*, **70**, 235410.

6. Zhao, S. J., Wang, S. Q., Cheng, D. Y., and Ye, H. Q. (2001) Three distinctive melting mechanisms in isolated nanoparticles, *J. Phys. Chem. B*, **105**, 12857.

7. Aristov, Yu. I., Tokarev, M. M., Cacciola, G., and Rectuccia, G. (1997) Properties of the system "calcium chloride–water" confined in pores of the silica gel: Specific heat, thermal conductivity, *Rus. J Phys. Chem.*, **71**, pp. 391–394.

8. Freni, A., Sapienza, A., Glaznev, I. S., Aristov, Yu. I., and Restuccia, G. (2012) Testing of a compact adsorbent bed based on the selective water sorbent "silica modified by calcium nitrate," *Int. J. Refrig.*, **35**, pp. 518–524.

9. Aristov, Yu. I., Tokarev, M. M., Cacciola, G., and Restuccia, G. (1996) Selective water sorbents for multiple applications: 1. $CaCl_2$ confined in mesopores of the silica gel: Sorption properties, *React. Kinet. Cat. Lett.*, **59**, pp. 325–334.

10. Bessergenev, V. G., Kovalevskaya, Yu. A., Paukov, I. E., Starikov, M. A., Opperman, H., and Reichelt, W. (1992) Thermodynamic properties of $MnMoO_4$ and $Mn_2Mo_3O_8$, *J. Chem. Thermodyn.*, **24**, pp. 85–98.

11. Kelley, K. K. and Moore, G. E. (1943) The specific heat at low temperatures of anhydrous chlorides of calcium, iron, magnesium and manganese, *J. Amer. Chem. Soc.*, **65**, pp. 1264–1267.

12. Nanver, L. K., Weyer, G., and Deutch, B. I. (1980) Precipitation of β-tin in Sn implanted silicon, *Phys. Status Solidi A.*, **61**, pp. K29–K34.

13. Meisingset, K. K. and Grønvold, F. (1986) Thermodynamic properties and phase transitions of salt hydrates between 270 and 400 K. IV. $CaCl_2 \cdot 6H_2O$, $CaCl_2 \cdot 4H_2O$, $CaCl_2 \cdot 2H_2O$, and $FeCl_3 \cdot 6H_2O$, *J. Chem. Thermodyn.*, **18**, pp. 159–173.

14. Kolokolov, D. I., Stepanov, A. G., Glaznev, I. S., Aristov, Yu. I., and Jobic, H. (2008) Water dynamics in bulk and dispersed in silica $CaCl_2$ hydrates

studied by ^{2}H NMR, *J. Phys. Chem. C*, **112**, pp. 12853–12860.

15. Kolokolov, D. I., Stepanov, A. G., Glaznev, I. S., Aristov, Yu. I., Plazanet, M., and Jobic, H. (2009) Water dynamics in bulk and dispersed in silica CaCl$_2$ hydrates studied by neutron scattering methods, *Micropor. Mesopor. Mater.*, **125**, pp. 46–50.

16. Plazanet, M., Glaznev, I. S., Stepanov, A. G., Aristov, Yu. I., and Jobic, H. (2006) Dynamics of hydrated water in CaCl$_2$ complexes, *Chem. Phys. Lett.*, **419**, pp. 111–114.

17. Johnson, G. K., Flotow, H. E., O'Hare, P. A. G., and Wise, W. S. (1982) Thermodynamic studies of zeolites: Analcime and dehydrated analcime, *Amer. Mineral.*, **67**, pp. 736–748.

18. Paukov, I. E., Moroz, N. K., Kovalevskaya, Yu. A., and Belitsky, I. A. (2002) Low-temperature thermodynamic properties of disordered zeolites of the natrolite group, *Phys. Chem. Miner.*, **29**, pp. 300–306.

19. Aristov, Yu. I. (2012) Adsorptive transformation of heat: Principles of construction of adsorbents database, *Appl. Therm. Eng.*, **42**, pp. 18–24.

20. Chakraborty, A., Saha, B. B., Ng, K. C., Koyama, S., and Srinivasan, K. (2009) Theoretical insight of physical adsorption for a single-component adsorbent + adsorbate system: I. Thermodynamic property surfaces, *Langmuir*, **25**, pp. 2204–2211.

21. Tanashev, Yu. Yu. and Aristov, Yu. I. (2000) Thermal conductivity of a silica gel + calcium chloride system: The effect of adsorbed water, *J. Eng. Phys. Thermophys.*, **73**, pp. 876–884.

22. Freni, A., Tokarev, M. M., Okunev, A. G., Restuccia, G., and Aristov, Yu. I. (2002) Thermal conductivity of selective water sorbents under the working conditions of a sorption chiller, *Appl. Therm. Eng.*, **22**, pp. 1631–1642.

23. Carslaw, H. S. and Jaeger, J. C. (1988) *Conduction of Heat in Solids*, 2nd edn. (Clarendon Press, Oxford).

24. Tanashev, Yu. Yu., Krainov, A. V., and Aristov, Yu. I. (2013) Thermal conductivity of composites "a porous matrix modified by an inorganic salt": Influence of sorbed water, *Appl. Therm. Eng.*, **61**, pp. 401–407.

25. Erich Pietsch, E. H. (Ed.) (1957) *Gmelins Handbuch der Anorganischen Chemie. Calcium Teil B: Lieferung 2* (Verlag Chemie, GmbH).

26. Mirzaev, Sh. M., Yakubov, Yu. N., Akhmedov, A. A., Boltaev, S. A., and Shodiev, O. K. (1996) Experimental study of the temperature dependence of the CaCl$_2$, SrCl$_2$, CaCl$_2 \cdot 6$H$_2$O and SrCl$_2 \cdot 6$H$_2$O absorbent heat conductivity, *Appl. Solar Energy*, **32**, pp. 65–70.

27. Magomedov, U. B. (1993) Heat conductivity of aqueous solutions of salts at high pressures, temperatures and concentrations, *Rus. Teplophizika Visokih Temperatur.*, **31**, pp. 504–507.

28. Harriott, P. (1975) Thermal conductivity of catalyst pellets and other porous particles, Part I, Review of models and published results, *Chem. Eng. J.*, **10**, pp. 65–71.

29. Zarichniak, Yu. P. and Novikov, V. V. (1978) The effective conductivity of heterogeneous systems with a disordered structure, *J. Eng. Thermophys.*, **34**, pp. 648–655.

30. Luikov, A. V., Shashkov, A. G., Vasiliev, L. L., and Fraiman, Yu. E. (1968) Thermal conductivity of porous systems, *Int. J. Heat Mass Transfer.*, **11**, pp. 117–140.

31. Bjurström, H., Karawacki, E., and Carlsson, B. (1984) Thermal conductivity of a microporous particulate medium: Moist silica gel, *Int. J. Heat Mass Transfer.*, **27**, pp. 2025–2036.

32. Aristov, Yu. I., Restuccia, G., Cacciola, G., and Parmon, V. N. (2002) A family of new working materials for solid sorption air conditioning systems, *Appl. Therm. Eng.*, **22**, pp. 191–204.

Chapter 11

Melting–Solidification of Salt Solution/Hydrates in Pores

The main method of synthesizing composites "salt in porous matrix" (CSPMs) is impregnation of the host matrix with an aqueous salt solution with subsequent evaporation of water [1, 2]. The solution is concentrated, and the salt (or its hydrate) precipitates inside the matrix pores. While the salt absorbs water vapor, an inverse process occurs, namely, salt hydrates and an aqueous solution are formed inside the matrix pores. Therefore, the physicochemical properties of the salt hydrates and solutions confined to the matrix pores are of great interest for understanding and optimizing the synthesis and performance of CSPMs.

One may expect that these properties can differ from those in the bulk. For confined hydrates, it is due to the division of the material into tiny pieces with an enormous increase in surface energy [3]. The confined solutions may be affected by the strong potential of the pore wall, ion adsorption, wetting phenomenon, or just by the restricted pore geometry [4].

In this chapter, we consider the processes of melting and solidification inside the CSPM pores, including solubility diagrams, depression of melting point, solution subcooling, etc.

Nanocomposite Sorbents for Multiple Applications
Yu. I. Aristov
Copyright © 2020 Jenny Stanford Publishing Pte. Ltd.
ISBN 978-981-4267-50-2 (Hardcover), 978-981-4303-15-6 (eBook)
www.jennystanford.com

11.1 "CaCl$_2$–Water" System inside Silica Gel Pores

It is well known that a substance's confinement inside small pores may affect its properties, leading to a melting behavior different from that in the bulk [5]. The main effects include a depression of its melting temperature (T_m) and enhancement of the solubility of tiny particles. The depression has been established in the literature for many single-component systems, such as water, organic liquids, metal particles, etc. [6–8]. Very large effects are known for 5 nm gold particles: ΔT_m higher than 500°C was reported for this system [9]. The melting point depression was first theoretically analyzed in Refs. [3, 10] by considering the contribution of surface energies of solid and liquid phases. A thermodynamic analysis predicts that the melting point depression $\Delta T_m(d) = T_m(\infty) - T_m(d)$ for a small crystal with a typical size d is determined by the following equation:

$$\Delta T_m(d) = 4\sigma T_m(\infty)/(d\,\Delta H\,\rho) \sim 1/d \qquad (11.1)$$

where $T_m(\infty)$ is the normal (bulk) melting point, σ is the surface energy of the solid–liquid interface, ΔH is the bulk fusion enthalpy, and ρ is the solid density.

The dynamic approach was suggested by Lindemann [11], who assumed that a crystal melts when the root-mean square deviation of its atoms exceeds the average interatomic distance. This concept was further developed in Ref. [12], and a modification of Eq. (11.1) was suggested, which gives important correction at very small crystal size. In this book, we use the basic Eq. (11.1), which is quite sufficient for our analysis.

Two commercial silica gels with well-defined pore structures were used as porous hosts, namely, a mesoporous silica gel KSK and a microporous silica gel KSM (Table 6.1). Their pores were filled with an aqueous CaCl$_2$ solution at a fixed degree of pore filling γ. The salt concentration C in the solution was varied from 0 to 48 wt.%; the latter corresponds to an aqueous CaCl$_2$ solution saturated at $T = 29$°C (Fig. 6.1).

11.1.1 Solidification–Melting Phase Diagram

A differential scanning calorimetry (DSC)-111 "SETARAM" unit was used for detecting the phase transition and measuring the heat capacity of $CaCl_2/SiO_2$ (KSK) in the temperature range of 170–320 K. Before the measurements, the samples were rapidly cooled down to −150°C [13]. The thermograms for the cooling and heating modes were recorded under the dry nitrogen flow. The melting temperature was detected either as the temperature corresponding to the maximum of a DSC peak or by a standard "on-set" method. Other details of these experiments can be found in Ref. [13].

The degree of pore filling, γ, appears to be a crucial parameter in melting/solidification in the silica gel pores. For instance, at $\gamma \leq 0.95$, no DSC signal is detected at all (see Sec. 11.1.2). At the degree of pore filling of 0.96 or more, a typical thermogram of the system $CaCl_2/H_2O/SiO_2$ (KSK) has two endothermic peaks (Fig. 11.1). The symmetric low-temperature peak appears at 214 K (the on-set temperature is equal to 205 K). Its position does not change with the solution concentration. At $CaCl_2$ concentrations lower than 30 wt.%, this peak may not appear due to the supercooling effect, which will be specially discussed below. The position of the second asymmetric peak strongly depends on the solution concentration.

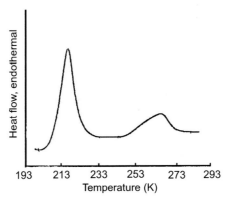

Figure 11.1 DSC thermogram of the $CaCl_2$ aqueous solution (C = 37 wt.%) confined to the pores of the KSK silica gel. The heating rate is 5 K/min. Reprinted by permission from Springer Nature from Ref. [13], Copyright 1997.

For interpreting this thermogram, it is useful to consider a solubility diagram for the bulk "$CaCl_2$–H_2O" system, which is of a simple eutectic type (Figs. 6.1 and 11.2). For salt concentrations lower than that at the eutectic point (30.3 wt.%), the solubility line defines the conditions at which ice crystals start to form (the "ice line"). For higher concentrations, the "crystallization line" defines the conditions at which salt hydrates or anhydrous salt crystallize from the solution. The low-temperature peak corresponds to the melting of the eutectic mixture of ice and $CaCl_2 \cdot 6H_2O$ at C = 37 wt.% and T = −55°C (point A in Fig. 11.2). At higher T (segment AB), a $CaCl_2$ hexahydrate melts, which shows itself as a wide asymmetric peak that determines the border of the solubility line (point B).

Analyzing the position of the second peak in the thermograms of the $CaCl_2/H_2O$/KSK system, a solubility diagram of the solution confined to the KSK pores can be plotted (Fig. 11.2). It appears to belong to the same simple eutectic type and lies below the diagram of the bulk solution. It is worthy to note that the phase diagrams for both the bulk and dispersed solutions do not follow the Schreder–Le Chatelier equation [14], which indicates a strong deviation of these solutions from an ideal binary solution.

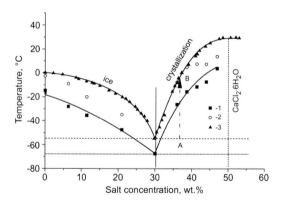

Figure 11.2 "Solidification–melting" phase diagram for the binary H_2O–$CaCl_2$ mixture: in the dispersed state (in the KSK pores) — 1 (on-set), 2 (peak); in the bulk — 3. Below the horizontal lines is the field of the eutectic mixture. Reprinted by permission from Springer Nature from Ref. [13], Copyright 1997.

The melting temperature inside the pores is certainly lower by some 15°C (on-set base) over the whole concentration range from pure water to the highest stoichiometric crystalline hydrate ($CaCl_2 \cdot 6H_2O$). For instance, for pure water in the KSK pores, we have obtained ΔT_m = 14.8°C with the on-set method and ΔT_m = 2.6°C from the peak position. The former value is likely to characterize water melting in the narrowest pores, whereas the latter corresponds to the largest ones. The shift in melting temperature in average pores lies in between the two and can be briefly estimated as ΔT_m = (14.8 + 2.6)/2 = 8.7°C. Our estimation by Eq. (11.1) gives ΔT_m= 9.0°C for 15 nm pores. A more detailed thermodynamic analysis [6] suggested a melting temperature depression equal to 9.3°C for water in such pores (see Eq. (19) in Ref. [6]).

Interestingly, during the heating run, a wide exothermic peak was observed at $T \approx -90$°C (not presented). It may correspond to the crystallization of the supercooled salt solution or to its glassy state. The rapid cooling of the sample results in supercooling of the solution or its transformation to an amorphous solid state. Indeed, no crystallization peak is observed during the cooling of the sample. The supercooling, which is well known for the bulk $CaCl_2$ aqueous solution [15], indicates that the probability for a critical nucleus to appear is low enough and under fast cooling, the solution transforms to a non-crystalline low-temperature state (supercooled liquid or amorphous solid). Additional information about this state of the confined $CaCl_2$–water system is obtained by dynamic mechanical thermal analysis (DMTA) and nuclear magnetic resonance (NMR) (see Sec. 11.1.3). The sample heating is much slower than cooling, and the system is transformed to the crystalline state, which is accompanied by heat release.

The same experimental methodology is applied to study the same solution in the small pores of KSM silica gel (d = 3.5 nm), but no DSC signal is detected. Assuming the $1/d$-dependence (Eq. (11.1)) valid for these pores, we estimate the melting temperature of water in the KSM pores: T_m = –38°C. So the melting point of water or low concentrated $CaCl_2$ solutions in the KSM pores should be within the tested temperature range $T > -100$°C. The absence of the solidification–melting transitions in the KSM pores can be explained as follows: (1) Equation (11.1) is not valid in so small pores, and the actual T_m is lower than –100°C; (2) the fusion enthalpy in small pores is strongly reduced below its bulk value [6, 16], making the melting peak undetectable; (3) the system is vitrified; (4) water (or solution) in small pores is strongly supercooled [7, 17], and its solidification temperature is lower than −100°C. Let us discuss these reasons in more details:

1. One should expect that Eq. (11.1) is not valid for very small pores because it predicts $\Delta T_m \to \infty$ at $d \to 0$. Hence, this equation has to be corrected at very small d [12].

2. The reduction in the fusion enthalpy (ΔH_f) for water confined to the KSK pores was found in Ref. [13]: ΔH_f = 265 kJ/g, which is significantly lower than the fusion heat in the bulk (326 kJ/g). It is interesting to note that the former enthalpy is close to that calculated in Ref. [6] for 15 nm water aggregates: ΔH_f = 262 kJ/g. Since ΔH_f diminishes at low temperature, an additional decrease $\Delta(\Delta H_f)$ is expected in the KSM pores, where the melting temperature $T_m(\text{KSM}) < T_m(\text{KSK})$:

$$\Delta(\Delta H_f) = \int_{T_m(\infty)}^{T_m(d)} [C_p^{\text{liq}}(T) - C_p^{\text{sol}}(T)]dT < 0$$

where C_p^{liq} and C_p^{sol} are the specific heats for liquid and solid states, respectively. Hence, the tentative decrease in ΔH_f at $T = (−50 \div −100)$°C may be significant and a reason for not detecting a peak of the solution melting in the KSM pores.

3. An evidence for possible vitrification of the solution in the KSM pores is obtained by a DMTA study [13] (see Sec. 11.1.3).

4. Strong supercooling may be one more reason for the absence of exo- and endo-effects for liquids dispersed in the micropores. Supercooling is observed even for a bulk aqueous solution of $CaCl_2$: the solidification exo-peak is shifted downward compared to the melting point by some 10°C. Hence, the formation of a critical germ is a rare event, which determines the rate of the whole solidification process. The restricted pore geometry of the KSK silica gel enhances the supercooling effect: the solidification temperature decreases by 10–30°C compared to the bulk system [13] (see also Sec. 11.1.3).

The very interesting consequence of supercooling in this system is the threshold effect of pore filling on the solution's solidification, which is reported in the next section.

11.1.2 Effect of Pore Filling

We have already mentioned earlier that the degree of pore filling, γ, is a crucial parameter for melting/solidification processes in the KSK pores. Here we detail this statement by considering the melting of a CaCl$_2$ eutectic solution (C = 30.3 wt.%) at various degrees of KSK pore filling. These experiments were performed by Dr. G. Di Marco (Messina, Italy).

If $\gamma > 1$, a typical thermogram of the system "eutectic solution of CaCl$_2$/KSK" contains two calorimetric peaks (Fig. 11.3): (1) a wide endothermic peak at −75°C (**I**) and (2) a sharp endothermic peak at −55°C (**II**).

The intensity of the latter peak increases at the larger degree of filling (Fig. 11.3), and its position corresponds to the melting of a bulk CaCl$_2$ eutectic solution (T_m = −55°C) [13, 15]. Indeed, at $\gamma \geq 1$, a portion of the solution is located outside the pores, occupies the space between the silica grains (Fig. 11.4), and demonstrates melting behavior similar to the bulk solution. When the value of γ approaches 1.0, peak **II** gets smaller and disappears at $\gamma < 1.0$ (Fig. 11.3).

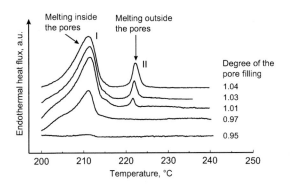

Figure 11.3 Thermograms of the system "CaCl$_2$ eutectic solution confined to the silica gel pores" at various pore filling degrees.

The endothermic peak (**I**) at T = −75°C corresponds to the melting of a CaCl$_2$ eutectic solution inside the silica mesopores. The depression of the melting point for the confined eutectic solution is in line with the results of Sec. 11.1.1 and Ref. [13]. At $\gamma > 1.0$, the signal intensity does not depend on the pore filling, while at

$0.95 < \gamma < 1.0$, a gradual decrease in peak **II** was observed, which was much larger than the decrease in the γ value. At last at $\gamma \leq 0.95$, this melting peak completely disappeared (Fig. 11.3). At the same time, no exothermic solidification peak was observed at $\gamma \leq 0.95$. This indicates that the pore filling degree has a kind of threshold effect on the solidification–melting processes in the confined $CaCl_2$ eutectic solution.

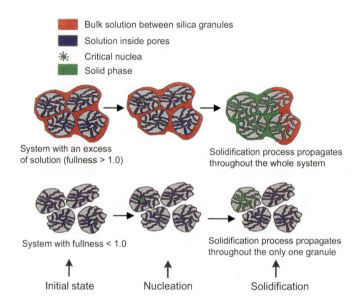

Figure 11.4 Schematics of the solution solidification for connected (upper) and disconnected (lower) clusters.

Why neither solidification nor melting of the confined solution was observed at $\gamma \leq 0.95$? As mentioned earlier, the crystallization of the bulk $CaCl_2$ solution during cooling is hindered due to the slow formation of a critical germ, which is a rate-controlling process. As soon as the germ is formed in a local zone (Fig. 11.4), the nucleation process starts and covers the neighboring regions of the solution. If the solution forms a coupled cluster, the germ formation initiates a posterior propagation of the crystallization front through the whole solution. This is the case of $\gamma > 1.0$, when the pore space inside the silica grains is completely filled by the solution and the solution excess forms a liquid film on the external grain surface, thus creating a coupled solution cluster (the upper part of Fig. 11.4). As a result, the solution freezes and an appropriate solidification peak is detected [13]. During heating, the process of solution melting is observed: at first inside the pores and then outside the pores (peaks **I** and **II** in Fig. 11.3).

On the contrary, if the silica grains are not connected by the liquid film, the formation of a critical germ in one of the grains results in the freezing of only that single grain but not of the whole system (the lower part of Fig. 11.4). Neither exothermic nor endothermic effects are detected during the sample cooling and heating, respectively. To avoid the pore filling effect, all the DSC measurements described in Sec. 11.1.1 were performed at $\gamma > 1.0$.

Thus, the solution's confinement to the silica gel pores strongly affects the nucleation process at low temperatures. Possible reasons for this effect may be a considerable disordering of the solution structure in the vicinity of the pore surface or enhancement of the solution viscosity in pores [7, 18].

11.1.3 Supercooling and Vitrification of Salt Solutions in Silica Micropores

If the solution inside the pores is unable to crystallize, it remains liquid or glassy at low temperature. The DSC method has very low sensitivity to investigate the state of the salt solution inside the pores at conditions when crystallization does not take place. Therefore, to study supercooling and vitrification of aqueous $CaCl_2$ solutions in the pores of KSK and KSM silica gels, a dynamic thermomechanical analysis (DTMA) [19] was used. This method is two–three orders of magnitude more sensitive than the common DSC technique.

The internal friction (tan δ) and dynamic module (E') were measured at the frequencies of 1, 3, and 10 Hz by means of a Rheometric Scientific dynamic mechanical analyzer. The KSK grains were impregnated with either an excess ($\gamma > 1.0$) or a deficiency ($\gamma < 0.95$) of the eutectic $CaCl_2$ solution. The samples were pressed into a mold and protected with a tin foil to obtain a rod with the size suited for mechanical measurements ($1.5 \times 11.3 \times 10$ mm^3). More details can be found elsewhere [13].

At the excess of salt solution ($\gamma > 1.0$), the internal friction peak at $-105°C$ (Fig. 11.5a) and the corresponding inflection in the module shift to higher temperatures when the frequency is increased (Fig. 11.6). Such a behavior is typical for the thermally activated relaxation and may be tentatively associated with the glass transition of the solution in the silica pores, like the relaxation of a metastable vitreous state [20]. Probably, this transition is responsible for the exothermic DSC peak observed at −110°C. Further heating of the sample leads to a rise in the dynamic modulus at $T = -75°C$. It correlates with the DSC endothermic peak observed at the same temperature and caused by the solution crystallization inside the KSK pores. Indeed, if the solution inside the silica granules becomes solid, the mechanical strength of the whole sample increases. The peak of internal friction at −50°C and the relative module drop (Fig. 11.5a) are not sensitive to the frequency change, which is typical for the first-order transitions. In fact, at this temperature, melting of the salt solution located outside the pores takes place as revealed by the DSC method (Fig. 11.3).

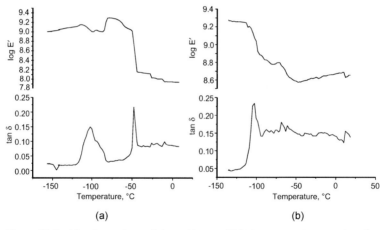

Figure 11.5 The dynamic module and internal friction versus temperature for the eutectic $CaCl_2$ solution in the KSK pores at $\gamma > 1$ (a) and $\gamma < 1$ (b).

Quite different behavior was observed at the degree of pore filling $\gamma < 0.95$ (Fig. 11.5b). A peak of the internal friction at $T = -105°C$ is accompanied by a drop of the dynamic modulus. Its position depends on the frequency [13]; hence, this peak corresponds to a second-order transition, e.g., relaxation of the salt solution inside the pores similar to that observed at $\gamma > 1.0$ (Fig. 11.5a). Further heating leads to decreasing the dynamic modulus due to a gradual softening of the vitreous solution. No crystallization is observed, which agrees with the data of DSC measurements reported in Sec. 11.1.1.

Two peaks of the internal friction are observed for the eutectic CaCl$_2$ solution that fills the KSM grains (the average pore size d = 3.5 nm) with some excess, $\gamma > 1.0$ (Fig. 11.6b). The low-temperature peak and the corresponding drop of the dynamic modulus, observed at $T \approx -105°C$, are temperature dependent, thus associating with the relaxation of the salt solution inside the micropores. The peak at c.a. −55°C is due to the melting of the solution outside the pores.

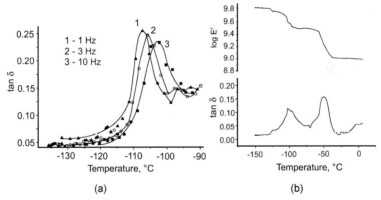

Figure 11.6 (a) Internal friction as a function of temperature for the CaCl$_2$/H$_2$O/KSK at various frequencies; (b) dynamic module and internal friction versus temperature for the eutectic CaCl$_2$ solution in the KSM pores at $\gamma > 1$ [2].

All the data obtained by DSC and DTMA techniques paint the following picture of the phase transitions in the silica pores at $\gamma > 1$. After fast cooling of the system (eutectic CaCl$_2$ solution)/(silica gel) down to $(-150 \div -200)°C$, the solution transforms to a metastable state, namely, a supercooled solution or a glass. Relaxation of this state is observed at $(-100 \div -110)°C$ as a second-order transition. At $T \approx -90°C$, this solution undergoes crystallization in the KSK pores. Subsequent melting is observed at $T = -75°C$ inside the KSK mesopores and at $T = -55°C$ outside the KSK and KSM pores.

Significantly different is the picture in the case of incomplete pore filling with the salt solution ($\gamma < 0.95$). The low-temperature peak and the corresponding reduction in the dynamic modulus, observed at $T \approx -105°C$, are temperature dependent, thus associating with relaxation/softening of a supercooled or vitreous state of the salt solution inside the pores. The first-order phase transitions are not observed at $\gamma < 0.95$.

Further information about the state of $CaCl_2$ solutions confined into the silica gel pores was obtained by a ^2H NMR technique [21]. It allows studying the mobility of water molecules in CaC_2 solution inside the pores at low temperature, which gives useful information about the solution phase transitions. The ^2H NMR tests are presented and discussed in Chapter 12.

11.2 Other Salts and Matrices

Here the processes of solidification/melting of other "salt–water" systems confined to various porous matrices are considered.

11.2.1 LiBr Solution in Pores of Silica Gel KSK

A DSC-111 "SETARAM" unit was used for studying the phase transitions and measurement of the heat capacity of $LiBr/SiO_2$(KSK) in the temperature range of $170 \div 320$ K. Before the measurements, the samples were cooled down to $-100°C$ [22]. DSC thermograms were recorded at the heating rate 1 K/min. The melting temperature was detected either as the temperature corresponding to the maximum of a DSC peak or by a standard "on-set" method.

The solubility diagram for a bulk "LiBr–H_2O" system is of a simple eutectic type (Fig. 11.7) [23]. At the eutectic point A (39.1 wt.%, $-67.5°C$), a mixture of ice and $LiBr \cdot 5H_2O$ melts. A low-temperature "solidification–melting" phase diagram of this system in the confined state was measured at the salt concentration of 0–48 wt.%. This diagram lies below the diagram of the bulk solution by 15–30°C (Fig. 11.7), which is similar to the system $CaCl_2/SiO_2$ (KSK) considered above. In particular, an eutectic solution melts inside the pores at $-91.2°C$.

This system is also similar to the system $CaCl_2/SiO_2$ (KSK) in several other aspects [22]:

- Strong supercooling of the solution inside the silica mesopores,

- The absence of the melting peak at the salt concentration lower than the eutectic one,

- Reduction in the melting enthalpy in the confined state.

Other Salts and Matrices | 239

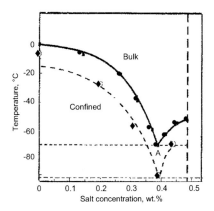

Figure 11.7 "Solidification–melting" phase diagram for binary mixture H₂O–LiBr: in the dispersed state (in the KSK pores) [22] and in bulk. Below the horizontal lines is the field of the eutectic mixture.

11.2.2 Various Hydrates of CaCl₂ in Pores of SBA-15

A new composite "CaCl₂ in the pores of SBA-15" was prepared and studied in Ref. [24], keeping in mind its application for adsorptive heat transformation [25]. The main advantage of the SBA matrix is its almost uniform pore structure [26]. Narrow pore size distribution of this host leads to the more narrow size distribution of the salt particles confined to its pores. Here we consider melting of various CaCl₂ hydrates (n = 2, 4, and 6) in the SBA pores.

The composite CaCl₂/SBA-15 was synthesized by impregnating the pores of the initial SBA-15 (diameter 7.5 nm) with an aqueous solution of calcium chloride of the salt concentration 50 wt.% as described in Ref. [24]. The salt content was 43 wt.%.

The DSC tests were performed using a calorimeter DSC404C Pegasus Netzsch. Samples of typical weight 25.0 ± 1.5 mg were placed inside the standard aluminum pans of 40 ml volume located in the measuring cell of the calorimeter. The pan was tightly covered to avoid water evaporation during heating runs. Nitrogen was passed through the measuring cell with a flow rate of 30 cm³/min. The typical temperature program was as follows: cooling with the constant rate 3 K/min down to T_c = −20 to −40°C and maintaining

Melting–Solidification of Salt Solution/Hydrates in Pores

the sample at this temperature for 10 min, then heating with a constant rate of 3 K/min up to T_h = 65–220°C (see Table 11.1).

Table 11.1 The cooling T_c, heating T_h, on-set T_{onset} temperatures and the specific enthalpy of melting, ΔH_m

Sample	T_c (°C)	T_h (°C)	T_{onset} (°C)	ΔH_m (J/g)
n = 2.0 bulk	25	220	176.1	87.2 [27]
n = 2.03 disperse	25	220	92.1/140.2	7.9/22.9*
	—	—	—	15.8/45.8**
n = 3.94 bulk	25	120	38.4	134.1 [27]
n = 3.98 disperse	−50	75	30.0	44.4*
	—	—	—	80.1**
	−30	—	30.7	50.6**
	−20	—	31.2	46.3**
n = 6.0 bulk	−30	65	30.4	198 [27]
n = 5.94 disperse	−40	65	21.4	85.9*
	—	—	—	128.8**

*calculated per 1 g of the composite
**calculated per 1 g of the hydrate
Source: Reprinted from Ref. [24], Copyright 2010, with permission from Elsevier.

Three such repetitive cycles were performed. DSC curves obtained were analyzed with "on-set" temperature T_{onset}, which determines a starting temperature of melting. The enthalpy of melting was obtained by integrating the DSC thermograms. The peak area was proportional to the change in enthalpy ΔH_m (J/g), and appropriate calibration factor was determined in a separate experiment.

The DSC thermograms of the confined hydrates $CaCl_2 \cdot nH_2O$ (n = 2, 4, and 6) exhibit two peaks, which are well resolved only for dihydrate (Fig. 11.8a). Both peaks correspond to a melting point of the appropriate hydrates in the pores and appear at a lower temperature relative to the corresponding bulk hydrate (Fig. 11.8 and Table 11.1), which agrees with theoretical predictions [3, 5, 11]. Two peaks observed in Fig. 11.8 may indicate that the confined salt/hydrate exhibits two different sizes inside the SBA-15 pores. Probably, smaller particles are positioned in the mesopores inside the SBA grains, while the larger ones occupy the space between these grains. The melting heat ΔH_m in the confined state is smaller than in bulk (Table 11.1) as predicted in Ref. [6].

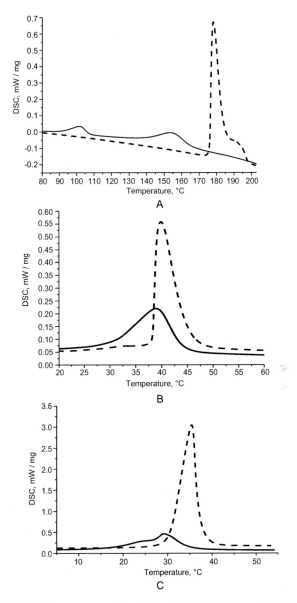

Figure 11.8 DSC thermograms of hydrates $CaCl_2 \cdot nH_2O$ with $n = 2$ (A), 4 (B), and 6 (C) in bulk (dashed lines) and confined (solid lines) states. Reprinted from Ref. [24], Copyright 2010, with permission from Elsevier.

11.3 Overview

In this chapter, we have considered processes of melting/solidification of salt hydrates and solution confined to the pores of various meso- and microporous matrices. The smaller the matrix pores, the stronger the properties of the dispersed substance, e.g., solubility diagram, melting temperature, solution subcooling, etc., differ from those in bulk. In sum, the combined DSC and DTMA study of the low-temperature melting/solidification of the salt solutions/hydrates confined to the silica meso- and micropores clearly showed the following:

- The degree γ of pore filling strongly affects the mentioned processes. Quite different scenarios are revealed when (a) the silica gel grains are completely filled with the solution and a part of the solution is located outside the grains ($\gamma > 1$), and (b) the pores are incompletely filled with the solution. No solidification/melting is observed at $\gamma < 0.95$. This is due to the destroyed connectivity of the liquid phase in the system, so that formation of the critical germ cannot lead to solidification in the whole solution (Fig. 11.4).
- The low-temperature "solidification–melting" phase diagram of the "salt/water" system was measured in a wide concentration range. The diagram lies below the diagram of the bulk solution by 15–30 K. This depression was analyzed on the base of the Gibbs–Thompson equation that accounts for contributions of the surface energy in the dispersed state.

Several other peculiarities of melting and solidification in this system are also reported and discussed. All the phenomena discussed earlier show that size-dependent effects may come in many varieties and manifest themselves in supercooling, vitrification, crystallization, and melting behavior in a way different from common bulk aqueous salt solutions. Besides the fundamental interest, the data obtained could be of importance in many commercial areas such as refrigeration, accumulation of low-temperature heat, frost prevention in building materials, etc.

References

1. Aristov, Yu. I., Tokarev, M. M., Cacciola, G., and Restuccia, G. (1996) Selective water sorbents for multiple applications: 1. $CaCl_2$ confined in mesopores of the silica gel: Sorption properties, *React. Kinet. Cat. Lett.*, **59**, pp. 325–334.

2. Tokarev, M. M. (2003) PhD thesis, Boreskov Institute of Catalysis, Novosibirsk.

3. Gibbs, J. W. (1928) *Collected Works*, New York.

4. Bellissent-Funel, M.-C. (2003) Status of experiments probing the dynamics of water in confinement, *Eur. Phys. J. E,* **12**, pp. 83–92.

5. Defay, R., Prigogine, I., Bellemans, A., and Everett, D. H. (1966) *Surface Tension and Adsorption* (Wiley, New York).

6. Brun, M., Lallemand, A., Quinson, J.-F., and Eyraud, C. (1977) A new method for simultaneous determination of the size and shape of pores: The thermoporosimetry, *Thermochim. Acta*, **21**, pp. 59–88.

7. Mu, R. and Malhotra, V. M. (1991) Effect of surface and physical confinement on the phase transitions of cyclohexane in porous silica, *Phys. Rev., B,* **44**, pp. 4296–4303.

8. Jackson, C. L. and McKenna, G. B. (1990) The melting behaviour of organic materials confined in porous solids, *J. Chem. Phys.*, **93**, pp. 9002–9011.

9. Buffat, P. and Borel, J. P. (1976) Size effects on the melting temperature of gold particles, *Phys. Rev. A,* **13**, pp. 2287–2298.

10. Thomson (Lord Kelvin) (1871) LX. On the equilibrium of vapour at a curved surface of liquid, *Philos. Magazine*, **42**, p. 448.

11. Lindemann, F. A. (1910) The calculation of molecular vibration frequencies, *Phyz. Z.*, **11**, pp. 609–613.

12. Shi, F. G. (1994) Size-dependent thermal vibrations and melting in nano-crystals, *J. Mater. Res.*, **9**, pp. 1307–1313.

13. Aristov, Yu. I., Di Marco, G., Tokarev, M. M., and Parmon, V. N. (1997) Selective water sorbents for multiple applications: 3. $CaCl_2$ solution confined in micro- and mesoporous silica gels: Pore size effect on the "solidification–melting" diagram, *React. Kinet. Cat. Lett.*, **61**, pp. 147–154.

14. Kondratiev, S. N. (Ed.) (1978) *Physical Chemistry* (Vishaya Shkola, Moscow) (in Russian), p. 196.

15. Kirk-Othmer. (1992) *Encyclopedia of Chemical Engineering*, 4th ed., vol. 4 (Wiley, New York).

16. Jackson, C. L. and McKenna, G. B. (1990) The melting behavior of organic materials confined in porous solid, *J. Chem. Phys.*, **93**, pp. 9002–9011.

17. Warnock, J., Awschalom, D., and Shafer, M. (1986) Geometrical supercooling of liquids in porous glass, *Phys. Rev. Lett.*, **57**, pp. 1753–1756.

18. Kaerger, J., Pfeifer, H., and Heink, W. (1988) Principles and application of self-diffusion measurements by nuclear magnetic resonance, *Adv. Magn. Reson.*, **12**, pp. 1–89.

19. Menard, K. P. (1999) *Dynamic Mechanical Analysis: A Practical Introduction* (CRC Press).

20. Gradin, P., Howgate, P. G., Selden, R., and Brown, R. (1989) In: *Comprehensive Polymer Science*, G. Allen and J. Bevington (Eds.) (Pergamon, New York).

21. Kolokolov, D. I., Glaznev, I. S., Aristov, Yu. I., Stepanov, A. G., and Jobic, H. (2008) Water dynamics in bulk and dispersed in silica $CaCl_2$ hydrates studied by ^2H NMR, *J. Phys. Chem. C*, **112**, pp. 12853–12860.

22. Shevel'kov, V. V. (1996) Bd. Thesis, Novosibirsk State University.

23. Kirgintsev, A. N., Trushnikova, L., and Lavrentieva, V. G. (1972) *Solubility of Inorganic Salts in Water* (Chemistry, Leningrad) (in Russian), p. 589.

24. Ponomarenko, I. V., Glaznev, I. S., Gubar, A. V., Aristov, Yu. I., and Kirik, S. D. (2010) Synthesis and water sorption properties of a new composite "$CaCl_2$ confined to SBA-15 pores," *Micropor. Mesopor. Mater.*, **129**, pp. 243–250.

25. Glaznev, I. S., Ponomarenko, I. V., Kirik, S. D., and Aristov, Yu. I. (2011) Composites $CaCl_2$/SBA-15 for adsorptive transformation of low temperature heat: Pore size effect, *Int. J. Refrig.*, **34**, pp. 1244–1250.

26. Schueth, F., Sing, K. S., and Weitkamp, J. (Eds.) (2002) *Handbook of Porous Solids*, vol. 2 (Wiley-VCH).

27. *Kirk-Othmer Encyclopedia of Chemical Technology*, 4th edition, v. 4, p. 413.

Chapter 12

Molecular Dynamics of Sorbed Water

As demonstrated in the previous chapters, the sorption properties of a salt and its crystalline hydrates confined to the pores of matrices with middle and small pores can significantly differ from their bulk properties. For instance, the formation of dihydrates of calcium chloride and calcium nitrate in the silica gel pores occurs at much lower vapor pressure than in the bulk (see Secs. 5.2.3 and 5.3.3). This effect has been discussed within the bounds of the two hypothesis:

1. Increased contribution of the reagents surface energy as compared with that of the products (Sec. 6.6.1);
2. Enhanced water mobility in the dispersed hydrate as compared with the bulk one (Sec. 10.1.1).

These assumptions are similar to the thermodynamic [1, 2] and dynamic [3] approaches suggested for analyzing the depression of the melting point of small particles (Sec. 11.1).

This chapter addresses the study of changes in the mobility of the water linked with the confined salt. To compare it with the mobility in bulk hydrates, we use methods that are very sensitive to the proton mobility: ^1H NMR, ^2H NMR, and neutron scattering.

Nanocomposite Sorbents for Multiple Applications
Yu. I. Aristov
Copyright © 2020 Jenny Stanford Publishing Pte. Ltd.
ISBN 978-981-4267-50-2 (Hardcover), 978-981-4303-15-6 (eBook)
www.jennystanford.com

12.1 ¹H NMR Spectroscopy

The ¹H NMR method is a powerful tool for studying the state of adsorbed water, in particular, for determining proton mobility, constants of the dipole–dipole interaction in a water molecule and between water molecules [4]. It is also widely used for investigation of water in hydrated crystals [5]. Here we report results of the ¹H NMR study of $CaCl_2$ hydrates confined to the silica gel pores of various sizes from 3.5 to 34 nm [6]. The samples were prepared by impregnating various silica gels (Table 12.1) with a 40% aqueous solution of $CaCl_2$ and subsequent drying at 150°C for 6 h.

Table 12.1 Characteristics of the silica gels used as hosts and the salt content in the composites studied

N	Silica type	D_{pore} (nm)	S_{BET} (m²/g)	V_{pore} (cm²/g)	C (wt.%)
1	KSM	3.5	600	0.3	13.8
2	Davisil 635	7.5	490	0.95	33.4
3	Davisil 645	12	300	1.15	38.7
4	KSK	15	350	1.0	32.0
5	C3-15	34	140	1.1	37.6

Source: Reprinted by permission from Springer Nature from Ref. [6], Copyright 1998.

12.1.1 Water in CaCl₂ Hydrates Confined to Silica Pores of Middle Size

A typical ¹H NMR spectrum of a silica gel–$CaCl_2$–xH_2O system at room temperature is a superposition of a narrow single line and a broad doublet (Fig. 12.1). The single line corresponds to a "liquid-like" water, while the doublet is typical for a "solid-like" water. For instance, similar doublets were observed for bulk $CaCl_2 \cdot 2H_2O$ and $CaCl_2 \cdot 6H_2O$ [7]. It is worthy to note that the shape of NMR signal reflects the degree of water mobility rather than a real phase composition of the hydrated water. For instance, the narrow single line is observed even at n = 0.27–0.33 (Table 12.2) when no liquid phase is present in the system (see also Sec. 12.2).

The fraction q of hydration water that demonstrates a "solid-like" behavior can be estimated from the integral intensity of the broad component in the NMR spectrum. Quantitative analysis of this component was performed by using Pake's model [5]. The varied parameters are as follows:

1. The relative intensity q of the broad component.
2. The constant α of dipole–dipole interaction of protons in the water molecule.
3. The parameter β describing the interaction of protons of the neighboring water molecules.

The constant of dipole–dipole proton interaction in the confined hydrate can be found from the shape of the NMR spectrum. Its value $\alpha = \alpha_o(1 - 3<\Theta^2>)$ depends on the constant of dipole–dipole interaction in the bulk hydrate $\alpha_o = 23$ kHz and the average amplitude of water molecule librations $<\Theta^2>$. In the pores of 7.5–34 nm size, the constant α is smaller than 23 kHz (Fig. 12.1b), which gives the value of $<\Theta^2> = 0.04 \div 0.08$ rad^2. In the narrow range between $3 < n < 4$, the broad doublet disappears, which agrees with the fact that the tetrahydrate CaCl$_2$·4H$_2$O melts at $T \leq 28°C$ inside the pores of 15 nm size or smaller. The doublet observed at $2 < n < 4$ is attributed to the large crystals of the dihydrate CaCl$_2$·2H$_2$O, which remain solid and coexist in equilibrium with the salt solution (see Sec. 5.3).

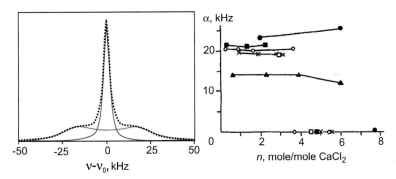

Figure 12.1 (a) Typical ^{1}H NMR spectrum of "silica gel–CaCl$_2$–xH$_2$O" system: dotted line—experiment; solid line—theory; (b) dependence of the dipole–dipole interaction constant α on the water content n in the pore of different sizes: ▲—3.5 nm, ■—7.5 nm, ○—12 nm, ×—15 nm, □—34 nm, and ●—bulk hydrates ($n = 2$ and 6). Reprinted by permission from Springer Nature from Ref. [6], Copyright 1998.

The larger librations amplitude in the dispersed state indicates that the water molecules are more mobile in the confined hydrates. This finding is in line with the enhanced ability of the confined CaCl$_2$

248 | *Molecular Dynamics of Sorbed Water*

to absorb water vapor, as discussed in Sec. 5.3. Indeed, a larger entropy of water molecules in the confined hydrate makes water sorption more profitable as compared with a more rigid bulk hydrate. As a result, the equilibrium pressure over the confined hydrate becomes lower than in the bulk. This qualitative analysis is in good agreement with the entropy changes during the $CaCl_2$ hydration estimated from the low-temperature heat capacity tests (Chapter 10).

12.1.2 Water in CaCl$_2$ Hydrates Confined to Silica Pores of Small Size

The NMR spectrum of water molecules of the "$CaCl_2$–xH_2O" system in the KSM silica gel with 3.5 nm pores is much narrower than that in the matrices with larger pores. The relative intensity q of the broad component is estimated to be lower than 24% (Table 12.2); however, this component was recorded up to the hydration degree $n = 6$ (Fig. 12.1b). No plateau corresponding to any salt hydrate was observed in the sorption isotherm of this system, and the sorption equilibrium was bivariant at any uptake [8]. Probably, the broad doublet is attributed to a hydrated state of $CaCl_2$ with low water mobility.

Table 12.2 The relative intensity q of the broad component (%) at various hydration degree n and pore size

3.5 nm		7.5 nm		12 nm		15 nm		34 nm	
n	q	n	q	n	q	n	q	n	q
0.62	24	0.33	53	0.27	71	0.93	58	—	—
2.3	23	1.35	57	0.93	65	1.9	34	2.9	45
3.9	9	2.26	46	1.64	69	2.9	36	4.5	0
6	24	4.8	0	3.7	63	3.1	30	—	—
7.7	0	—	—	3.7	0	5	0	—	—

Source: Reprinted by permission from Springer Nature from Ref. [6], Copyright 1998.

In the pores of 3.5 nm size, the constant α is much smaller (14 kHz) than that in the bulk (Fig. 12.1b), and the librations amplitude $<\Theta^2> = 0.15$ rad^2 indicates greater librations disordering of water molecules compared to that in the other mesoporous silica gels. This is in line with the higher vapor pressure over the "$CaCl_2$–xH_2O" system confined to the KSM pores [8].

12.2 ^{2}H NMR Spectroscopy

^{2}H NMR spectroscopy has been shown to be a powerful technique for probing the dynamics of water molecules [9, 10]. Moreover, ^{2}H NMR also provides information on the structure of crystalline hydrates [11]. The line shape of ^{2}H NMR signals, being completely defined by intramolecular quadrupole interaction, is especially sensitive to the nature of molecular motion and to its rate. Spin–lattice (T_1) and spin–spin (T_2) relaxation times also bring information on the energy and the rate of the different inter- and intramolecular motions. In the following sections, we report the results of ^{2}H NMR study on the dynamic behavior of water in deuterated analogues of $CaCl_2$ hydrates— $CaCl_2 \cdot 2D_2O$, $CaCl_2 \cdot 4D_2O$, and $CaCl_2 \cdot 6D_2O$—in both bulk and dispersed states.

Anhydrous $CaCl_2$ salt of chemical-grade purity, mesoporous KSK silica, and deuterated water (99% ^{2}H isotope enrichment) were used without further purification. The bulk hydrates $CaCl_2 \cdot 2D_2O$ and $CaCl_2 \cdot 4D_2O$ were prepared by exposing the granules of $CaCl_2$ (0.1–0.2 mm size) to surrounding heavy water vapor at fixed P_{D2O}. The $CaCl_2 \cdot 6D_2O$ was prepared by re-crystallization from a saturated solution in D_2O. Dispersed $CaCl_2 \cdot nD_2O$ ($n = 2, 4$, and 6) hydrates were prepared by impregnating the silica KSK with a saturated solution of $CaCl_2$.

The ^{2}H NMR experiments were performed at 61.432 MHz on a Bruker Avance-400 spectrometer, using a high power probe with 5 mm horizontal solenoid coil. The temperature of the samples was controlled with a flow of nitrogen gas stabilized with a variable-temperature unit (Bruker model BVT-3000) with a precision of c.a. 1 K. More experimental details can be found elsewhere [12].

12.2.1 ^{2}H NMR spectra of $CaCl_2 \cdot nD_2O$ Hydrates in Bulk and Confined States

Analysis of ^{2}H NMR spectra of the bulk and confined hydrates $CaCl_2 \cdot nD_2O$ has shown that all the studied hydrates exhibit similar temperature dependence of ^{2}H NMR line shape. However, a quantitative difference is found between the bulk and dispersed hydrates. A similar comment can be made for the temperature dependences of ^{2}H NMR spin–lattice (T_1) and spin–spin (T_2) relaxation times. Therefore, the main features of the common pattern can be demonstrated through detailed consideration of only one of the hydrates, e.g., by a detailed analysis of ^{2}H NMR spectra for CaCl$_2$·6D$_2$O hydrate in bulk and dispersed states.

The temperature dependences of ^{2}H NMR spectra for the bulk and dispersed CaCl$_2$·6D$_2$O hydrate (Fig. 12.2) can be conditionally divided into three regions. In the *low*-temperature region, the spectra exhibit ^{2}H NMR broad powder pattern, typical for solid or immobile water. At some certain temperature, a narrow peak of "liquid-like" water arising from the mobile (rapidly and isotropically reorienting) water molecules appears at the center of the "solid-like" spectrum. The region where the narrow "liquid-like" and the broad "solid-like" signals coexist characterizes the *intermediate*-temperature region of spectra variation. At temperatures above T_{NMR}, the experimental spectrum consists only of a single isotropic signal with a Lorentzian line shape. The value of T_{NMR} marks the beginning of the *high*-temperature region of spectra variation. The dynamics of water and the structure of the hydrates in the bulk state can be derived from the analysis of ^{2}H NMR line shape. In the *low*-temperature region, the bulk and dispersed hydrates show almost identical ^{2}H NMR line shapes with similar principal features: Each spectrum represents a superposition of two Pake-type powder patterns.

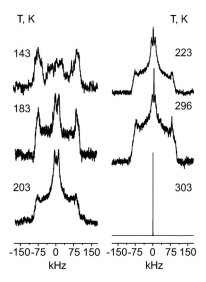

Figure 12.2 Temperature dependence of the ^{2}H NMR spectrum line shape of bulk CaCl$_2$·6D$_2$O hydrate. Reprinted with permission from Ref. [12], Copyright 2008, American Chemical Society.

Simulation of the spectra based on the classical theory for ^{2}H NMR solid-state spectra line shape [13] shows that one of the signals has a large quadrupled constant $Q_1 \sim 250$ kHz and a small asymmetry parameter $\eta_1 \sim 0.1$, which are typical for hydrogen-bonded water molecules in the crystalline hydrates [11, 14] or ice [9, 15]. In other words, the spectra are typical for water, static on NMR timescale. If we denote τ_C as the characteristic time of the molecular motion, then $\tau_C \gg \tau_{NMR}$, where $\tau_{NMR} = Q^{-1} \sim 4 \times 10^{-6}$ s for static water. The other signal has a smaller quadruple constant $Q_2 \sim 125$ kHz and a much larger asymmetry parameter $\eta_2 \sim 0.85$. It is likely to correspond to water involved in some fast anisotropic motion ($\tau_C \ll \tau_{NMR}$). In the *low*-temperature region, hydrates are purely in a crystalline state, and the only reasonable type of motion not prohibited by the hydrates lattice symmetry is the twofold flips about the bisector of the angle D-O-D = 2θ. Such behavior is typical for crystalline hydrates [10, 14, 15].

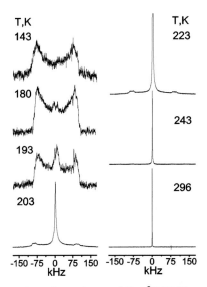

Figure 12.3 Temperature dependence of the ^{2}H NMR spectrum line shape of confined CaCl$_2$·6D$_2$O hydrate. Reprinted with permission from Ref. [12], Copyright 2008, American Chemical Society.

The ^{2}H NMR spectrum of the water molecules rapidly flipping by 180° around the bisecting angle is simulated [12] using the formalism elaborated by Spiess [16] and Wittebort [17]. Simulations of spectra

Molecular Dynamics of Sorbed Water

line shape show a perfect agreement with the experimental data for all hydrates [12]. The ^{2}H NMR simulating parameters of the spectra are given in Table 12.3 for all studied hydrates. While the parameter θ provides structural information on the hydrogen-bonded water molecules for bulk and dispersed hydrates, the quadrupole constant Q_1 for static water can also yield information on the length of the O-D..Cl hydrogen bond. An estimation of hydrogen bond length can be made based on semi-empirical dependence between the quadrupole constant and the length of hydrogen bond, found in Ref. [11]:

$$Q = 310 - 3 \times \frac{190.6}{r^3} \qquad (12.1)$$

where Q is the observed quadrupole constant in kHz and r is the length of hydrogen bond in nm.

Comparison of the simulated parameters θ and $r_{D..Cl}$ (Table 12.3) for bulk and dispersed hydrates shows that they have very close values. Thus, the local structures of bulk and dispersed hydrates are almost identical in the low-temperature region.

Table 12.3 ^{2}H NMR spectra line shape simulation results for CaCl$_2$ · nD$_2$O hydrates in the low-temperature region

	Bulk hydrate			Dispersed hydrate		
	$n = 2$	$n = 4$	$n = 6$	$n = 2$	$n = 4$	$n = 6$
Q_1 (kHz)	250	250	250	260	240	240
η_1	0.08	0.09	0.09	0.09	0.12	0.16
Q_2	125	125	125	130	120	120
η_2	0.85	0.83	0.83	0.84	0.84	0.84
θ (°)	53.2	53.2	53.2	53.2	53.2	53.2
$r_{D..Cl}$* (nm)	0.21	0.20	0.21	0.22	0.21	0.20
$r_{D..Cl}$** (nm)	0.22	0.23	0.22			

*Data taken from Ref. [12], with permission of American Chemical Society, Copyright 2008.
**Taken from literature X-ray and neutron diffraction data.

Within the *intermediate*-temperature region, the intensity of the "liquid-like" isotropic signal of the bulk hydrate increases rapidly with temperature. It is clear that the "liquid-like" signal represents water in a *mobile* state, characterized by fast molecular reorientation with $\tau_C \ll \tau_{NMR}$, which completely averages the quadrupolar solid-state spectral features. It is reasonable to assign the "liquid-like" signal to the melted hydrate. The intermediate-temperature region ends at temperature T_{NMR} when the solid-state signal completely disappears and the spectrum is represented by the single "liquid-like" signal. This temperature is the temperature at which the bulk hydrates completely melt as confirmed by the data in Table 12.4.

Table 12.4 Characteristic temperatures for bulk and dispersed $CaCl_2 \cdot nD_2O$ hydrates

	Bulk hydrate			Dispersed hydrate		
	$n = 2$	$n = 4$	$n = 6$	$n = 2$	$n = 4$	$n = 6$
T_{NMR} (K)	453	318	303	323	263	233
T_C (K)	—	—	—	—	296	285
T_{melt} (K)	450	313	301	365, 401	301	290

Source: Reprinted with permission from Ref. [12], Copyright 2008, American Chemical Society.

The temperature behavior of the NMR line shape for the bulk and dispersed hydrates is similar. However, the dispersed hydrate is characterized by a broader intermediate-temperature region in which the solid and melted hydrates coexist, and by an essentially lower value of T_{NMR} (by 55–130°C!) at which the broad line disappears. This is one more confirmation that the hydrated water is more mobile in the confined state as compared with that in the bulk.

As opposed to the bulk hydrates, the temperature T_{NMR} cannot be associated with the hydrate melting as it is significantly lower (by 40–70°C) than the melting temperature of appropriate hydrates inside the silica gel pores (Table 12.4). Hence, the narrow single line represents water in a *mobile* state rather than reflects its phase composition. Further information about the water mobility is obtained by the analysis of T_1 and T_2 relaxation times.

12.2.2 ^{2}H NMR T_1 and T_2 Relaxation Times Analysis

Figure 12.4 shows the temperature dependences of ^{2}H NMR spin–lattice (T_1) and spin–spin (T_2) relaxation times for the "liquid-like" signal of both bulk and confined $CaCl_2 \cdot nD_2O$ hydrates in the intermediate- and high-temperature regions. The solid curves drawn

254 *Molecular Dynamics of Sorbed Water*

through the data points are theoretical fits according to the models developed for the motion of water molecules [12].

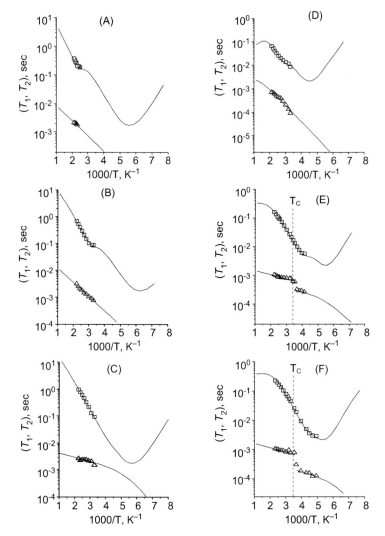

Figure 12.4 Temperature dependences of T_1 (□) and T_2 (△) for the bulk (A, B, C) and confined (D, E, F) $CaCl_2 \cdot 2D_2O$ (A,D), $CaCl_2 \cdot 4D_2O$ (B,E), and $CaCl_2 \cdot 6D_2O$ (C,F) hydrates. Reprinted with permission from Ref. [12], Copyright 2008, American Chemical Society.

It is clear from the data in Fig. 12.4 that at least two different types of motion are present in both bulk and dispersed hydrates. A fast motion ($\tau_C \sim \omega_0^{-1} = 2.6 \times 10^{-9}$ s, $\omega_0 = 61.43$ MHz) with higher activation barrier governs the T_1 temperature dependence. A slower one ($\tau_C \sim 10^{-6}$ to 10^{-7} s) with a smaller activation barrier defines the T_2 temperature dependence. The qualitative difference between the bulk and dispersed hydrates lies in a narrow-temperature interval of rapid T_2 growth (leap at T_C) observed in two of the dispersed hydrates (marked by a vertical line on Fig. 12.4(E, F)). Comparison with the DSC data (Chapter 11) shows that the leap temperature T_C is close to the melting temperature of the confined hydrate (Table 12.4). Hence, as opposed to the temperature T_{NMR}, the leap in the temperature dependence $T_2(T)$ indicates a true melting process inside the pores. The slopes of T_2 dependence before and after this leap are visually identical. It should be noted that there is no noticeable leap in the T_2 temperature dependence for the dispersed $CaCl_2 \cdot 2D_2O$ hydrate.

Theoretical fits to the T_1, T_2 temperature dependences (Fig. 12.4) are made based on a physically reasonable picture of the motion of water molecules in the melted $CaCl_2$ hydrates as described in Ref. [12]. Both the bulk and confined hydrates exhibit three types of molecular motion. Two of them represent fast motions: internal $180°$ flips with the characteristic time of 10^{-10} to 10^{-11} s, local motions by jumps between two neighboring water positions or precession of the water molecule around some arbitrary axis, which are also performed within the time of 10^{-10} to 10^{-11} s. The third slow isotropic reorientation is performed on the time of 10^{-6} to 10^{-7} s for both the bulk and dispersed hydrates. In the dispersed hydrates, the water molecules reorient isotropically one order of magnitude faster in the temperature range of 230–490 K, which means the water is more mobile in the dispersed hydrates.

Thus, the analysis of ^2H NMR data shows a general resemblance between the bulk and dispersed $CaCl_2 \cdot nD_2O$ hydrates. They have similar environments in the solid state as it follows from close values of the length of hydrogen O-$D..Cl$ bonds between the water molecule's hydrogen atoms and chlorine atoms. Water molecules are either immobile on the NMR timescale or they exhibit fast two-site jumps around the C_2 axis in the solid state. The intramolecular motion of water depends only on the local environment for all hydrates, which is mainly governed by the hydrogen bonding of water molecules.

For the bulk hydrates, the temperature of transformation of the "solid-like" spectrum to the "liquid-like" T_{NMR} is in good correspondence with the melting points T_{melt} for the bulk hydrates, $T_{NMR} \approx T_{melt}$ (see Table 12.4). In contrast to the bulk hydrates, the transformation of the "solid-like" to "liquid-like" signal occurs in a relatively broad temperature range for the dispersed hydrates. Further, the temperature T_{NMR} is 55–130°C lower than that for the bulk hydrates. This means that inside the pores, water molecules exhibit isotropic reorientation at much lower temperature; therefore, the water molecules are much more mobile in the dispersed hydrates.

Further information about the molecular motions of the hydrated water in the bulk and confined states was obtained by neutron scattering techniques [18].

12.3 Neutron Scattering

Neutron scattering is a powerful spectroscopic method used to measure motions of atoms. Inelastic neutron scattering observes the change in the energy of the neutron as it scatters from a sample and can be used to probe a wide variety of different physical phenomena such as the motions of atoms, the rotational modes of molecules, sound modes and molecular vibrations, etc. [19].

Neutron scattering experiments were performed at the Institut Laue-Langevin (Grenoble, France) using two different instruments: the time-of-flight spectrometer IN5 allowed to measure QENS spectra and low-energy excitations, whereas high-energy vibrational modes were observed on the beryllium-filter spectrometer IN1-BeF (see Ref. [18] for more details).

12.3.1 QENS Domain

A comparison between QENS spectra obtained at the same experimental conditions for the bulk and dispersed $CaCl_2 \cdot 2H_2O$ is shown in Fig. 12.5. The spectrum measured for bulk $CaCl_2 \cdot 2H_2O$ corresponds to the instrumental resolution, which implies that water motions are slower than 10^{-9} s. On the other hand, large quasi-elastic broadening is obtained for $CaCl_2 \cdot 2H_2O$ dispersed in silica gel (Fig. 12.5b), which reveals that translational and rotational modes of water occur on the timescale of 10^{-9} to 10^{-12} s.

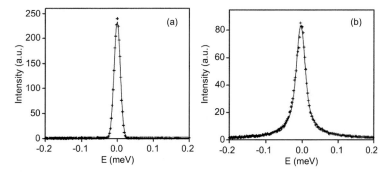

Figure 12.5 Comparison between experimental (crosses) and fitted (solid lines) QENS spectra obtained for CaCl$_2$·2H$_2$O in bulk (a) and dispersed in silica (b), at the same temperature 440 K, and the same wave vector transfer, Q = 0.97 Å^{-1} (λ = 10 Å). Reprinted from Ref. [18], Copyright 2009, with permission from Elsevier.

The spectra obtained at 350 K for the different dispersed CaCl$_2$ hydrates are reported in Fig. 12.6. The increase in the quasi-elastic broadening upon increasing the water concentration indicates that water dynamics becomes faster. The scattering from the samples containing water can be analyzed only in terms of hydrogen motions, because of the large incoherent cross section of this atom. The QENS spectra contain information on both rotational and translational motions of water. The measured intensities are governed by the incoherent scattering functions corresponding to these two motions. Assuming that they are uncoupled, the total scattering function is the convolution of the individual scattering functions for translation and rotation. The information obtained from ^{2}H NMR spectroscopy was utilized to model the rotations or local motions as flips by 180° observed for water by ^{2}H NMR [12].

Convolution of all these local motions by a long-range translation and by the instrumental resolution gives excellent fits to the experimental spectra (Figs. 12.5b and 12.6). The fast flips are found on the timescale of 10^{-10} to 10^{-11} s, which is in the same range of values as in the ^{2}H NMR study. Whatever the loading or the temperature, the flipping motion is found to be about one order of magnitude faster than the jumps. The water self-diffusivity can be obtained from a half-width at the half-maximum of the QENS spectra that corresponds to a long-range translation: $D = <r^2>/6\tau_0$ where $<r^2>$ is the mean-square jump length and τ_0 is the residence time [12]. The self-diffusivity of water in various hydrated states of the dispersed salt is reported in Fig. 12.7 versus $1/T$. The figure demonstrates that the larger the water content, the higher its diffusivity. However, even for $n = 9$, the values stay below the diffusivities of pure water. Therefore, the presence of the salt reduces the mobility of water in dispersed hydrates.

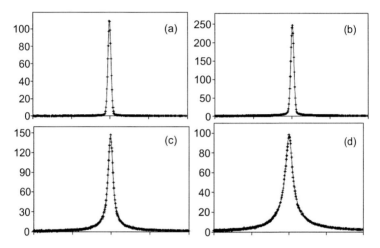

Figure 12.6 QENS spectra obtained for $CaCl_2 \cdot nH_2O$ hydrates dispersed in silica (a) $n = 1/3$, (b) $n = 2$, (c) $n = 4$, (d) $n = 6$. $T = 350$ K, $Q = 0.97$ Å^{-1}. The solid lines correspond to fits to the experimental data (+). All the spectra were measured with $\lambda = 10$ Å. Reprinted from Ref. [18], Copyright 2009, with permission from Elsevier.

The self-diffusivity obtained by QENS at the highest water loading, $D_{QENS} = 8.6 \times 10^{-10}$ m^{2}s^{-1}, is very close to the value derived by the PFG NMR method on the same sample, $D_{NMR} = 7.3 \times 10^{-10}$ m^2/s [20], both being measured at 30°C. The typical distance of the water diffusion measured by the QENS technique is very short, 1–30 Å, while much larger displacements are probed by the PFG NMR method, of the order of a micrometer. Nevertheless, the diffusion mechanisms at the short and long times seem to be identical.

For diffusivity in a porous solid, the equation $D_{pore} = \varepsilon D_{bulk}/\chi$ is valid [21] where D_{bulk} is the diffusivity in homogeneous infinite space, ε is the solid porosity, and χ is the tortuosity of the pores. The former takes into account the fact that the solid skeleton is not permeable to diffusing molecules and their transport occurs only through the empty space of the pores. The latter (χ) is an empirical parameter that takes into account, first of all, that the pores are not straight but quite twisting. For diffusion of water over distance L shorter than the average pore size d, the pore tortuosity can be neglected and $\chi = 1$. This is the case for the QENS measurements: $L = (0.1 - 3)$ nm $<< d \approx 15$ nm. Therefore, at $n = 9$, the water diffusivity in pores can be estimated from the simplified equation $D_{pore} = \varepsilon D_{bulk}$, where the water self-diffusivity in the bulk $CaCl_2$ solution $D_{bulk} = 1.2 \times 10^{-9}$ m^2/s [22]: $D_{pore} = 8.7 \times 10^{-10}$ m^2/s, which is very close to D_{QENS}. For the NMR tests [20], the impact of pore twisting was essential; hence, the tortuosity can be estimated as $\chi = \varepsilon D_{bulk}/D_{pore} = D_{QENS}/D_{NMR} = 1.19$ in surprising agreement with the data of Ref. [20]. The fact that the self-diffusivities obtained by the two techniques are consistent means that the mechanism of water diffusion in the dispersed hydrates is the same over a very wide range of length scale.

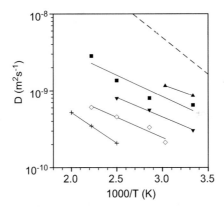

Figure 12.7 Arrhenius plot of the self-diffusion coefficients of water in $CaCl_2 \cdot nH_2O$ hydrates dispersed in silica $n = 1/3$ (+), $n = 2$ (◇), $n = 4$ (▼), $n = 6$ (■), and $n = 9$ (▲); compared with pure water (dotted line). Reprinted from Ref. [18], Copyright 2009, with permission from Elsevier.

The evolution of the "$CaCl_2$–silica KSK–water" system at increasing water content can be summarized as follows:

- $n = 0$. Pure calcium chloride occupies 20% of the silica pore volume. It is suggested that the interaction between salt molecules is stronger than that between the salt and silica, so that the confined salt forms separated nanocrystals on the silica surface rather than a continuous surface layer.

- n = 1/3. The lowest hydrate contains 1 molecule of water per 3 salt molecules and occupies more than 20% of the pore volume. Its assembling inside the silica pores is likely to replicate that of the dry salt with appropriate swelling of salt nanocrystals.
- n = 2. The crystalline hydrate $CaCl_2 \cdot 2H_2O$ is detected in this composite by XRD technique. It is stable over a wide temperature range, and its melting point is 365–400 K (Chapter 11). This hydrate occupies 31.5% of the pore volume, and partial blocking of the pores of silica is possible [23].
- n = 4. This solid hydrate would occupy more than 45.5% of the silica pore volume. The solid hydrate $CaCl_2 \cdot 4H_2O$ has not ever been detected in this composite by the XRD technique. The melting point of this confined hydrate is c.a. 300 K [12]; hence under these scattering tests, it is liquid. It is quite probable that it wets the hydrophilic surface of silica, which results in the formation of a connected liquid layer on the surface. The thickness of the layer can be estimated as $d \approx V_{hs}/S$, where V_{hs} is the volume occupied by the hydrated salt (Table 12.5), S = 172 m^2/g is the silica surface area (both are calculated per 1 g of the composite); $d \approx 1.9$ nm.

Table 12.5 Density ρ of the bulk salt, hydrates, and solution, volume V_{hs} occupied by the hydrated salt in the pores (per 1 g of the composite), appropriate volume fraction δ and the effective layer thickness d

n	ρ (g/cm^3)	V_{hs} (cm^3/g)	δ	d (nm)
0	2.51	0.134	0.20*	—
0.33	—	—	> 0.20*	—
2	2.14	0.21	0.315*	—
4	1.84	0.32	0.455*	1.9
6	1.71	0.39	0.586*	2.3
9	1.38	0.60	0.906*	—

*Calculated from the density of bulk salt, hydrates or solution
Source: Ref. [18].

- n = 6. The melting point of this hydrate in the silica KSK pores is 280–290 K; hence, it is a liquid as well, with the layer thickness $d \approx 2.3$ nm.

- $n = 9$. The salt solution with the concentration of 40.6 mas.% occupies the entire volume of silica pores (Table 12.5), and the water transport is a molecular diffusion in this confined solution.

This model, in whole, is quite consistent with the neutron scattering results, which, in addition, contribute to specifying new interesting details on assembling the hydrated salt on the silica surface. It also agrees well with our previous results on the equilibrium and dynamics of water sorption on the composite $CaCl_2$/silica gel KSK (SWS-1L) [23].

References

1. Gibbs, J. W. (1928) *Collected Works* (New York).
2. Thomson (Lord Kelvin) LX. On the equilibrium of vapour at a curved surface of liquid, (1871) *Philos. Magazine*, **42**, p. 448.
3. Lindemann, F. A. (1910) The calculation of molecular vibration frequencies, *Phyz. Z.*, **11**, pp. 609–613.
4. Hougardy, J., Stone, W. E. E., and Fripiat, J. J. (1976) NMR study of adsorbed water. I. Molecular orientation and protonic motions in the two-layer hydrate of a Na vermiculite, *J. Chem. Phys.*, **64**, pp. 3840–3852.
5. Pake, G. E. (1948) Nuclear resonance absorption in hydrated crystals: Fine structure of proton line, *J. Chem. Phys.*, **16**, pp. 327–337.
6. Tokarev, M. M., Kozlova, S. G., Gabuda, S. P., and Aristov, Yu. I. (1998) ^{1}H NMR in nanocrystalls $CaCl_2 \cdot H_2O$ and water sorption isobars in the system $CaCl_2$–silica gel, *J. Struct. Chem.*, **39**, pp. 212–216.
7. Yano, S. (1959) Proton magnetic resonance in hydrates of halogen compounds of Mg, Ca, Sr and Ba, *J. Phys. Soc. Jpn.*, **14**, pp. 942–954.
8. Aristov, Yu. I., Tokarev, M. M., Cacciola, G., and Restuccia, G. (1996) Selective water sorbents for multiple applications: 2. $CaCl_2$ confined in micropores of the silica gel: Sorption properties, *React. Kinet. Cat. Lett.*, **59**, pp. 335–342.
9. Benesi, A. J., Grutzeck, M. W., O'Hare, B., and Phair, J. W. (2004) Room temperature solid surface water with tetrahedral jumps of ^{2}H nuclei detected in ^2H_2O-hydrated porous silicates, *J. Phys. Chem. B*, **108**, pp. 17783–17790.
10. Stepanov, A. G., Shegai, T. O., Luzgin, M. V., Essayem, N., and Jobic, H. (2003) Deuterium solid-state NMR study of the dynamic behavior of

deuterons and water molecules in solid $D_3PW_{12}O_{40}$, *J. Phys. Chem. B,* **107**, pp. 12438–12443.

11. Soda, G. and Chiba, T. (1969) Deuteron magnetic resonance study of cupric sulfate pentahydrate, *J. Chem. Phys.,* **50**, pp. 439–455.

12. Kolokolov, D. I., Stepanov, A. G., Glaznev, I. S., Aristov, Yu. I., and Jobic, H. (2008) Water dynamics in bulk and dispersed in silica $CaCl_2$ hydrates studied by ^2H NMR, *J. Phys. Chem. C,* **112**, pp. 12853–12860.

13. Abragam, A. (1961) *The Principles of Nuclear Magnetism* (Oxford University Press, Oxford).

14. Long, J. R., Ebelhaeuser, R., and Griffin, R. G. (1997) ^2H NMR line shapes and spin–lattice relaxation in $Ba(ClO_3)_2 \cdot 2H_2O$, *J. Phys. Chem. A,* **101**, pp. 988–994.

15. Wittebort, R. J., Usha, M. G., Ruben, D. J., Wemmer, D. E., and Pines, A. (1988) Observation of molecular reorientation in ice by proton and deuterium magnetic resonance, *J. Am. Chem. Soc.,* **110**, pp. 5668–5671.

16. Spiess, H. W. (1978) Rotation of molecules and nuclear spin relaxation. In: *NMR Basic Principles and Progress*, Diehl, P., Fluck, E., and Kosfeld, R. (Eds.) (Springer-Verlag, New York), **15**; p. 55.

17. Wittebort, R. J., Olejniczak, E. T., and Griffin, R. G. (1987) Analysis of ^2H nuclear magnetic resonance lineshapes in anisotropic media, *J. Chem. Phys.,* **86**, pp. 5411–5417.

18. Kolokolov, D. I., Stepanov, A. G., Glaznev, I. S., Aristov, Yu. I., Plazanet, M., and Jobic, H. (2009) Water dynamics in bulk and dispersed in silica $CaCl_2$ hydrates studied by neutron scattering methods, *Micropor. Mesopor. Mater.,* **125**, pp. 46–50.

19. Lovesey, S. W. (1984). *Theory of Neutron Scattering from Condensed Matter, Vol. 1: Neutron Scattering* (Clarendon Press, Oxford).

20. Aristov, Yu. I., Glaznev, I. S., Gordeeva, L. G., Koptyug, I. V., Ilyina, L. Yu., Kärger, J., Krause, C., and Dawoud, B. (2006) *NATO-ASI Series, Series II,* Vol. 219 (Springer), p. 553.

21. Ruthven, D. M. (1984) *Principles of Adsorption and Adsorption Processes* (Wiley, New York).

22. Baron, M. M., Kvyat, E. I., Podgornaya, E. A., Ponomareva, A. M., Ravdel, A. A., and Timofeeva, Z. (1974) *Handbook of Physico-Chemical Data* (Khimiya, Leningrad) (Chemistry), p. 183.

23. Aristov, Yu. I., Glaznev, I. S., Freni, A., and Restuccia, G. (2006) Kinetics of water sorption on SWS-1L (calcium chloride confined to mesoporous silica gel): Influence of grain size and temperature, *Chem. Eng. Sci.,* **61**, pp. 1453–1458.

Chapter 13

Sorption Dynamics: An Individual Composite Grain

This chapter addresses the dynamics of vapor sorption on new composites "salt in porous matrix" (CSPMs). An improvement in the dynamics of vapor sorption by a bulk salt was one of the motivations to divide the salt into tiny pieces by its confinement to the matrix pores (see Sec. 1.6). Indeed, the reaction between the bulk salt and vapor may be very slow because

- It needs an intense reorganization of the salt crystalline structure [1];
- A crystalline solvate phase is formed on the surface of the salt, and further reaction requires diffusion of vapor through the solvate layer, which could be a very slow process [2].

Moreover, there can be a pronounced "pressure–temperature" region near equilibrium where reaction (1.1) in bulk is inhibited [1]. This results in a noticeable hysteresis between synthesis and decomposition reactions, which is likely to have a dynamic nature. In this chapter, we consider the dynamics of vapor sorption by CSPMs for the two cases:

1. The sorption is initiated by a small deviation from the equilibrium, e.g., by a little change in the sorbate concentration (pressure) over the sorbent at a constant temperature.

Nanocomposite Sorbents for Multiple Applications
Yu. I. Aristov
Copyright © 2020 Jenny Stanford Publishing Pte. Ltd.
ISBN 978-981-4267-50-2 (Hardcover), 978-981-4303-15-6 (eBook)
www.jennystanford.com

2. The pressure/temperature deviation is large enough, as it takes place in numerous practical technologies, such as adsorptive heat transformation, regeneration of desiccants, etc.

First, we compare the dynamics of water sorption by calcium chloride in bulk and confined to the host matrices with meso- and macropores (silica KSK and expanded vermiculite, respectively) to demonstrate that the confinement of salt to small pores can indeed significantly accelerate the sorption rate. Then we consider the kinetics of quasi-equilibrium water sorption by individual grains of the composites "CaCl$_2$/silica gel" (SWS-1L) [3–5] and its dependence on the grain size, salt content, and temperature. It was performed by the classical isothermal differential step (IDS) method [6] briefly described in Sec. 2.6.2. We analyze contributions of diffusion and chemical reaction, transport through the gas and liquid phases, enhancement of the pore tortuosity due to the salt presence, etc. Besides, ^{1}H NMR (nuclear magnetic resonance) micro-imaging [7] was used to visualize the spatial distribution of the sorbed water in an individual composite grain [8, 9].

13.1 Brief Comparison of Water Sorption Dynamics by Bulk and Confined Calcium Chloride

Before the detailed kinetic study, we performed a simple comparison of water sorption dynamics by calcium chloride in bulk and confined to the host matrix with mesopores (silica gel KSK). Both samples were prepared as grains of 0.2–0.25 mm size. The bulk salt was mixed with quartz powder to avoid its aggregation. The reaction studied was $CaCl_2 \cdot H_2O + H_2O = CaCl_2 \cdot 2H_2O$. At the same conditions, sorption by the composite was found to be much faster: the equilibrium was reached for 2 min instead of 60–70 h in the bulk (the only reaction of the bulk salt is displayed in Fig. 13.1a). Moreover, an S-shaped kinetic curve with an induction period of 2–3 h, typical for gas–solid reactions [2], transformed into a convex curve more common for adsorption processes controlled by gas diffusion. The latter mechanism was confirmed by numerous IDS tests reported in the following sections.

Isothermal Dynamics of Water Sorption in CSPMs | 265

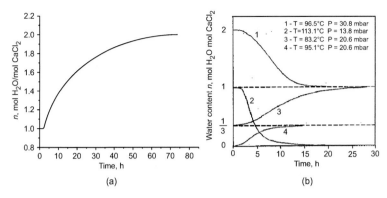

Figure 13.1 Kinetics of various hydration reactions in bulk: (a) T = 48.0°C, P = 8.7 mbar; (b) in the vermiculite pores. Figure (b) reprinted by permission from Springer Nature from Ref. [10], Copyright 2000.

An intermediate pattern was observed for the composite $CaCl_2$–vermiculite (SWS-1V): the S-shaped kinetic curves with the induction period of several hours and complete transformation within 5–20 h (Fig. 13.1b) were recorded for transitions between various $CaCl_2$ hydrates confined to the expanded vermiculite with macropores of 1–10 μm size [10].

The confined $CaCl_2$ solution also absorbs vapor faster than the bulk one. It was revealed by measuring, under the same conditions described in Ref. [11], the characteristic sorption times: 210 min (bulk), 50 min (SWS-1L, 15 nm pores), and 16 min (SWS-1S, 3.5 nm pores).

Hence, the salt dispersion inside the matrix's pores is an efficient tool to accelerate the salt–vapor reaction and make the sorption time typical for adsorption processes rather than for gas–solid reactions.

13.2 Isothermal Dynamics of Water Sorption in CSPMs

For more detailed study of water sorption kinetics and measuring the effective diffusivity, the well-known IDS method was used. Theoretical foundations of this method can be found elsewhere [6, 12] (see also Sec. 2.6.2).

13.2.1 Composite CaCl$_2$/Silica Gel (SWS-1L)

The kinetic curves of the water sorption on loose SWS-1L grains were measured by the IDS method at $T = 33$–$69°C$ and $P(H_2O) = 8$–70 mbar over water uptake range 0–0.47 g/g for the three grain sizes R_p (0.355–0.425, 0.71–0.85, and 1.2–1.4 mm). The water adsorption was initiated by a small pressure jump. The evolution of the sample weight was measured by a CAHN microbalance under isothermal external conditions (see Ref. [3] for further details). For reducing thermal effects, the adsorbent grains were mixed with small pieces of copper wire (weight ~ 85 mg). These precautions increased the heat capacity of the measuring cell and made the sample more isothermal.

Temporal variation of the sample weight is shown in Fig. 13.2. For the initial region of the uptake curves, $m_t/m_\infty \sim t^{0.5}$ or $m_t/m_\infty = A \cdot t^{0.5}$ (Fig. 13.2a). This behavior is typical for adsorption processes that are controlled by the diffusion of sorptive inside the adsorbent grain [6, 12] and is quite different from the kinetics of gas–solid reactions [2]. It means that the water sorption in SWS-1L is not controlled anymore by the interaction between the salt and water molecules typical for bulk salts [2]. This is also confirmed by much shorter sorption time for the composite sorbent (1–10 min, instead of 1–50 h for bulk CaCl$_2$).

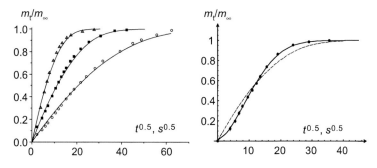

Figure 13.2 Uptake curves of the water sorption by SWS-1L at $T = 59.0°C$: (a) At various grain size: △ — 0.355–0.425 mm, ■ — 0.71–0.85 mm, ○ — 1.2–1.4 mm, $P = 20.2$ mbar [3]. Solid lines are calculated by Eq. (2.28); (b) Grains of 0.355–0.425 mm, $P = 40.6$ mbar; ● — experiment, solid line — non-isothermal model of Ref. [13], dashed line — Eq. (2.28). Figure (a) reprinted from Ref. [3], Copyright 2006, with permission from Elsevier.

Measuring the slope $A = (2S/V)(D_{ap}\, t/\pi)^{0.5}$ allows the apparent diffusivity D_{ap} to be calculated. This diffusivity describes the propagation of the adsorption front toward the grain center. If substitute this diffusivity to Eq. (2.28), conformity is observed between the experimental and theoretical uptake curves (solid lines in Fig. 13.2a). For large grains of 0.71–0.85 mm and 1.2–1.4 mm, the apparent diffusivity does not depend on the grain size R_p and is equal to $(2.6 \pm 0.4)\times10^{-11}$ m²/s. For the smallest grains, the value of the apparent diffusivity reduces, probably due to the heating of grains. Indeed, for such grains, the adsorption rate is sharply increased and the heat sink to the surrounding is not sufficient to dissipate it [13]. To reduce these thermal effects, an advanced methodology was suggested in Ref. [4].

The contribution of thermal effects can be accounted for in the frame of the non-isothermal model developed by Ruthven et al. [13]. Theoretical uptake curves for the smallest grains, calculated with the apparent diffusivity earlier obtained for grains with $2R_p \geq 0.71$ mm, well agree with the experimental data (Fig. 13.2b).

The pore diffusivity D_e, which characterizes replacement of a single water molecule, may be calculated from the relation $D_e = D_{ap}$ $[\varepsilon + (1 - \varepsilon)K]/\varepsilon$, where ε is the grain porosity and $K = K(c)$ is the local slope of the adsorption isotherm [12]. The latter was numerically calculated from the experimental sorption isotherm (Fig. 13.3). For instance, the effective pore diffusivity D_e at $T = 59°C$ is found to be $(3.3 \pm 0.8) \times 10^{-7}$ m²/s. This value can be compared with the Knudsen diffusivity for cylindrical pores of radius $r_p = 7.5$ nm [12]: $D_{kn} = 9700\, r_p\, (T/M)^{0.5} = 3.1 \times 10^{-6}$ m²/s. The ratio D_{kn}/D_e formally gives the ratio $(\varepsilon/\chi) \approx 10$ and allows determination of the pore tortuosity $\chi \approx 5$, which is larger than a typical value for pure silica gels $(\chi = 2–3)$ [12]. This extra tortuosity may be caused by the salt's presence inside the pore, e.g., by the blockage of a pore due to the salt.

For further analysis, three composites with a lower salt content (between 6 and 19 wt.%) were prepared and studied in Ref. [5] (see Table 13.1). It appears that the water pore diffusivity does not depend on the salt content (not presented). Hence, the pore's tortuosity increases from (2–3) to c.a. 5 at the lowest $CaCl_2$ content of 6 wt.%. Probably, even a little amount of the salt is sufficient to block the shortest diffusion routes and to slow down vapor diffusion significantly. This hypothesis is in good agreement with the texture data of the composites tested. Table 13.1 shows that the reduction in the composite pore volume is larger than the volume occupied by the salt, and the degree of the pore blockage does not depend on the salt content. In this table, V_p is the experimental specific pore volume of the composite, V_{sil} is the experimental pore volume per 1 g of the silica gel, $V^*_p = 1.16 \text{ cm}^3/\text{g} - m_{CaCl_2}/\rho_{CaCl_2}$ is the calculated pore volume of the composite per 1 g of the silica gel, the pore blockage degree $= (V^*_p - V_p)/V^*_p$.

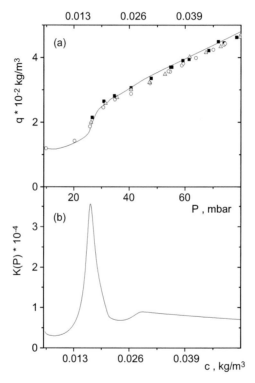

Figure 13.3 (a) Water adsorption isotherm of SWS-1L measured at $T = 59°C$; (b) Coefficient $K(P)$ numerically calculated from the above isotherm. Reprinted from Ref. [3], Copyright 2006, with permission from Elsevier.

Another reason for the decrease in the sorption rate may be strong interaction of water molecules with the dispersed salt, which increases the waiting time of the molecule on the surface, thus decreasing the frequency of jumps and, hence, diffusivity. In the latter case, the waiting time τ_w should be an exponential function of temperature, $\tau_w \sim \exp(-B/T)$ [14], which should lead to the appropriate temperature dependence of D_e.

Table 13.1 Texture parameters of the CaCl$_2$/silica gel composites with various salt contents

Sample	V_p (cm^3/g)	V_{sil} (cm^3/g_{sil})	V^*_p (cm^3/r$_{sil}$)	Blockage degree
Silica gel	1.16	1.16	1.16	0
CaCl$_2$ (6%)/silica	0.72	0.76	1.14	0.33
CaCl$_2$ (13%)/silica	0.68	0.78	1.10	0.29
CaCl$_2$ (19%)/silica	0.53	0.65	1.07	0.39

The D_e measurements in the temperature range 33–69°C reveal no dependence $D_e(T)$ (not presented) [3]. This may be considered additional confirmation of the Knudsen diffusion as the main transport mechanism, at least, at relatively low water uptakes. More complicated analysis of the sorption dynamics is required at large uptakes when the essential fraction of the matrix pores is filled with the salt solution [4] as considered in Sec. 13.2.2.

Thus, the isothermal water sorption by SWS-1L is much faster than by the bulk calcium chloride. The kinetics may be described by the Fickian diffusion model, and the Knudsen diffusion is deemed to be a mass transport mechanism. No manifestation of the chemical reaction between the salt and water vapor is recorded. Hence, dividing the salt (CaCl$_2$) into tiny pieces by its confinement to the silica mesopores, indeed, results in acceleration of the slow chemical reaction itself, so that the vapor transport becomes a rate-limiting process.

It is important that at large uptakes, the "salt/matrix" system becomes multi-phase and consists of solid (matrix, salt, salt hydrates), liquid (salt solution), and gas (vapor) phases. Hence, the sorption dynamics cannot be considered anymore in the frame of common (IDS) models of gas/vapor transport in a porous medium, and the transport through the liquid phase should be taken into account as well [15, 16]. Preliminary analysis of the isothermal kinetics of water sorption in CSPMs that are partially filled with the salt solution is considered in the next section [4, 17].

13.2.2 Enhancement of Vapor Transport in Partially Saturated CSPM Pores

At comparatively small amounts ($n < 4$–5) of water sorbed by the "$CaCl_2$/silica gel" composite [4], the process is primarily controlled by water vapor transport in the pore space, which occurs mainly via Knudsen diffusion (see above). As the amount of sorbed water increases, the sorption rate and the apparent diffusivity D_{ap} grow (not presented) [4]. This indicates the appearance of another channel for water transport in the pores. This is quite likely in view of the change in the phase composition of the system during water sorption. As discussed in Chapter 5, the first portions of water are sorbed via the reaction $CaCl_2 + nH_2O = CaCl_2 \cdot nH_2O$, which yields solid salt hydrates with n = 1/3, 1, and 2 [18]. At later stages, the salt dihydrate adds two more water molecules, thus yielding the tetrahydrate in the pores that melts at T = 30°C (Chapter 11). Hence, at $n > 2$, there is a salt solution (melt) in the pores rather than the solid hydrated salt. The liquid phase may be arranged differently in the pore space. Some of the arrangement variants are shown in Fig. 13.4 [4]:

- The solution is localized in narrow waists (Fig. 13.4a) and blocks the Knudsen diffusion of vapor through them. This lengthens the diffusion path, which is equivalent to an increase in the tortuosity factor.
- The solution spreads over the surface (Fig. 13.4b), forming a connected cluster through which water molecules can diffuse continuously. The vapor phase also forms a connected cluster, so the diffusion fluxes through the two phases are independent and additive (see below).
- The vapor domains do not form a connected cluster and are separated by solution domains (Fig. 13.4c). The transport of water molecules includes their transfer across the vapor/solution interface. A similar situation takes place when solution domains are separated by vapor domains.
- The entire pore space is filled with the salt solution (Fig. 13.4d). In this case, water diffuses only through the liquid phase.

Isothermal Dynamics of Water Sorption in CSPMs | **271**

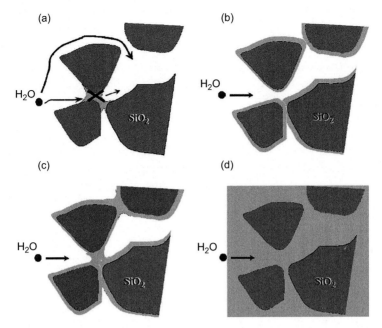

Figure 13.4 Variants of the arrangement of the salt and its solution in matrix pores (see discussion in the main text). Reprinted by permission from Springer Nature from Ref. [4], Copyright 2011.

It is expected that with an increasing amount of sorbed water, the gas phase transport will diminish and the liquid phase transport will increase. Under the assumption that water is transported simultaneously and independently over the connected clusters of the vapor and liquid phases (Fig. 13.4b) and that the pores have the simple shape of a cylinder (Fig. 13.5a), the total diffusion flux J_Σ can be represented as the sum of the corresponding fluxes:

$$\begin{aligned} J_\Sigma &= J_g + J_l = -D_g S_g \frac{dC_g}{dz} - D_l S_l \frac{dC_l}{dz} \\ &= -\left[D_g(S_g/S) + D_l(S_l/S) \left(\frac{dC_l}{dz} \Big/ \frac{dC_g}{dz} \right) \right] S \frac{dC_g}{dz} \\ &= -\left[D_g(S_g/S) + D_l(S_l/S) \frac{dC_l}{dC_g} \right] S \frac{dC_g}{dz} \\ &= -D_\Sigma S \frac{dC_g}{dz} \end{aligned} \qquad (13.1)$$

where $D_\Sigma = D_g(S_g/S) + D_l(S_l/S)\dfrac{dC_l}{dC_g} = D_g(S_g/S) + D_l(S_l/S)K$

Here, z is the coordinate along the pore (Fig. 13.5a), S_g and S_l are the mass transfer areas of the gas and liquid phases, S is the total mass transfer area ($S = S_g + S_l$), C_g and C_l are the water concentrations in the gas and liquid phases, and $K = dC_l/dC_g$ is the local slope of the sorption isotherm (Fig. 13.5b). Equation (13.1) is consistent with the analytical expressions suggested in the literature for describing the additive contributions from different mechanisms of water transport in a porous medium [15, 16, 19–21], which assumed the fluxes to be parallel and linearly depend on the concentration gradients.

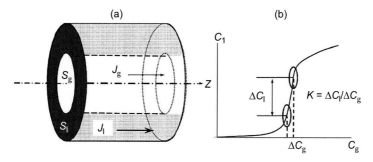

Figure 13.5 (a) Diffusion fluxes in the gas and liquid phases and (b) determination of the coefficient K. Reprinted by permission from Springer Nature from Ref. [4], Copyright 2011.

According to Eq. (13.1), transport through the liquid phase intensifies with increasing S_l and K, and this leads to an increase in the total flux if $D_g(S_g/S) + D_l(S_l/S) > D_g$ or $D_l K > D_g$ [16]. The vapor diffusivity can be estimated for the Knudsen transport mechanism as $D_g = 1.8 \times 10^{-6}$ m^2/s. We shall assume that the diffusivity of water in the aqueous solution of calcium chloride in the pores is equal to the diffusivity of water in the massive calcium chloride solution. The latter depends only slightly on the salt concentration in the 0.05–3.0 M range and is $(1.2 \pm 0.1) \times 10^{-9}$ m^2/s [22]. At $K > D_g/D_l = 1.5 \times 10^3$, the presence of the liquid phase will increase the diffusion flux. The K value for the composite CaCl$_2$/silica depends on the water uptake and varies between 3×10^3 and 35×10^3 (see Fig. 13.3b).

Therefore, the contribution from the liquid phase diffusion should be taken into account. Apparently, the real contribution of this channel is smaller than estimated above because the actual distribution of solution and vapor domains significantly differs from the simplified model presented in Fig. 13.5a. It is likely that vapor domains in the pores are separated by solution domains (Fig. 13.4d). This slows down the water transport because water diffusion includes transfer across the vapor–solution interface, and the escape of a water molecule from the solution phase into the vapor phase requires thermal activation [23].

The contribution from diffusion through the liquid phase is especially important for adsorbents that have large isotherm slope K. For CSPMs, the K value may reach 10^4–10^5, which is 1–2 orders of magnitude larger than for an unmodified porous matrix (e.g., silica gel) [16]. This is due to two circumstances:

1. These composites sorb 5–7 times more water.
2. The salt introduced into the pores sorbs a large amount of water within a narrow range of water concentrations in the gas phase (step-like sorption isotherm).

Thus, water diffusion through the liquid phase plays a much more significant role in the CSPM than it does in the host matrix.

13.2.3 Other Composites

Isothermal kinetics of water sorption was studied by the IDS method for several other CSPMs, namely, $Ca(NO_3)_2$/silica gel [24, 25], $CaCl_2$/alumina [5], $MgSO_4$/silica gel [17], and $LiNO_3$/silica gel [26]. For all these systems, the process of water sorption by the salt accelerates if the salt is confined to the silica pores and the shape of the uptake curve changes from S-shaped to convex as shown in Fig. 13.6. The initial part of the uptake curve can be described by the Fickian diffusion model, which enables calculating the effective water diffusivity D_{ap} and that in pores, D_e. The latter value is found to be close to the Knudsen diffusivity in those pores.

For all these sorbents, the sorption rate does not depend on the salt's nature and content, but depends on the matrix's pore structure, which confirms the dominant role of vapor diffusion rather than the vapor–salt reaction.

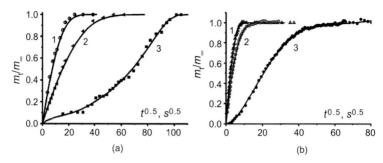

Figure 13.6 Kinetic curves of water sorption by (a) Ca(NO$_3$)$_2$/silica gel at T = 70°C (1) and 50°C (2) and the pure salt (3) (fraction 0.71–0.85 mm) [25]; (b) LiNO$_3$/silica gel at T = 35°C, the silica pore size 9 nm (1) and 15 nm (2) as well as the pure salt (3), fraction 0.4–0.5 mm [26]. Figure (a) reprinted from Ref. [25], Copyright 2009, with permission from Elsevier.

13.3 NMR Imaging Study of Water Sorption

Another method used for studying quasi-equilibrium water sorption by CSPMs was ^{1}H NMR micro-imaging. This method has been proven to be a useful tool for a non-destructive determination of spatial and temporal variations of liquid contents in a variety of porous materials, rapid enough to follow the drying and sorption processes in real time. In the imaging experiments, the ^{1}H NMR signal is not proportional to the local water uptake, so a preliminary calibration of NMR signal was done [8].

13.3.1 Sorption Profiles in Individual CSPM Pellet

The ^{1}H NMR micro-imaging technique was applied to study water vapor sorption by individual cylindrical silica gel and alumina pellets impregnated with hygroscopic salts. The two-dimensional (2D) images or the 1D profiles of the sorbed water distribution are detected sequentially to monitor the transport of water within the pellets in real time in the course of the sorption process.

^{1}H NMR micro-imaging experiments were performed at 300 MHz on an Avance NMR spectrometer equipped with the micro-imaging accessory. A cylindrical alumina (⌀ 3.6 mm) or silica gel (⌀ 6 mm) pellet containing CaCl$_2$ was positioned in the cylindrical glass cell residing in the probe of the NMR instrument. An air stream at room temperature was saturated with water vapor up to a required relative humidity and then passed along the pellet surface in the direction of the pellet axis (see details in Ref. [8]).

Figure 13.7a visualizes the water vapor sorption process by a cylindrical silica gel pellet with a uniform distribution of $CaCl_2$. The 2D imaging patterns demonstrate a nearly perfect cylindrical symmetry of the process despite possible presence of the local defects in the pellet. Therefore, a 1D radial profile of water content is sufficient to fully characterize the distribution of water within the pellet. The 1D profiles of water content along the diameter of the cylindrical pellet for the same silica gel sample are shown in Fig. 13.7b. It takes much less time to acquire a 1D profile (ca. 35 s) as compared to a 2D one (ca. 700 s); therefore, the 1D profile is a better representation of an instantaneous distribution of water at a given time.

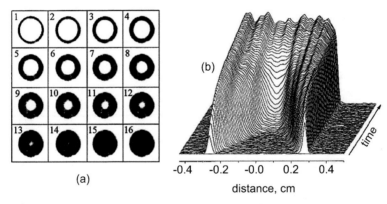

Figure 13.7 Two-dimensional images (a) and one-dimensional profiles (b) of the water sorbed by a cylindrical silica gel pellet containing $CaCl_2$ with uniform distribution of the salt. Air flow 390 L/h, relative air humidity 55%. Reprinted with permission from Ref. [8], Copyright 2000, American Chemical Society.

Figure 13.8a shows that for the alumina pellet filled with $CaCl_2$, the sorption rate is limited by the penetration of vapor into a dry region in the interior of the pellet, and the propagation of the sorbed water is characterized by a sharp front slowly moving toward the pellet center. The slowing down of water transport in the initially dry grain is due to the fact that shortly after the initiation of the sorption process, the volume adjacent to the surface of the pellet becomes almost fully saturated with the salt solution (Figs. 13.7 and 13.8).

As a result, the inner pellet regions do not contribute significantly to the efficiency of sorption, while the entire process is rate-limited by the vapor transport through the solution layer near the external surface of the pellet. This situation tentatively corresponds to cases (c) and (d) in Fig. 13.4.

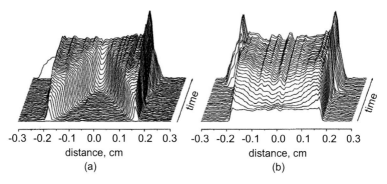

Figure 13.8 One-dimensional water content profiles detected in the course of the water sorption by a cylindrical alumina pellet (∅ 3.6 mm). Accumulation profile time was 38.5 s; every other profile is shown; flow 310 L/h, RH = 100%. The initial water contents of the pellet were (a) 0% and (b) 56%. Reprinted with permission from Ref. [8], Copyright 2000, American Chemical Society.

The propagation of the sorbed water front through the pellet that has already been partially saturated with water is much faster (Fig. 13.8b). In the latter case, water transport through the connected liquid phase is more efficient than that through the gas phase in the dry pellet. This probably corresponds to the case of parallel vapor fluxes through both liquid and gas phases (Fig. 13.4b).

In sum, the ^{1}H NMR micro-imaging technique allowed establishing the basic features of water transport within the individual grains containing $CaCl_2$ upon water vapor sorption. The experimental results demonstrate that soon after the initiation of the sorption process, the regions of the pellet adjacent to the surface become fully saturated and, therefore, the salt in the inner parts of the pellet does not contribute to the sorption efficiency. Furthermore, due to the presence of the salt in the inner regions of the pellet, the penetration of water into the dry areas becomes the bottleneck of the sorption process. Slow water diffusion through the solution layer near the grain's external surface was considered in Ref. [27] as a limiting step of air drying process by the composites $CaCl_2$/alumina and $CaCl_2$/(porous carbon Sibunit). The proposed non-stationary model interpreted a gradual decrease in the drying rate that was experimentally observed, as a result of monotonic thickening of the solution layer near the external grain surface. The model fits well the experimental data for both sorbents if the effective diffusivity at the limiting step is close to the water diffusivity in an aqueous calcium chloride solution.

References

1. Andersson, J. Y. (1986) *Kinetic and Mechanistic Studies of Reactions between Water Vapour and Some Solid Sorbents* (Department of Physical Chemistry, the Royal Institute of Technology, Stockholm, Sweden), S-100 44, pp. 27–44.

2. Galwey, A. and Brown, M. (1999) *Thermal Decomposition of Ionic Solids* (Elsevier Science, Amsterdam).

3. Aristov, Yu. I., Glaznev, I. S., Freni, A., and Restuccia, G. (2006) Kinetics of water sorption on SWS-1L (calcium chloride confined to mesoporous silica gel): Influence of grain size and temperature, *Chem. Eng. Sci.*, **61**, pp. 1453–1458.

4. Ovoshchnikov, D. S., Glaznev, I. S., and Aristov, Yu. I. (2011) Water sorption by the calcium chloride/silica gel composite: The accelerating effect of the salt solution present in the pores. *Kinet. Catal.*, **52**, pp. 620–628.

5. Glaznev, I. S. (2006) Water sorption dynamics in individual grain and beds of sorbents $CaCl_2$/silica and $CaCl_2$/alumina, PhD Thesis, Boreskov Institute of Catalysis.

6. Crank, J. (1975) *Mathematics of Diffusion* (Oxford University Press, London).

7. Blumich, B. (2000) *NMR Imaging of Materials* (Oxford University Press, Inc., New York).

8. Koptyug, I. V., Khitrina, L. Yu., Aristov, Yu. I., Tokarev, M. M., Iskakov, K. T., Parmon, V. N., and Sagdeev, R. Z. (2000) An ^{1}H NMR microimaging study of water vapor sorption by individual porous pellets, *J. Phys. Chem. B*, **104**, pp. 1695–1700.

9. Aristov, Yu. I., Koptyug, I. V., Glaznev, I. S., Gordeeva, L. G., Tokarev, M. M., and Ilyina, L. Yu. (2002) ^{1}H NMR microimaging for studying the water transport in an adsorption heat pump, *Proc. Int. Conf. Sorption Heat Pumps*, September 23–27, 2002, Shanghai, China, pp. 619–624.

10. Aristov, Yu. I., Restuccia, G., Tokarev, M. M., Buerger, H.-D., and Freni, A. (2000) Selective water sorbents for multiple applications. 11. $CaCl_2$ confined to expanded vermiculite, *React. Kinet. Cat. Lett.*, **71**, pp. 377–384.

11. Tokarev, M. M. and Aristov, Yu. I. (1997) Selective water sorbents for multiple applications. 4. $CaCl_2$ confined in the silica gel pores: Sorption/desorption kinetics, *React. Kinet. Cat. Lett.*, **62**, pp. 143–150.

12. Ruthven, D. M. (1984) *Principles of Adsorption and Adsorption Processes* (Wiley, New York).

13. Ruthven, D. M., Lee, L., and Yucel, H. (1981) Kinetics of nonisothermal sorption: Systems with bed diffusion control, *AIChE J.*, **27**, pp. 654–660.

14. De Boer, J. H. (1953) *The Dynamical Character of Adsorption* (Oxford University Press, London).

15. Churaev, N. V. (2000) *Liquid and Vapor Flows in Porous Bodies: Surface Phenomena* (Taylor & Francis Group).

16. Ho, C. K. and Webb, S. W. (2006) *Gas Transport in Porous Media* (Springer, The Netherlands), p. 443.

17. Ovoshchnikov, D. S. (2009) Dynamics of water sorption initiated by a drop of temperature (jump of pressure), Ms. Thesis, Boreskov Institute of Catalysis.

18. Aristov, Yu. I., Tokarev, M. M., Cacciola, G., and Restuccia, G. (1996) Selective water sorbents for multiple applications: 1. $CaCl_2$ confined in mesopores of the silica gel: Sorption properties, *React. Kinet. Cat. Lett.*, **59**, pp. 325–334.

19. Deryagin, B. V., Churaev, N. V., and Muller, V. M. (1987) *Poverkhnostnye Sily (Surface Forces)* (Nauka, Moscow) (in Russian).

20. Churaev, N. V. 1990 *Physical Chemistry of Mass Transfer in Porous Bodies* (Nauka, Moscow, p. 272) (in Russian).

21. Jury, W. A., Russo, D., Streile, G., and El Abd, H. (1990) Evaluation of volatilization by organic chemicals residing below the soil surface, *Water Resour. Res.*, **26**, pp. 13–20.

22. Harned, H. S. and Parker, H. W. (1955) The diffusion coefficient of calcium chloride in dilute and moderately dilute solutions at $25°$, *J. Am. Chem. Soc.*, **77**, pp. 265–266.

23. Valiullin, R. R., Skirda, V. D., Stapf, S., and Kimmich, R. (1997) Molecular exchange processes in partially filled porous glass as seen with NMR diffusometry, *Phys. Rev. E*, **55**, pp. 2664–2671.

24. Simonova, I. A. (2009) Composite water sorbents $Ca(NO_3)_2$/silica and LiNO3/silica, PhD Thesis, Boreskov Institute of Catalysis.

25. Simonova, I. A., Freni, A., Restuccia, G., and Aristov, Yu. I. (2009) Water sorption on the composite "silica modified by calcium nitrate," *Micropor. Mesopor. Mater.*, **122**, pp. 223–228.

26. Simonova, I. A. and Aristov, Yu. I. (2008) Composite sorbents "lithium nitrate in silica" for transformation of low temperature heat, *Altern. Energy Ecol.*, **11**, pp. 95–99.

27. Ostrovskii, N. M., Chumakova, N. A., Bukhavtsova, N., Vernikovskaya, N. V., and Aristov, Yu. I. (2007) Modeling of the limiting step of water sorption by composite sorbents of the "calcium chloride in porous matrix" type, *Russ. Theor. Found. Chem. Eng.*, **41**, pp. 83–90.

Chapter 14

Sorption Dynamics: A Composite Bed

For many adsorption technologies, the adsorbent material is used in more realistic configuration than a single grain, namely, a bed of loose or consolidated grains. In this chapter, we consider the sorption dynamics for SWS-1L [1] and SWS-1A [2] layers prepared with a binder. In this case, the pressure jump that causes the adsorption/desorption was large enough. The effects of grain size, the binder and salt contents are analyzed to succeed an intent transition of water sorption process from the intra- to intergrain mass transfer kinetic modes.

Two modern techniques are used to visualize the distribution of sorbed water inside the bed and its temporal evolution: nuclear magnetic resonance (NMR) micro-imaging [3] and gamma ray microscopy [4].

14.1 Water Sorption Profiles Measured by NMR Method

This method is a useful tool for a non-destructive determination of spatial and temporal evolution of adsorbed water in porous solids [3]. It is rapid enough to follow the drying and sorption processes in real time [5]. This method is already described in Chapter 13, where it has been used to detect one-dimensional (1D) profiles of

Nanocomposite Sorbents for Multiple Applications
Yu. I. Aristov
Copyright © 2020 Jenny Stanford Publishing Pte. Ltd.
ISBN 978-981-4267-50-2 (Hardcover), 978-981-4303-15-6 (eBook)
www.jennystanford.com

the sorbed water distribution in individual composite "salt in porous matrix" (CSPM) pellets [5, 6].

14.1.1 Experimental Methodology: Synthesis of Consolidated Layers and NMR Measurements

Grains of silica gels KSK and KSM (of pore size 15 and 6 nm, respectively) or alumina A1 were impregnated with a 38 wt.% $CaCl_2$ aqueous solution and then dried at 150°C (analogies of SWS-1L, SWS-1S, and SWS-1A, respectively). The consolidated bed was prepared as a cylindrical tablet of 16 mm diameter and 4–8 mm thickness made of the primary grains of CSPMs mixed with a binder (pseudo-boehmite). Details of the synthesis can be found elsewhere [1, 2].

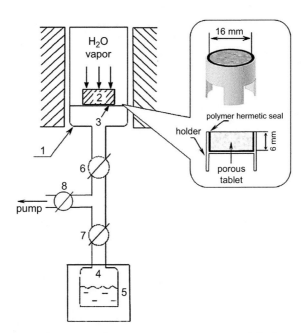

Figure 14.1 Schematics of NMR micro-imaging tests: (1) glass cell with CSPM sample (positioned inside the NMR micro-imaging probe); (2) sample; (3) sample holder; (4) evaporator; (5) thermostat; (6–8) valves. Reprinted by permission from Springer Nature from Ref. [1], Copyright 2006.

The sample was positioned inside the measuring cell of a Bruker Avance DRX-300 NMR spectrometer (Fig. 14.1). The sample cell was initially evacuated and then connected to an evaporator to maintain a constant vapor pressure of 7 mbar. Only the upper flat surface of the tablet was accessible for vapor. Other details of the measurements can be found elsewhere [1, 2]. One-dimensional distribution of the sorbed water was detected as a function of time at various combinations of the changeable parameters: the binder content (B = 0–30 wt.%), the salt content (C = 2.5–20 wt.%), and the size of the primary grains (0.04–0.056 mm or 0.25–0.5 mm).

14.1.2 Effect of Binder Content

If no binder is used, water vapor is sorbed uniformly over the bed of loose SWS-1L grains (Fig. 14.2a, profiles 4–8). It means that the vapor pressure is constant over the bed (and equal to that above the sorbent bed), and the sorption kinetics is determined by vapor diffusion inside the individual composite grains, that is, by the intra-grain diffusion (scenario **d** in Fig. 3.9).

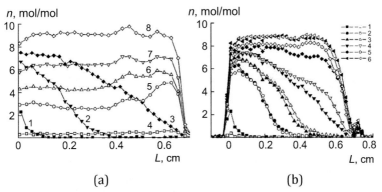

Figure 14.2 (a) Sorbed water profiles for an SWS-1L bed with B = 30 wt.% (1–3) at sorption times 11 min (1), 1 h 37 min (2), and 5 h 57 min (3); a bed of loose SWS-1L grains (4–8) at sorption times 11 min (4), 37 min (5), 54 min (6), 1 h 37 min (7), and 5 h 52 min (8); (b) Sorbed water profiles for consolidated bed of the CaCl$_2$/silica composites with different pore size R_{meso} = 3.0 nm (dark symbols) and 7.5 nm (light symbols) at sorption times 10 min 50 s (1), 1 h 37 min (2), 3 h 4 min (3), 5 h 57 min (4), 13 h 11 min (5), and 17 h 30 min (6). B = 30 wt.%, fraction 0.25–0.5 mm. Reprinted by permission from Springer Nature from Ref. [1], Copyright 2006.

If the binder is added, the water sorption front is formed and moves inside the bed (Fig. 14.2a, 1–3) [1]. Hence, the sorption process is limited by the vapor diffusion in the bed (the intergrain diffusion—scenario **b** in Fig. 3.9). It is no wonder that in this case, the water sorption dynamics is almost independent of the grain pore size (Fig. 14.2b).

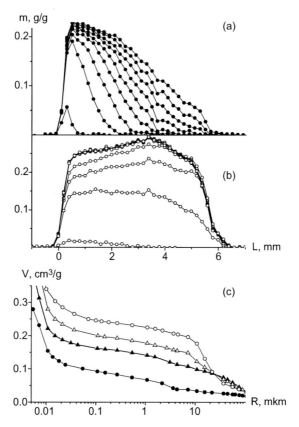

Figure 14.3 Effect of the binder content on the sorbed water profiles (a, b) and the pore volume distribution (c) in the CaCl$_2$/alumina beds: B = 20 (•), 10 (▷), 5 (▶), and 2.5 (○) wt.%. C = 22–25 wt.%, the alumina fraction 0.25–0.5 mm. The profiles were measured every 86 min. Reprinted from Ref. [2], Copyright 2010, with permission from Elsevier.

One-dimensional spatial profiles of water sorbed in the SWS-1A bed prepared with various binder contents show that at large content B = 20 wt.%, a sorption front is formed, which slows down as it propagates into the sample (Fig. 14.3a). Such a pattern argues for a dominant role of the intergrain diffusional resistance (scenario b). The displacement of the sorption front with time $x(t)$ during the 2 h $\leq t \leq$ 10 h time interval is proportional to $t^{0.5}$ [2], which is characteristic of diffusion type processes.

From the slope of the straight line in x^2 versus t coordinates, the effective diffusivity of water in the layer is estimated as $x^2/t = (3.6 \pm 0.1) \times 10^{-10}$ m^2/s at $B = 20$ wt.%. It takes ca. 10 h for water to reach the rear surface of the 6 mm thick layer.

Reduction in the binder content down to 2.5 wt.% leads to an increase in the volume of the transport pores (Fig. 14.3c) and to a corresponding acceleration of vapor transport in the bed. The water sorption occurs uniformly over the whole bed volume, hence the intragrain vapor diffusion becomes a rate-limiting process (scenario **d** in Fig. 3.9). At 10 wt.% $\geq B \geq$ 5 wt.%, an intermediate behavior between inter- and intragrain transfer resistances is observed (not presented) [2].

Thus, these results clearly demonstrate that variation in the sole parameter, the binder content, permits a crossover between the limiting diffusional regimes of water transport to be achieved.

14.1.3 Effect of Primary Grain Size

The size of primary grains is another parameter for tuning transport properties of the consolidated beds as it affects both intra- and intergrain diffusional resistances (see Sec. 3.3). For larger primary grains (0.25–0.5 mm), the diffusional resistance in the grain is larger than in the layer, and a uniform profile of the sorbed water is observed over the whole bed (Fig. 14.4a). The average uptake is determined by the water sorption in individual grains, so that 50% of the equilibrium uptake is reached in 3 h. In the layer composed of small grains, the sorption slows down due to a dramatic reduction in the volume of transport pores (Fig. 14.4c). Intergrain resistance becomes dominant, and the profile pattern with a sorption front is observed (Fig. 14.4b). This front separates the bed's volume adjacent to the external surface that is saturated with the salt solution, from the bed's inner volume, which is almost empty of the sorbed water. This pattern is similar to that reported for a single SWS pellet in Sec. 13.3.1 (Fig. 13.8a).

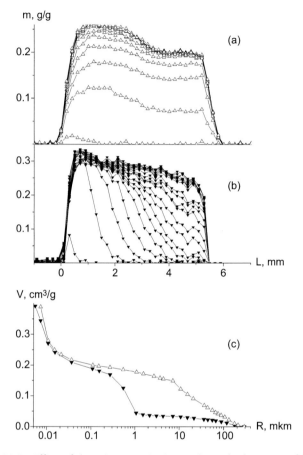

Figure 14.4 Effect of the primary grain size on the sorbed water profiles (a, b) and the pore volume distribution (c): △ — 0.04–0.056 mm, ▼ — 0.25–0.5 mm. C = 22–24 wt.%, B = 5 wt.%. The profiles were measured every 86 min. Reprinted from Ref. [2], Copyright 2010, with permission from Elsevier.

The displacement of the adsorption front with time $x(t)$ at 2 h $\leq t \leq$ 8 h is proportional to $t^{0.5}$. The slope of the straight line $x^2(t)$ gives the effective diffusivity $D_{ap} = (4.5 \pm 0.2) \times 10^{-10}$ m^2/s, which determines the sorption rate in the bed composed of small alumina grains. If the bed porosity ε and tortuosity χ ($\varepsilon/\chi \approx 0.2$–0.4) are taken into account, the water diffusivity inherent in the confined solution can be estimated as (1–2) × 10^{-9} m^2/s. This value is close to the water diffusivity in a bulk CaCl$_2$ solution in the concentration range of 0.05–3 M, (1.19 ± 0.09) × 10^{-9} m^2/s [7]. This confirms the assumption that the rate of water sorption is limited by the diffusion through the layer near the external surface that is saturated with the salt solution.

The observed crossover between the limiting diffusional resistances confirms the significant effect that the primary grain size has on the water diffusion in consolidated layers. Indeed, at constant B and C, a variation in the grain size permits fine-tuning of the relative contributions of intra- and intergrain diffusional resistances and optimization of water transport in the CSPM bed.

14.1.4 Effect of Salt Content

The effect of salt content is more complex and depends on the binder content. At $B \leq 5$ wt.%, the salt has a modest effect on the pore size distribution (Fig. 14.5c). Therefore, at all salt contents, the water transport is determined by the intragrain diffusion, and the concentration of the sorbed water is homogeneous and increases gradually in time (not presented) [2]. At $B = 20$ wt.%, the average size and volume of the transport macropores significantly decrease (Fig. 14.5c), which slows down the water sorption. Yet the intragrain diffusion resistance remains dominant at $C = 5$ wt.%, which results in the quasi-uniform concentration profiles (Fig. 14.5b). At large $C = 24$ wt.%, the layer becomes dense enough, and the opposite behavior is observed: A distinct adsorption front is developed that propagates inside the bed with a gradually falling velocity (Fig. 14.5a). An intermediate behavior is observed at $C = 14$ wt.% (not presented) [2]. Hence, the salt content is another parameter that can be varied to achieve the crossover between the regimes of intra- and intergrain diffusion, but this is possible only if the initial layer, composed of alumina and pseudo-boehmite, is dense enough.

14.2 Water Sorption Profiles Measured by Gamma Ray Microscopy

As already mentioned in Sec. 13.3, the ^{1}H NMR signal is not proportional to the local water uptake. It significantly reduced at $n <$ 4 because water molecules are part of solid hydrates $CaCl_2 \cdot 0.33H_2O$, $CaCl_2 \cdot H_2O$, and $CaCl_2 \cdot 2H_2O$ (Fig. 6.1) and have low mobility. At these low uptakes, the profiles of the sorbed water were measured by gamma ray microscopy [8], the sensitivity of which does not depend on the phase state of sorbed water [4].

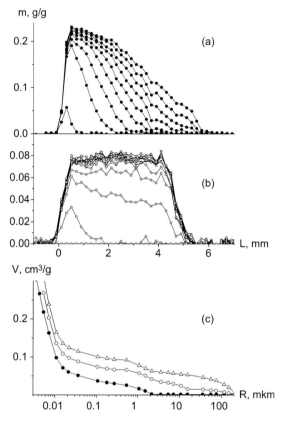

Figure 14.5 Effect of the salt content on the sorbed water profiles (a, b) and the pore volume distribution (c): ● — 24 wt.% (a), ○ — 5 wt.% (b), △ — 0 wt.%. C = 22–24 wt.%, B = 20 wt.%. The profiles were measured every 86 min. Reprinted from Ref. [2], Copyright 2010, with permission from Elsevier.

The profiles were detected at a relative pressure P_{H_2O}/P_0 of 0.32, 0.14, and 0.07; the corresponding equilibrium uptakes are 6, 4, and 2 mol/mol. The slope of the straight line $x^2(t)$ gives the effective diffusivity $(1.9 \pm 0.2) \times 10^{-10}$ m^2/s at $P_{H_2O}/P_0 = 0.32$ (Fig. 14.6d). This value is close to that measured by NMR micro-imaging $(3.6$–$4.5) \times 10^{-10}$ m^2/s and is consistent with the mechanism of water diffusion through the layer saturated with the salt solution. At

$P_{H_2O}/P_0 = 0.14$, the diffusivity goes down to $(0.7 \pm 0.1) \times 10^{-10}$ m²/s (Fig. 14.6d), probably due to the slow reaction between the salt and vapor.

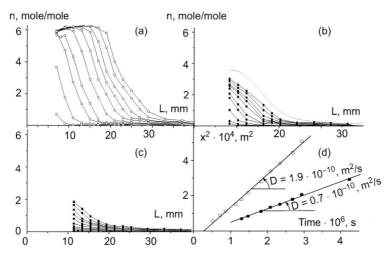

Figure 14.6 The sorbed water profiles in the CaCl$_2$/alumina bed at P_{H_2O}/P_0 = 0.32 (a), 0.14 (b), and 0.07 (c). B = 20 wt.%, C = 28 wt.%. The profiles were measured every 75 h. Dashed profile was recorded after 49 days [8].

14.3 Summary

To summarize this section, CSPMs can be prepared as a bed of loose grains or grains consolidated with the binder (pseudo-boehmite). The bed texture may be intently changed by varying the amount of the binder and salt as well as the size of the primary matrix grains. These modifications may result in the transition between the two limiting transport mechanisms, namely, intergrain and intragrain vapor diffusions. For the former regime, a sharp sorption front is formed that propagates into the bed with the displacement proportional to $t^{0.5}$. Analysis of the front dynamics allows estimation of the effective diffusivity of water in the bed, which is equal to $(1.9–4.5) \times 10^{-10}$ m²/s depending on the porous structure of the layer. This value is quite close to the water diffusivity in the aqueous salt solution, which is formed inside the matrix pores. In the case of intragrain diffusion, the concentration of the sorbed water is uniform over the entire layer, so that near-rectangular profiles are observed. In any case, the sorption dynamics is dictated by vapor diffusion rather than by the vapor–salt reaction, which confirms the fact that the dispersion of the salt inside the matrix's pores essentially accelerates the salt's interaction with water vapor.

These results can be used to formulate practical recommendations to optimize the primary grain size as well as the salt and binder contents for improving the dynamic operation of CSPM beds in various practical applications.

References

1. Aristov, Yu. I., Koptyug, I. V., Gordeeva, L. G., Il'ina, L. Yu., and Glaznev, I. S. (2006) Dynamics of water vapor sorption in a $CaCl_2$/silica gel/binder bed: The effect of the bed pore structure, *Kinet. Catal.*, **47**, pp. 776–781.

2. Glaznev, I. S., Koptyug, I. V., and Aristov, Yu. I. (2010) A compact layer of alumina modified by $CaCl_2$: The influence of composition and porous structure on water transport, *Micropor. Mesopor. Mater.*, **131**, pp. 358–365.

3. Blumich, B. (2000) *NMR Imaging of Materials* (Oxford University Press, Inc., New York).

4. Nizovtsev, M. I., Stankus, S. V., Sterlyagov, A. N., Terekhov, V. I., and Khairulin, R. A. (2008) Determination of moisture diffusivity in porous materials using gamma-method, *Int. J. Heat Mass Transfer*, **51**, pp. 4161–4167.

5. Koptyug, I. V., Khitrina, L. Yu., Aristov, Yu. I., Tokarev, M. M., Iskakov, K. T., Parmon, V. N., and Sagdeev, R. Z. (2000) An ^{1}H NMR microimaging study of water vapor sorption by individual porous pellets, *J. Phys. Chem. B*, **104**, pp. 1695–1700.

6. Aristov, Yu. I., Koptyug, I. V., Glaznev, I. S., Gordeeva, L. G., Tokarev, M. M., and Ilyina, L.Yu. (2002) ^{1}H NMR microimaging for studying the water transport in an adsorption heat pump, *Proc. Int. Conf. Sorption Heat Pumps*, September 23–27, 2002, Shanghai, China, pp. 619–624.

7. Baron, M. M., Kvyat, E. I., Podgornaya, E. A., Ponomareva, A. M., Ravdel A. A., and Timofeeva, Z. N. (1974) *Handbook of Physico-Chemical Data* (Khimiya, Leningrad) (Chemistry), p. 183.

8. Glaznev, I. S. (2006) Water sorption dynamics in individual grain and beds of sorbents $CaCl_2$/silica and $CaCl_2$/alumina, Ph.D Thesis, Boreskov Institute of Catalysis.

Chapter 15

Isobaric Sorption Dynamics: A Temperature Initiation

In this chapter, we consider a special case when adsorption is initiated by a drop in adsorbent temperature as opposed to the common initiation by a jump in vapor pressure over the adsorbent (see Chapters 13 and 14). This way of adsorption initiation is realized in temperature-driven units for adsorptive heat transformation (AHT). This emerging technology was suggested for utilization and storage of low-temperature heat (see Refs. [1–3] and Chapter 16). AHT is one of the most promising applications of composites "salt in porous matrix" (CSPMs), because adsorbents optimal for AHT should have S-shaped isobars and exchange large mass of adsorbate (see Refs. [3–5] and Sec. 3.2.2). A survey of CSPM dynamics in AHT can be found elsewhere [6, 7].

The CSPM sorption dynamics under typical conditions of AHT cycles is reported and analyzed. It is important that the adsorption dynamics initiated by the temperature drop can significantly differ from that initiated by the common pressure jump (see Ref. [8] and Sec. 15.3).

First, the large temperature jump (LTJ) method [9] is briefly introduced. It is, essentially, a new experimental approach that closely imitates the specific conditions of isobaric adsorption stage of a temperature-driven AHT cycle. Then we report recent results

---

Nanocomposite Sorbents for Multiple Applications
Yu. I. Aristov
Copyright © 2020 Jenny Stanford Publishing Pte. Ltd.
ISBN 978-981-4267-50-2 (Hardcover), 978-981-4303-15-6 (eBook)
www.jennystanford.com

obtained by the LTJ method, which demonstrate how the dynamics of sorption on CSPMs depends on the grain size, salt content, isobar shape, driving temperature difference (DTD), and the presence of residual air. Water, methanol, ethanol, and ammonia are used as sorptives. Their sorption dynamics for both mono- and multilayer configurations of the CSPM beds is reported.

15.1 The Large Temperature Jump Method

The LTJ method was suggested to study the dynamics of sorptive adsorption/desorption under laboratory conditions that nearly repeat the isobaric stages of a real AHT cycle [9]. An adsorbent is placed on a metal support that imitates a HEx fin. The support is quickly cooled/heated by a thermal carrier circulating under the support. The vapor pressure over the adsorbent is maintained almost constant by connecting the measuring cell to a large vapor vessel that moderates the pressure change.

An experimental setup contains three main compartments: measuring cell, vapor vessel, and evaporator with a liquid refrigerant (Fig. 15.1). The adsorbent (mass < 500 mg) is placed on a metal support, whose temperature may be adjusted with an accuracy of ±0.1°C using a heat carrier circuit coupled by a three-way valve (3WV) to either circulating thermal bath 1 or bath 2. The constant-volume vapor vessel as well as all connecting pipelines are maintained at a fixed temperature by using an air bath oven. The refrigerant vapor is generated by the evaporator with a cooling jacket. The evaporator temperature is managed by circulating thermal bath 2 through the valve V_{ev}. The sorbate uptake is calculated from the pressure evolution $P(t)$ in the apparatus of a constant volume [9]. For that reason, this version of the method is called volumetric LTJ (V-LTJ). The vapor pressure is measured by an absolute pressure transducer MKS Baratron® type 626A (accuracy ±0.01 mbar). More experimental details can be found elsewhere [9–11].

The main intrinsic limitation of the V-LTJ method is that only a small mass of adsorbent (≤0.5 g) can be tested, so only simplest flat-layer configurations of adsorbent bed can be studied. This is quite far from realistic configurations applied in modern AHT units.

To overcome this obstacle, a gravimetric version of the LTJ method (G-LTJ) has recently been suggested in Ref. [12]. In the G-LTJ method, a representative piece of real HEx is filled with an adsorbent (Ad-HEx) and directly placed on the balance cell. The weight of Ad-HEx is continuously monitored while adsorption/desorption proceeds. This allows AHT dynamics to be studied under more realistic conditions; however, it needs sophisticated weighting cell and careful experimentation to avoid the influence of flux fluctuations of the heat carrier fluid.

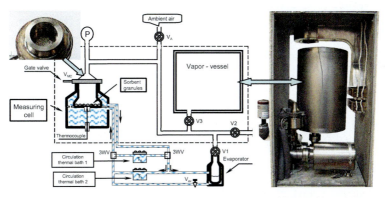

Figure 15.1 Schematic and real view of the volumetric LTJ setup. Reprinted from Ref. [10], Copyright 2008, with permission from Elsevier.

A calorimetric version of the LTJ method (C-LTJ) was recently suggested in Ref. [13] and applied in Ref. [14]. Presently, the C-LTJ method has restricted application and still needs further validation. A very new (T-LTJ) version was developed in Ref. [15]. The latter is based on direct measurement of the temperature difference ΔT of a heat carrier at the inlet and outlet of the tested HEx fragment after a fast drop/jump in the inlet temperature. The T-LTJ tightly repeats the procedure used in real HExs for transformation and storage of low-temperature heat.

The V-LTJ method is still the most developed and informative procedure. It has been widely used to study how the dynamics of adsorptive cooling cycle depends on the properties of CSPMs, namely, isobar shape [16], salt content [11], grain size [11, 17], as well as on the process organization (DTD [11, 18], presence of residual air [17], sorption and desorption times [19]).

15.2 Sorption Dynamics: Monolayer of Loose CSPM Grains

This section concerns the simplest adsorbent bed: a monolayer of loose CSPM grains.

15.2.1 Exponential Kinetics

Typical evolution of the temperature of the metal plate, the vapor pressure over the sample, and the calculated water uptake are shown in Fig. 15.2 for the composite $CaCl_2$/silica (SWS-1L). The heating (cooling) rate strongly depends on the adsorber construction, features of carrier circuit, 3WV, etc., and should be maximized to simplify further mathematical analysis. In our setup, the plate reached its final temperature after c.a. 10 s (Fig. 15.2a). So the sorption takes place at the almost constant temperature of the plate equal to its final temperature. The pressure reduction due to vapor sorption is slower and lasts some 10 min (Fig. 15.2b).

For both adsorption and desorption runs, an exponential evolution of the uptake w on time is found (see Fig. 15.2c for the adsorption run):

$$q(t) = q(0) \pm \Delta q \exp(-t/\tau) \tag{15.1}$$

This equation is quite universal: It is valid for almost all tested sorbents and sorbates up to 70–80% of the final conversion [7, 9, 11, 16, 18, 20]. The evolution of the tail may be slower than exponential. Equation (15.1) describes the adsorption/desorption dynamics by a *single* characteristic time τ that permits a stupendous simplification of the analysis and characterization of AHT dynamics as well as easy quantitative comparison of various adsorbents, boundary conditions, etc.

It is really amazing that quite complex and strongly coupled process, which includes adsorption, chemical reaction between the salt and water vapor as well as simultaneous heat and mass transfer, ultimately results in the simple exponential dependence of the sorption uptake/release on time. This kinetic law, probably, reflects the universal regularity of nature that the rate of relaxation to equilibrium is proportional to the deviation from the equilibrium. For our particular case, it means that a linear driving force (LDF) model [21] can be satisfactorily applied to analyze the major part of isobaric AHT stages. However, contrary to the common LDF model, the formally derivable rate constant $K = 1/\tau$ has to be independent of temperature [7] instead of the exponential dependence $K(T) = 15 D_{ap}(T)/R_p^2 \sim \exp(-E_a/RT)$ (see Sec. 2.6.2.3) still widely used for modeling AHT units [22]. The latter formula was originally introduced by Glueckauf [21] to simplify a numerical analysis of the sorption dynamics in chromatographic columns. The operating conditions in evacuated AHT units are quite different from those in flow-through columns; hence, the feasibility of the LDF model for AHT needs supplementary analysis and clarification [7, 8].

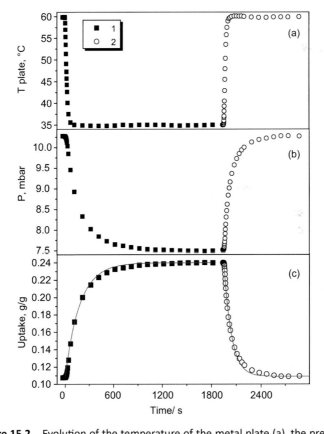

Figure 15.2 Evolution of the temperature of the metal plate (a), the pressure of pure water vapor over the SWS-1L sample (b), and the water uptake (c) for cooling (1) and heating (2) runs for grains of 0.8–0.9 mm size. Reprinted from Ref. [10], Copyright 2008, with permission from Elsevier.

15.2.2 Effect of Isobar Shape

One of the most important among the adsorbent properties that strongly affect the dynamic performance of AHT appears to be the shape of the sorption isobar. It was predicted in Ref. [23] and experimentally confirmed in Ref. [16]. Okunev et al. [23] numerically calculated the dynamics of water sorption on a single SWS-1L grain initiated by a drop in plate temperature from T_a to T_b and compared it with the dynamics of desorption initiated by a symmetric temperature jump from T_b to T_a at the same pressure. They found that the sorption dynamics is closely linked with equilibrium properties of the adsorbent, in particular, with the shape (convex, concave, or linear) of the segment of water sorption isobar between the initial and final temperatures. Under the symmetric drop/jump, desorption is faster than adsorption for a concave segment and vice versa for a convex one, while for a linar isobar the sorption and desorption curves coincide [23].

For experimental verification of this theoretical finding, the adsorption/desorption dynamics was studied by the LTJ method for a monolayer of loose SWS-1L grains [16] because this composite has both concave and convex segments of the sorption isobar (e.g., 1–2 and 2–3 in Fig. 15.3, respectively). When the metal plate is subjected to a temperature jump from 35°C to 60°C, it initiates the desorption process. A part of the sorbed water exits to the gas phase, and the system passes from the initial point 1 (35°C, 7.8 mbar) to the final point 2′ (60°C, 12.2 mbar). During the symmetric temperature drop (60 → 35°C), the composite sorbs vapor and returns to its initial state at point 1. Figure 15.4 shows that the desorption run is, indeed, faster than the reverse sorption run in complete agreement with the predictions of Ref. [23].

The qualitative explanation of such behavior is done by considering a coupled heat and mass transfer in the "adsorbent grain–metal plate" system [23]. For the 2′→1 run (drop 60 → 35°C), at the beginning of sorption process ($t \approx 10$ s) (near point 2′), the driving force for heat transfer is maximal because the plate is already cold ($T \approx 35$°C), while the sorbent grains are still at the initial temperature $T \approx 60$°C. Therefore, the conditions for removing the heat of sorption are optimal. However, the release of sorption heat is poor because at point 2′, the derivative dw/dT (proportional to the slop of the tangent line B) is small (Fig. 15.3). Hence, the temperature reduction causes only little water sorption. On the contrary, when approaching point 1, the temperature difference between the plate and the grain is getting smaller, while the amount of the adsorption heat to be removed is increasing, replicating the gradual rise in the derivative dw/dT (tangent A). This permanent discordance between reducing the driving force of heat removal and increasing the release of sorption heat does not favor the sorption process and results in very slow arrival of the sorption equilibrium at longer times (Fig. 15.4a).

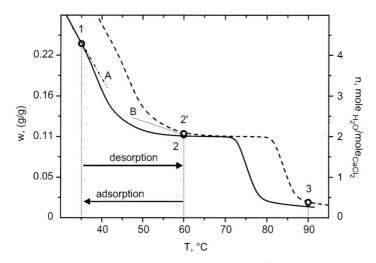

Figure 15.3 Water sorption isobars for SWS-1L at 7.8 mbar (solid line) and 12.2 mbar (dashed line). The initial and final values of temperature jumps are marked by points 1 — 35°C; 2, 2' — 60°C; 3 — 90°C. Reprinted from Ref. [16], Copyright 2009, with permission from Elsevier.

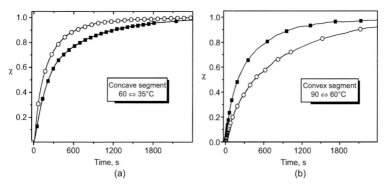

Figure 15.4 Experimental kinetic curves of water adsorption (solid squares) and desorption (open circles) for SWS-1L grains at different temperature jumps: (a) 35 ↔ 60°C, points 1 ↔ 2'; (b) 60 ↔ 90°C, points 2 ↔ 3 (Fig. 15.3). Reprinted from Ref. [16], Copyright 2009, with permission from Elsevier.

Much better correlation takes place during the desorption run $1 \to 2'$ $(60 \to 35°C)$. Indeed, in this case, the derivative dw/dT and, hence, the heat consumption due to water desorption are maximum at $t = 0$ (near point 1), which conforms to the most favorable conditions of heat transfer from the metal to the grain. At larger times, both fluxes are being gradually decreased. This conformity is the main reason why for the concave segment of the curve $w(T)$, the sorption process is slower than the desorption one as found in Refs. [16, 23].

Experimental testing of adsorption/desorption kinetics for a convex segment of the SWS-1L sorption isobar between points 2 $(60°C, 7.8\,mbar)$ and 3 $(90°C, 12.2\,mbar)$ (Fig. 15.3) is performed by the appropriate temperature drop/jump $(60 \leftrightarrow 90°C)$. In this case, the discordance of the heat fluxes takes place during the desorption process, which appears to be more slow than the adsorption one (Fig. 15.4b) and unambiguously proves the finding of Ref. [23].

Thus, a degree of correlation between the driving force for heat transfer and the derivative dw/dT, which defines the heat demand for desorption or heat release during adsorption, is a dominant factor for overall sorption dynamics in real AHT. Since this driving force gradually reduces with the process time, the derivative should decrease, too. Hence, d^2w/dT^2 has to be negative. It means that to shorten the isobaric stages, the isobar has to be concave at the desorption run, while convex at the adsorption run (Fig. 15.5). In the extreme case, these lines degenerate into appropriate stepwise isobars (Fig. 15.5): For the fastest desorption, the step has to be positioned at the initial desorption temperature $T_2 = T_{des}$, whereas for the fastest adsorption, it has to be positioned at the initial adsorption temperature T_4 [16].

Hence, the dynamics of single isobaric stages of AHT cycle can be significantly improved by optimizing the isobar shape of the adsorbent used as confirmed by estimations made in Ref. [24]. However, in actual practice, the gain is smaller because the same sorbent cannot ensure the ideal dynamics of both adsorption and desorption stages of the AHT cycle at once. Therefore, an isobar step has to be positioned between the initial and final stage temperatures to reach a compromise between adsorption and desorption rates and ensure the fastest cycle in whole [24].

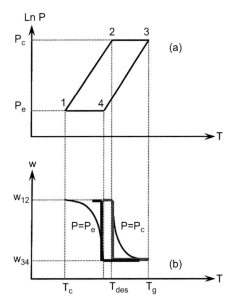

Figure 15.5 Basic AHT cycle (a) and dynamically optimal isobar shape (b). Adapted from Ref. [16], Copyright 2009, with permission from Elsevier.

Further analysis of adsorbents dynamically optimal for AHT cycles is performed in Chapter 16 (see Sec. 16.1.2.3), whereas the isobar shape optimal from dynamic point of view is analyzed by a numerical analysis of various model isobars in Refs. [6, 25].

15.2.3 Effect of Grain Size

The size R_g of adsorbent grains can simultaneously affect both heat and mass transfer in the adsorbent bed. As expected, adsorption is faster for smaller grains of SWS-1L (Fig. 15.6a) [7]. The shape of these kinetic curves is near-exponential over the range 0–90% of the final uptake, while the tail is longer than exponential. For larger grains, the ratio of characteristic sorption times $\tau_{0.5}(1.5 \text{ mm})/\tau_{0.5}(0.85 \text{ mm}) = 2.5$ is close to the ratio $(1.5 \text{ mm}/0.85 \text{ mm})^2 = 3.1$. That may indicate an essential contribution of intragrain vapor diffusion to the total adsorption rate. Strong acceleration of sorption process with the decrease in the grain size was also found for water adsorbent FAM-Z02 in Ref. [20].

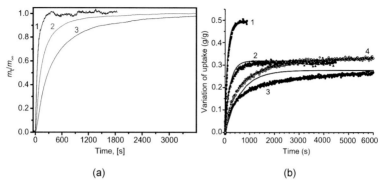

Figure 15.6 Grain size effect on the dynamics of (a) water sorption on SWS-1L: 1 — 0.35–0.42 mm, 2 — 0.8–0.9 mm, and 3 — 1.4–1.6 mm. Temperature drop 60 → 35°C, $P(H_2O)$ = 10.3 mbar [7]; (b) methanol sorption on LiCl (21 wt.%)/SiO$_2$ (1–3) and LiCl (15 wt.%)/SiO$_2$ (4): 0.4–0.5 (1), 0.8–0.9 (2) and 1.6–1.8 (3, 4) mm [11]. Lines indicate exponential fits. Temperature drop 50 → 35°C, $P(CH_3OH)$ = 60 mbar. Figure (a) reprinted from Ref. [7], Copyright 2009, and Figure (b) reprinted from Ref. [11], Copyright 2011, with permission from Elsevier.

The effect of grain size on methanol sorption by the composite LiCl/SiO$_2$ (KSK) was studied in Ref. [11]. Only for small sorbent grains, the kinetics is nearly exponential (Fig. 15.6b). For 1.6–1.8 mm grains, the time $\tau_{0.9}$ abnormally increases up to 1.0–1.5 h. Simultaneously, the uptake change reaches the equilibrium value of 0.50 g/g for 0.4–0.5 mm grains and just 0.30 g/g for 1.6–1.8 mm grains, so the conversion is essentially not completed for larger grains. This may be due to the following reason. Methanol sorption starts from the pores adjacent to the external grain surface. It leads to the formation of LiCl–methanol solution, so that a layer of this solution occupies the silica pores near the external grain surface. Because of this, the sorption rate gets limited by the slow methanol diffusion through this solution layer that may almost stop the process. Similar effect was observed by ^{1}H NMR micro-imaging technique for water sorption on a CaCl$_2$/silica single pellet (see Sec. 13.3.1 and Ref. [26]). The contribution of this blocking effect depends on the salt content in composites and will be further considered in the next section.

The near-exponential kinetic curves were also obtained by the LTJ method for sorption of ethanol on LiBr/SiO$_2$ [27] and ammonia on BaCl$_2$/vermiculite [28], the process being faster for smaller grains (not presented).

15.2.4 Effect of Salt Content

The salt is the main sorbing element of CSPMs, which may essentially affect the sorption dynamics, and its influence becomes apparent in different fashions:

- As shown in Sec. 13.2.1, even a little amount of salt is sufficient to block the shortest diffusion routes, increase the pore tortuosity, and significantly slow down the vapor diffusion. For the silica KSK, the degree of pore blockage does not depend on the salt content (from 6 to 19 wt.%).
- Another predictable effect is the longer time necessary for equilibrium setting in composites with a larger salt content [11]. For instance, the increase in the LiCl content from 13 to 21 wt.% results in a certain increase in the characteristic time of both adsorption and desorption from 44 to 70 s and from 15 to 32 s, respectively (see Fig. 15.7a for adsorption). In our opinion, the main reason is that a larger mass of methanol is exchanged (from 0.32 to 0.50 g/g). Interestingly, at short times, the absolute rate does not depend on the salt content. This indicated that the sorption dynamics is limited by heat and mass transport rather than by intrinsic interaction of methanol with the salt.

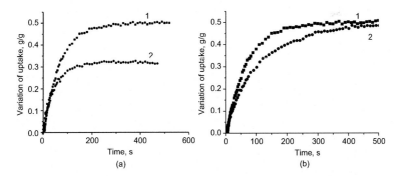

Figure 15.7 LTJ curves of methanol sorption on (a) LiCl/SiO$_2$ (Davisil): salt content 13 wt.% (1) and 21 wt.% (2); (b) LiCl (21 wt.%)/(silica gel) with the silica gel Davisil (1) and KSK (2). T-drop 50 → 35°C, P = 60 mbar. Figure (a) reprinted from Ref. [11], Copyright 2011, with permission from Elsevier.

- More sophisticated salt effects are related to the formation of salt solution inside the pores. The liquid phase may be located differently in the pore space (Fig. 13.4), which makes the salt effect on sorption dynamics quite complicated [29]. Strongly inhomogeneous distribution of the salt solution was observed in Ref. [26] by ^{1}H NMR micro-imaging: the solution occupies a region adjacent to the external grain surface and slows down vapor penetration to the grain interior. A similar rate reduction is revealed for methanol sorption on the LiCl/silica composite by the LTJ method [11]: the uptake reaches the equilibrium value of 0.50 g/g for grains of 0.4–0.5 mm size and much smaller uptake of 0.30 g/g for grains of 1.6–1.8 mm size; hence, the sorption is essentially not completed for larger grains.

That study revealed the effect to be more pronounced at larger salt content and smaller specific volume of the silica pores [11]: the dramatic reduction in rate happens when the salt solution occupies a very large portion of the pore space so that there is no sufficient empty space for continuous vapor diffusion. Indeed, at moderate uptake, the solution spreads over the surface (Fig. 13.4b), forming a connected cluster through which water molecules can diffuse continuously. The vapor phase also forms a connected cluster, so the diffusion fluxes through the two phases are independent and additive, as considered in Sec. 13.2.2 and Fig. 13.5. This profitable situation breaks at too large uptakes because the vapor domains do not form a connected cluster anymore and are separated by solution domains (Fig. 13.4c). The transport of methanol molecules includes their transfer across the vapor/solution interface and the escape of a molecule from the solution phase into the vapor phase, which requires thermal activation [29].

We compared the methanol sorption dynamics for the two composites LiCl (21 wt.%)/silica that differ solely by the specific pore volume of the host matrix: 1.0 cm^3/g of the silica KSK and 1.2 cm^3/g of the Davisil Gr646. The adsorption/desorption on the KSK-based composite is certainly slower (Fig. 15.7b), and this correlates with the empty pore volume V_{dif} available for the vapor diffusion. For the Davisil-based composite, $V_{dif} = 0.23$ cm^3/g $\gg 0.07$ cm^3/g for KSK. The latter residual volume appears to be not sufficient for efficient transport of methanol vapor. These examples demonstrate that

- Matrices with larger pore volume ensure better sorption dynamics;
- Moderate difference in the matrix pore volume may result in a drastic retardation of the sorption rate at large uptakes.

Hence, the salt content has to be properly selected to ensure an efficient vapor transport, which is possible until the empty pore volume is at least 20% of the total pore volume [11].

15.2.5 Effect of Cycle Boundary Temperatures

This effect was studied for methanol sorption on the monolayer configuration of the LiCl (21 wt.%)/silica composite [11]. The initial methanol pressure was P = 60 and 274 mbar for adsorption and desorption runs, respectively. To start desorption runs, the temperature of the metal support was jumped up from the initial temperature T_2 = 60°C to various final temperatures T_{des} = 70, 75, 80, 85, and 90°C. For adsorption runs, the temperature drop was from the initial temperature T_4 = 50°C down to various final T_{ads} = 30, 35, and 40°C. The evolution of temperature was exponential with a characteristic time of 15 s. More details of the experimental procedure can be found elsewhere [11].

The characteristic adsorption time strongly depends on the final adsorption temperature T_{ads} (Fig. 15.8). The process is quite fast at T_{ads} = 30 and 35°C, and the sorption constant $k = 1/\tau$ taken from the initial slop of kinetic curve is proportional to the DTD $\Delta T = T_{ads} - T_4$, $k = A\Delta T$ (the dashed line in Fig. 15.9). This probably indicates that, for large ΔT, the sorption rate is basically determined by the heat transfer between the sorbent and metal support.

At the maximal temperature, T_{ads} = 40°C, which means at a minimal DTD of 10°C, a drastic kinetic hindrance is observed, and the characteristic time τ increases up to 365 s (Fig. 15.8 and the symbol (*) in Fig. 15.9). Interestingly, at short sorption times, the process is faster (see the insert in Fig. 15.8), and the initial slope corresponds to $\tau = (120 \pm 10)$ s or $k = (8.3 \pm 0.7) \times 10^{-3}$ s^{-1}. This value satisfactorily follows the dashed line **A** in Fig. 15.9. Hence, 10–15% of the confined salt reacts with methanol very fast, so that the heat transfer is a rate-limiting process. Then, the sorption significantly decelerates and a crossover is observed to mode determined by either a poor mass transfer in the sorbent grain or slow reaction (1.3) between the salt and vapor. In the latter case, the DTD ΔT determines the transfer of heat, and the additional driving force ΔT_{reac} is necessary to initiate the gas–solid reaction proper. For the composite studied, it can be estimated from the intersection of line **B** with the axis $k = 0$ (Fig. 15.9), $\Delta T_{reac} = 8$–9 K [6].

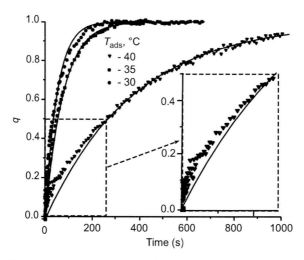

Figure 15.8 Dynamics of methanol adsorption at various boundary temperature T_{ads}. LiCl (21)/SiO$_2$-Gr646. 0.4–0.5 mm. $T_4 = 50°C$. Lines indicate exponential fits. Reprinted from Ref. [11], Copyright 2011, with permission from Elsevier.

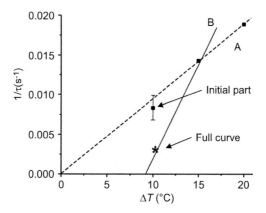

Figure 15.9 The adsorption constant as a function of the driving temperature difference $\Delta T = T_{ads} - T_4$. Reprinted from Ref. [11], Copyright 2011, with permission from Elsevier.

Thus, the sorption rate in the studied composites depends on the two driving forces. For the process to be fast, both should be sufficient as it is the case for the adsorption runs at T_{ads} = 30 and 35°C. At T_{ads} = 40°C, the driving force is small, and slow reaction between methanol and the salt controls the total sorption rate at q > 0.1. Similar non-monotonous effects of the final temperature T_{des} on the process dynamics are found for the desorption runs (not presented) [11]. These can be attributed to the peculiarities of the vapor–salt chemical reaction. Such a strong rate dependence on the driving force applied is quite typical for processes accompanied by a phase transformation, such as crystallization from supersaturated solutions or condensation of supersaturated vapor.

15.2.6 Effect of Layers Number *N*

All the previous kinetic data are measured for a monolayer (N = 1) of loose grains of various CSPMs. This configuration is poorly realistic for AHT from practical point of view as it results in too large parasitic heat losses due to inert thermal masses. Here we present results on studying the dynamics of methanol sorption for thicker layers (N = 2 and 4) [11].

As expected, both adsorption and desorption get slower for thicker layers (Fig. 15.10). The uptake curves are exponential, while the release curves show deviation from the exponential law, especially for N = 4 and long process time. The initial adsorption and desorption rates can be presented as a linear function of the ratio S/m = <heat transfer surface/sorbent mass> (not presented), which gives an indirect indication of a dominant role of the heat transfer from the metal support to the sorbent layer.

For the methanol sorption, the mass transfer resistance is also significant, so that sorption is always slower than desorption. It is important that even for the four-layer configuration, the total duration of the sorption and desorption runs is 10–15 min ($t_{0.9}$). It is quite acceptable for realizing dynamically efficient AHT cycles. On the other hand, to avoid the process retardation caused by slow non-exponential tails (Fig. 15.10b), it would be reasonable to restrict the duration of isobaric stages by the time $t_{0.8}$ or even less. This would allow avoiding a dramatic drop in the specific cooling power at long desorption times at the expense of a reasonable reduction in the cycle efficiency.

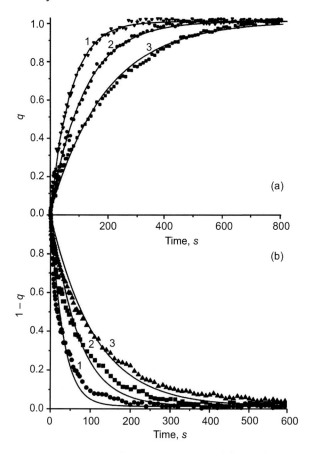

Figure 15.10 Kinetic curves of methanol sorption (a) and desorption (b) at various N = 1 (1), 2 (2), and 4 (3) and their exponential fits (lines). Reprinted from Ref. [11], Copyright 2011, with permission from Elsevier.

15.3 Summary

The CSPM sorption dynamics is studied under typical conditions of AHT cycles, that is, under the temperature initiation. The main findings of this study are as follows (for water, methanol, ethanol, and ammonia as sorbates):

- The initial part (60–90%) of sorption process follows exponential kinetics and can be characterized by a single characteristic time τ. This is valid for various composites, grain sizes, number of layers, and boundary conditions.
- A concave sorption isobar is profitable for desorption, while a convex one is profitable for sorption.
- The driving force for isobaric sorption is the temperature difference ΔT between the HEx and the sorbent bed. At small ΔT, the additional driving force may need to initiate the reaction between the salt and vapor. For LiCl/silica composite, the necessary supercooling is 8–9°C.
- For the multilayer case ($N > 1$), the sorption dynamics is invariant with respect to the ratio <heat transfer surface area/ sorbent mass>.

It is important that the adsorption dynamics initiated by the temperature drop can significantly differ from that initiated by the common pressure jump [8]. For instance, an appreciable difference in the character of the sorption front propagation inside the grain has recently been reported in Ref. [6]. Moreover, the T-initiation presupposes essential gradients of temperature between HEx fins and sorbent bed as well as inside the bed. This may result in the appropriate gradient of vapor pressure, if one assumes a local (P, T) equilibrium between the sorbent and vapor. Indeed, the equilibrium vapor pressure near colder sorbent sites/spots becomes lower than that near warmer sites. The diffusional flux F in the direction of temperature gradient can be evaluated as [8]

$$F = SD\frac{dC}{dx} = SD\frac{dC}{dT}\frac{dT}{dx} = SK_{ef}\frac{dT}{dx} \qquad (15.2)$$

where S is the mass transfer surface area, D is the vapor diffusivity, $K_{ef} = D(dC/dT)$ is the effective transfer parameter that depends on the equilibrium vapor concentration $C(T)$ over the adsorbent. Figure 5.11 shows dC/dT as a function of temperature calculated from the data of Ref. [30] at a fixed water uptake of 0.22 g/g for the composite sorbent SWS-1L. The common vapor diffusivity D moderately depends on temperature, namely, as $T^{1/2}$ and $T^{3/2}$ for the Knudsen and molecular diffusion mechanisms, respectively, whereas the $K_{ef}(T)$ dependence is almost exponential. The vapor phase contains the latent heat of sorption Q_{ads}; therefore, its transport delivers this heat to colder sites. This heat flux increases by a factor of 11 when the temperature rises from 35°C to 95°C (Fig. 15.11), which can be a reason of drastic enhancement of the total heat transfer coefficient that was experimentally revealed in Refs. [6, 8].

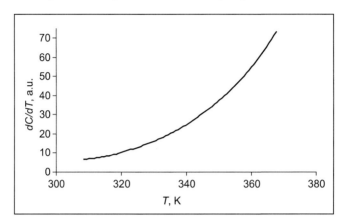

Figure 15.11 Temperature dependence of the derivative (dC/dT) for the composite sorbent SWS-1L at fixed water uptake of 0.22 g/g. Reprinted from Ref. [8], Copyright 2017, with permission from Elsevier.

This transport mechanism gives an indication of the tight link between heat and mass transfer processes in AHT units. This coupling is realized by water desorption at hotter sites, fast vapor transport through the interstitial voids, and subsequent adsorption at cooler sites. This mode of heat transfer can be referred to as the "heat pipe" effect, which is driven by the concentration gradient of water vapor imposed by the temperature non-homogeneity inside the CSPM bed.

References

1. Wang, R. Z. and Oliveira, R. G. (2006) Adsorption refrigeration: An efficient way to make good use of waste heat and solar energy, *Prog. Energy Combust. Sci.*, **32**, pp. 424–458.

2. Meunier, F. (2001) Adsorptive cooling: A clean technology, *Clean Prod. Processes*, **3**, pp. 8–20.

3. Aristov, Yu. I. (2008) Chemical and adsorption heat pumps: Cycle efficiency and boundary temperatures, *Rus. Theor. Found. Chem. Eng.*, **42**, pp. 873–881.

References

4. Aristov, Yu. I. (2007) Novel materials for adsorptive heat pumping and storage: Screening and nanotailoring of sorption properties (review), *J. Chem. Eng. Jpn.*, **40**, pp. 1241–1251.

5. Gordeeva, L. G. and Aristov, Yu. I. (2012) Composites "salt inside porous matrix" for adsorption heat transformation: A current state of the art and new trends, *Int. J. Low Carbon Tech.*, doi:10.1093/ijlct/cts050.

6. Aristov, Yu. I. (2013) Experimental and numerical study of adsorptive chiller dynamics: A loose grains configuration, *Appl. Therm. Eng.*, **61**, pp. 841–847.

7. Aristov, Yu. I. (2009) Optimal adsorbent for adsorptive heat transformers: Dynamic considerations, *Int. J. Refrig.*, **32**, pp. 675–686.

8. Aristov, Yu. I. (2017) Adsorptive transformation and storage of renewable heat: Review of current trends in adsorption dynamics, *Renew. Energy*, **110**, pp. 105–114.

9. Aristov, Yu. I., Dawoud, B., Glaznev, I., and Elyas, A. (2008) A new methodology of studying the dynamics of water sorption/desorption under real operating conditions of adsorption heat pumps: Experiment, *Int. J. Heat Mass Transfer*, **51**, pp. 4966–4972.

10. Glaznev, I. S. and Aristov, Yu. I. (2008) Kinetics of water adsorption on loose grains of SWS-1L under isobaric stages of adsorption heat pumps: The effect of residual air, *Int. J. Heat Mass Transfer*, **51**, pp. 5823–5827.

11. Gordeeva, L. G. and Aristov, Yu. I. (2011) Composite sorbent of methanol "LiCl in mesoporous silica gel" for adsorption cooling: Dynamic optimization, *Energy*, **36**, pp. 1273–1279.

12. Sapienza, A., Santamaria, S., Frazzica, A., Freni, A., and Aristov, Yu. I. (2014) Dynamic study of adsorbers by a new gravimetric version of the large temperature jump method, *Appl. Energy*, **113**, pp. 1244–1251.

13. Tierney, M., Ketteringham, L., Selwyn, R., and Saidani, H. (2016) Calorimetric measurements of the dynamics of a finned adsorbent: Early assessment of the activated carbon cloth–ethanol pair with prismatic aluminium fins, *Appl. Therm. Eng.*, **93**, pp. 1264–1272.

14. Tierney, M., Ketteringham, L., and Azri Mohd Nor, M. (2017) Performance of a finned activated carbon cloth–ethanol adsorption chiller, *Appl. Therm. Eng.*, **110**, pp. 949–961.

15. Tokarev, M. M. and Aristov, Yu. I. (2017) A new version of the large temperature jump method: The thermal response (T–LTJ), *Energy*, **140**, pp. 481–487.

16. Glaznev, I. S., Ovoshchnikov, D. S., and Aristov, Yu. I. (2009) Kinetics of water adsorption/desorption under isobaric stages of adsorption heat transformers: The effect of isobar shape, *Int. J. Heat Mass Transfer*, **52**, pp. 1774–1777.

17. Glaznev, I. S., Ovoshchnikov, D. S., and Aristov, Yu. I. (2010) Effect of residual gas on water adsorption dynamics under typical conditions of an adsorptive chiller, *Heat Transfer Eng. J.*, **31**, pp. 924–930.

18. Veselovskaya, J. V. and Tokarev, M. M. (2011) Novel ammonia sorbents "porous matrix modified by active salt" for adsorptive heat transformation: 4. Dynamics of quasi-isobaric ammonia sorption and desorption on $BaCl_2$/vermiculite, *Appl. Therm. Eng.*, **31**, pp. 566–572.

19. Aristov, Yu. I., Sapienza, A., Freni, A., Ovoschnikov, D. S., and Restuccia, G. (2012) Reallocation of adsorption and desorption times for optimizing the cooling cycle parameters, *Int. J. Refrig.*, **35**, pp. 525–531.

20. Glaznev, I. S. and Aristov, Yu. I. 2010. Dynamic aspects of adsorptive heat transformation. In: *Advances in Adsorption Technologies*, B. Saha and K. S. Ng (Eds.) (Nova Science Publishers, Singapore), pp. 107–163.

21. Glueckauf, E. (1955). Part 10. Formulae for diffusion into spheres and their application to chromatography, *Trans. Faraday Soc.*, **51**, pp. 1540–1551.

22. Yong, L. and Sumathy, K. (2002) Review of mathematical investigation on the closed adsorption heat pump and cooling systems, *Renew. Sustain. Energy Rev.*, **6**, pp. 305–338.

23. Okunev, B. N., Gromov, A. P., Heifets, L. I., and Aristov, Yu. I. (2008) A new methodology of studying the dynamics of water vapor sorption/desorption under real operating conditions of adsorption heat pumps: Modeling of coupled heat and mass transfer in a single adsorbent grain, *Int. J. Heat Mass Transfer*, **51**, pp. 246–252.

24. Okunev, B. N. and Aristov, Yu. I. (2014) Making adsorptive chillers faster by proper choice of adsorption isobar shape: Comparison with SWS-1L, *Energy*, **76**, pp. 400–405.

25. Okunev, B. N., Gromov, A. P., and Aristov, Yu. I. (2013) Modelling of isobaric stages of adsorption cooling cycle: An optimal shape of adsorption isobar, *Appl. Therm. Eng.*, **53**, pp. 89–95.

26. Koptyug, I. V., Khitrina, L. Yu., Aristov, Yu. I., Tokarev, M. M., Iskakov, K. T., Parmon, V. N., and Sagdeev, R. Z. (2000) An ^{1}H NMR microimaging study of water vapor sorption by individual porous pellets, *J. Phys. Chem. B*, **104**, pp. 1695–1700.

27. Gordeeva, L. G. and Aristov, Yu. I. (2009) Novel sorbents of ethanol "salt confined to porous matrix" for adsorptive cooling, *Proc. Int. Conf. Heat Powered Cycles (HPC-09)*, Berlin, Germany, September 7–9, 2009.

28. Veselovskaya, J. V., Tokarev, M. M., Grekova, A. D., and Gordeeva, L. G. (2012) Ammonia sorbents "porous matrix modified by active salt" for adsorptive heat transformation: 6. The ways of adsorption dynamics enhancement, *Appl. Therm. Eng.*, **37**, pp. 87–94.

29. Ovoshchnikov, D. S., Glaznev, I. S., and Aristov, Yu. I. (2011) Acceleration of the water sorption by composite "calcium chloride/silica gel" in the presence of salt solution inside the pores, *Russ. Kinetics Catalysis*, **52**, pp. 620–628.

30. Tokarev, M. M., Okunev, B. N., Safonov, M. S., Heifets, L. I., and Aristov, Yu. I. (2005) Approximation equations for describing the sorption equilibrium between water vapor and a $CaCl_2$-in-silica gel composite sorbent, *Rus. J. Phys. Chem.*, **79**, pp. 1490–1493.

Chapter 16

Adsorptive Transformation of Heat: Temperature-Driven Cycles

In Chapter 3, we gave a short introduction to adsorptive heat transformation (AHT). This technology has attracted an increasing interest as it allows utilization of low-temperature heat—wastes from industry, transport, dwellings, etc.—as well as new renewable heat sources [1]. As already mentioned in Sec. 3.2.2, two types of AHT cycles were suggested, which differ by the way of adsorbent regeneration (Fig. 3.3):

1. In temperature-driven (TD) cycles, an adsorbent is heated at a constant adsorptive pressure up to the temperature sufficient for desorption.
2. In pressure-driven (PD) cycles, the adsorptive pressure over the adsorbent is dropped at a constant temperature.

In this chapter, we consider applications of composites "salts in porous matrix" (CSPMs) in common TD AHT cycles. The development of these cycles was initiated by the 1973 oil crises, and the first, quite big (100–1000 kW), TD adsorptive chillers (ACs) appeared in the market as early as in 1986 [2, 3]. Various small chillers (less than 10–15 kW) were developed in the 21st century, and many of them have passed over from the prototype stage to small serial production [4].

Nanocomposite Sorbents for Multiple Applications
Yu. I. Aristov
Copyright © 2020 Jenny Stanford Publishing Pte. Ltd.
ISBN 978-981-4267-50-2 (Hardcover), 978-981-4303-15-6 (eBook)
www.jennystanford.com

Here we demonstrate that quite different adsorbents are needed for various TD AHT applications (freezing, ice making, air conditioning, heat pumping, etc.) under different climatic conditions. Therefore, CSPMs are deemed to be very promising for AHT as they allow intent and wide variation of sorption properties to amply fit changing requirements (Chapter 4). The current chapter is aimed at surveying the current state of the art in design, characterization, and application of CSPMs for TD AHT (see also Ref. [5] and references therein).

16.1 Thermodynamic Harmonization of Adsorbent and Working Conditions of TD AHT Cycle

An adsorbent is a key element of an AHT unit, and harmonization of its properties with the cycle's working conditions would significantly enhance the performance of AHT [6].

16.1.1 Various TD AHT Cycles

The basic 3T TD AHT cycle consists of two isosters and two isobars and is commonly presented on the Clapeyron diagram $\ln(P)$ versus $(-1/T)$ (Fig. 3.3a). Other usable ways to present this cycle are $w(T)$ and $T(S)$ diagrams, where S is the entropy change. Here we mainly consider the AHT cycle on the diagram $w(\Delta F)$, which is rarely used yet [7].

The Dubinin adsorption potential $\Delta F = -RT \ln(P/P_0)$ was introduced as a universal measure of the adsorbent's affinity for the adsorptive [8] (here P_0 is the saturated vapor pressure at temperature T). Fortunately, for the majority of adsorbents that are promising for AHT, there is a one-to-one correspondence between the equilibrium uptake w and the ΔF value [9]. In particular, it is valid for any adsorbent–adsorbate pair described by the Dubinin–Astakhov or Dubinin–Radushkevich equations.

The AHT cycle in the $w(\Delta F)$ presentation is very simple: It is just one line that represents both isobaric adsorption and desorption stages of a 3T TD AHT cycle (Fig. 16.1). If hysteresis is present, two different curves describe the adsorption and desorption runs.

Point A in Fig. 16.1 represents the boundary conditions at the end of the adsorption stage (point 1 in Fig. 3.3a). It corresponds to the adsorption potential $\Delta F_{min} = -RT_M \ln[P_L/P_o(T_M)]$, which characterizes the minimal adsorbent affinity necessary to bind an adsorptive at temperature $T_1 = T_M$ and pressure $P(T_L)$. This potential depends on the adsorptive nature as seen from Table 16.1, which displays the ΔF_{min} values for several TD AHT cycles selected as reference cycles in the key review [10]. Five basic applications, such as air conditioning (AC2), heat pumping (HP1, HP2), and deep freezing (DF1, DF2), were suggested in Ref. [10]. We have added one more air-conditioning cycle with $T_M = 15°C$ (AC1), which is also commonly in demand.

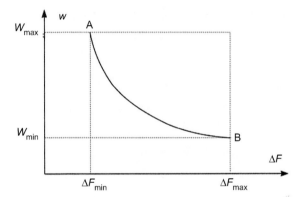

Figure 16.1 3T TD AHT cycle presented in the w versus ΔF coordinates. Reprinted from Ref. [7], Copyright 2014, with permission from Elsevier.

Table 16.1 ΔF_{min} value for the reference TD AHT cycles

			ΔF_{min} (kJ/mol)		
Cycle	T_L (°C)	T_M (°C)	H_2O	CH_3OH	NH_3
AC1	15	35–50	3.05–5.30	2.66–4.64	1.58–2.78
AC2	5	35–50	4.76–7.10	4.15–6.21	2.46–3.70
HP1	5	35–55	4.76–7.85	4.15–6.90	2.46–4.10
HP2	−15	35–55	9.02–12.4	7.53–10.4	4.43–6.20
DF1	−18	35–50	9.73–12.3	8.09–10.3	4.74–6.12
DF2	−50	35–50	18.6–21.6	15.2–17.8	8.90–10.5

Source: Reprinted from Ref. [7], Copyright 2014, with permission from Elsevier.

316 | *Adsorptive Transformation of Heat*

Point B in Fig. 16.1 represents the boundary conditions at the end of the desorption stage (point 3, Fig. 3.3a) and corresponds to the adsorption potential $\Delta F_{max} = -RT_H \ln[P_M/P_0(T_H)]$, which is defined by the conditions of heat rejection (T_M) and adsorbent regeneration (T_H). Adsorbate molecules that are bound to the adsorbent with an affinity lower than $\Delta F_{max} = -RT_H \ln[P_M/P_0(T_H)]$ can be desorbed during the regeneration stage 2–3. Those bound more strongly remain adsorbed and are not involved in the heat transformation cycle. The ΔF_{max} values associated with the regeneration conditions (P_M, T_H) are collected for $T_M = 35$ and 50°C, and $T_H = 60, 80, 100$°C in Table 16.2.

Table 16.2 ΔF_{max} value for the reference boundary conditions and selected values of T_H

T_M (°C)	35			50		
T_H (°C)	60	80	100	60	80	100
Water	3.50	6.25	8.90	1.32	3.95	6.47
Methanol	3.05	5.50	7.90	1.15	3.45	5.75
Ammonia	1.85	3.35	4.83	0.71	2.12	3.55

Source: Reprinted from Ref. [7], Copyright 2014, with permission from Elsevier.

The easiest conditions for air conditioning are $T_L = 15$°C and $T_M = 35$°C, which correspond to $\Delta F_{min} = 3.05, 2.66$, and 1.58 kJ/mol for water, methanol, and ammonia, respectively. The most severe AC conditions are $T_L = 5$°C and $T_M = 50$°C, which correspond to $\Delta F_{min} = 7.10, 6.21$, and 3.70 kJ/mol (Table 16.1). Thus, adsorbents promising for AC should have affinity for a working fluid lying between these limits. For the strongest adsorbent, the minimal regeneration temperature can be determined by Trouton's rule (Eq. (2.13)) as $T_{min} = 102$°C and 56°C for the weakest one, regardless of the nature of adsorbate (Table 16.3). These upper and lower boundary adsorbents are quite different, and all other adsorbents suitable for AC should lie between these two. A similar analysis can be done for heat pumping and deep freezing.

Table 16.1 and Fig. 16.2 show that various AHT applications need *quite different* adsorbents. AC2 asks for a moderate affinity and requires adsorbents very similar to HP1. The same is true for HP2 and DF1 but at a higher affinity level than for AC. Adsorbents for DF2 application require much greater affinity than those for other applications. AC1, AC2, and HP1 are the most commonly used applications that require adsorbents with affinity for water, methanol, and ammonia in the range of 3.0–7.9, 2.6–6.9, 1.5–4.1 kJ/mol, respectively (see Table 16.1).

Table 16.3 Minimal desorption temperature T_{min} and the Carnot efficiency COP^{Carnot}. The latter is calculated for the reference cycles assuming that $T_H = T_{min} + 20$ K

Cycle	T_L (°C)	T_M (°C)	T_{min} (°C)	COP^{Carnot}
AC 1	15	35–50	56–89	1.73–1.28
AC 2	5	35–50	68–102	1.37–1.12
HP 1	5	35–55	68–112	2.37–2.06
HP 2	–15	35–55	95–144	2.07–1.92
DF 1	–18	35–50	99–136	1.04–0.93
DF 2	–50	35–50	152–195	—

Source: Reprinted from Ref. [7], Copyright 2014, with permission from Elsevier.

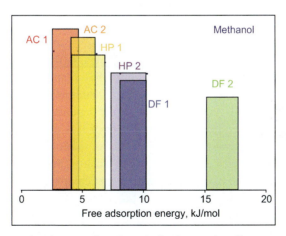

Figure 16.2 Classification of methanol adsorbents: the affinity at adsorption stage for various AHT applications. Reprinted from Ref. [7], Copyright 2014, with permission from Elsevier.

16.1.2 Adsorbent Optimal for TD AHT

This issue was briefly analyzed in Sec. 3.2.2. Here we consider it in more detail from both thermodynamic and dynamic points of view.

16.1.2.1 The first law efficiency

For an ideal 3T cycle (with zero thermal masses), the energy balance is given as

$$Q_c - Q_e - Q_d + Q_a = 0 \qquad (16.1)$$

(see definition of heats in Fig. 3.3a). The ideal first law efficiency or the coefficient of performance (COP) is then defined as

$$COP_c^{id} = Q_e/Q_d \qquad (16.2)$$

$$COP_h^{id} = (Q_c + Q_a)/Q_d = 1 + COP_c$$

for cooling and heating, respectively. For real AHT systems, heating of inert masses is also important, and the COP value is dependent on the adsorbate mass Δw exchanged in the cycle

$$COP_c(\Delta w) = Q_e\Delta w/[Q_d\Delta w + C(T_H - T_M)] = COP_c^{id} [\Delta w/(\Delta w + B)] \qquad (16.3)$$

where $B = C(T_H - T_M)/Q_d$, which is the ratio of the sensible heat of all inert masses to the latent heat necessary for desorption. To increase the first law efficiency, the exchanged mass $\Delta w = w_{max} - w_{min} = w(\Delta F_{min}) - w(\Delta F_{max})$ must be maximized, as is already pointed out in Sec. 3.2.2.

For well-designed AHT units, the COP approaches its maximum value (Q_e/Q_d) already at $\Delta w \geq (0.1–0.15)$ g/g and $(0.2–0.3)$ g/g for the exchange of water and methanol (ammonia), respectively. At larger Δw, the first law efficiency attains the maximum value (Q_e/Q_d) for "single effect" AHT cycles and is no longer dependent on Δw. Hence, to achieve the maximum first law efficiency, it is sufficient to use an adsorbent that exchanges the mentioned amount of a working fluid (or more) between the rich and weak isosters. Such an adsorbent may be considered harmonized with a particular cycle.

It is important to highlight that the maximum COP_c (0.7–0.9) for "single effect" AHT cycles is significantly lower than predicted for a reversible Carnot cycle with the same boundary temperatures [11] (see Table 16.3). For this reason, the second law efficiency should be considered.

16.1.2.2 The second law efficiency

For an ideal 3T TD AHT system, the energy balance (the first law, Eq. (16.1)) should be supplemented with the entropy balance (the second law)

$$- Q_c/T_M + Q_e/T_L + Q_d/T_H - Q_a/T_M = \Delta S \geq 0 \qquad (16.4)$$

If all the processes are completely reversible, the entropy generation is equal to zero ($\Delta S = 0$), and the Carnot COP can be calculated for cooling and heating [12, 13]:

$$\text{COP}^{\text{Carnot}} = (1/T_M - 1/T_H)/(1/T_L - 1/T_M) \qquad (16.5)$$

$$\text{COP}^{\text{Carnot}} = (1/T_L - 1/T_H)/(1/T_L - 1/T_M) \qquad (16.6)$$

The entropy is generated mainly during isobaric adsorption (4–1) and desorption (2–3) stages due to thermal coupling between the adsorber and a heat source/sink $\Delta T = T_H - T_{min}$ [11]. If the heat for regeneration is supplied directly at the minimal desorption temperature ($T_H = T_{min}$), no entropy is generated. As a result, the COP values calculated from the first and second laws are equal, so that the maximum (Carnot) efficiency is reached [14]. It imposes an appropriate requirement on the shape of the sorption isobar for an optimum sorbent: the rich and weak isosters must coincide, so that the cycle becomes degenerated, and the isobaric adsorption 4–1 and desorption 2–3 occur immediately and completely at the temperatures T_M and T_H (Fig. 16.3a), respectively [7]. The degenerated cycle can be realized *only* with an adsorbent characterized by a *mono-variant* equilibrium. The adsorbent has to sorb a large amount of adsorbate in a stepwise manner directly at T_M and P_L and completely desorb it at T_H and P_M (Fig. 16.3b). Hence, in theory, the adsorbent optimum for AHT should have a stepwise adsorption isobar. Real adsorbents, however, have S-shaped adsorption isobars (Fig. 16.3c) instead of strictly stepwise ones. Such adsorption equilibrium corresponds to isotherms of types IV, V, and VI in the IUPAC classification (Fig. 2.3). The second law harmonization of the adsorbent and the cycle means that the position of the step must coincide with the boundary temperatures $T_M = T_4$ at $P = P_L$ and $T_2 = T_H$ at $P = P_M$, as shown in Fig. 16.3b.

If the adsorbent is not optimal so that the adsorbate is removed at $T_H > T_{min}$, the efficiency becomes lower than the Carnot efficiency due to the thermal entropy production caused by the external thermal coupling as mentioned above [11]. The Carnot values calculated for the reference cycles at a moderate coupling of 20 K are displayed in Table 16.3. A key difference is found for AC1 at the boundary conditions $(T_L, T_M, T_H) = (15°C, 35°C, 76°C)$: $\text{COP}^{\text{Carnot}} - \text{COP}_c = 1.73 - 0.90 = 0.83$. This difference rapidly increases at larger

320 | *Adsorptive Transformation of Heat*

ΔT; therefore, to enhance the second law efficiency, the thermal coupling should be minimized. This suggests that a high second law efficiency may be attained by a proper choice of adsorbent without using efficient, but sophisticated methods of heat recovery in multi-bed [15] and thermal wave [16] systems, or "multi-effect" cycles in which waste heat from an upper cycle is used to drive a lower cycle [12].

16.1.2.3 The dynamic efficiency

As opposed to the thermodynamic efficiencies considered above, the adsorbent dynamic efficiency depends less on adsorbent inherent properties and more on a number of external factors [17]. Among these are the design of the "adsorber–heat exchanger (HEx)," efficiency of heat supply/rejection, pressure and temperature levels, etc. The final aim of the dynamic analysis is to maximize the specific AHT power, which can be defined for cooling as

$$\text{SCP} = Q_e/mt \qquad (16.7)$$

where m is the adsorbent mass and t is the cycle time.

The dynamic analysis may be simplified if the adsorption rate is controlled by the heat transfer between the adsorbent and the HEx fin. For a configuration with loose grains, it is a case for methanol and ammonia adsorption [18] because it occurs at a relatively high pressure of 0.1–10 bar. It is also often valid for water adsorption. Under the heat transfer control, the dynamic harmonization means that the most advantageous adsorption isobar is step-like with the largest possible ΔT between the step temperature and the HEx fin temperature (Fig. 16.3d).

Thus, step-like (sigmoid) adsorption isobars are advantageous from both thermodynamic and dynamic points of view. The dynamic requirement to maximize the average temperature difference between the adsorbent and the HEx fin is in contradiction with the thermodynamic requirement to minimize this difference [7, 17]. Thus, proper choice of the isobar step position greatly depends on which output parameter is to be maximized—COP or SCP. If both parameters are important, a reasonable compromise must be reached by an intelligent design of the Ad–HEx unit.

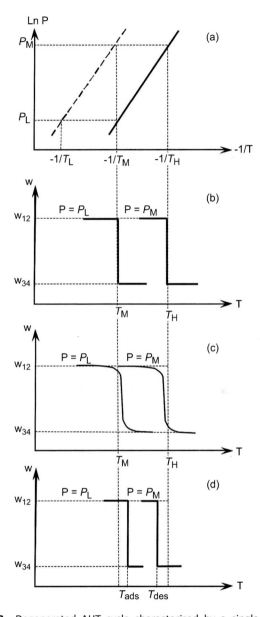

Figure 16.3 Degenerated AHT cycle characterized by a single adsorption/desorption isoster (a); corresponding step-like (b, d) and S-shaped (c) isobars of adsorption (at $P = P_L$) and desorption (at $P = P_M$) optimum from the thermodynamic (b, c) and dynamic (d) points of view. Reprinted from Ref. [7], Copyright 2014, with permission from Elsevier.

16.2 Adsorptive Cooling

Here we survey the application of CSPMs for adsorptive chillers, which is the most popular use of these composites. After the first presentation for ACs in Ref. [19], CSPMs were synthesized and tested in AC prototypes in many laboratories all over the world.

16.2.1 Composite Sorbents of Water

The first application of CSPMs in AHT prototypes was reported in Ref. [20] where the composite "CaCl$_2$/silica KSK" (SWS-1L) was tested in a lab-scale unit in ITAE-CNR, Messina, Italy (Fig. 16.4a).

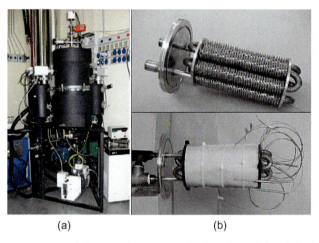

Figure 16.4 Views of the tested prototype of lab-scale AC (a) and the finned-tube adsorber without and with the sorbent (b). Figure (a) reprinted from Ref. [26], Copyright 2012, and Figure (b) reprinted from Ref. [20], Copyright 2004, with permission from Elsevier.

The prototype consists of a finned-tube HEx (Fig. 16.4b) connected to the evaporator and the condenser (Fig. 16.4). The HEx was filled with the sorbent (1.1 kg) in the form of irregular grains of 0.8–1.6 mm size prepared according to the procedure described in Ref. [21]. The experiments were carried out under the standard conditions of an AC cycle: the temperatures of the condenser, evaporator, and regeneration are 35–45°C, 5–10°C, and 80–110°C, respectively [20].

For the AC tested, the COP value is very sensitive to the condenser temperature and significantly reduces at $T_M = 40°C$ and $T_H < 0°C$ (Fig. 16.5a). The reason becomes clear from Fig. 16.5b where two AC cycles are plotted over the experimental SWS-1L adsorption/desorption isobars: A–B–C–D corresponds to $T_M = 35°C$, $T_L = 10°C$, and $T_H = 80°C$, while A*–B*–C*–D* corresponds to $T_M = 40°C$, $T_L = 10°C$, and $T_H = 80°C$. Increasing the condenser temperature by only 5°C results in the following:

- The mass of the exchanged water significantly reduced from 0.16 to just 0.06 g/g.
- The minimal desorption temperature increased from 61°C to 73°C.

Hence, no regeneration is possible at all at $T_H < 73°C$ and $COP_c = 0$, while at $T_H > 80$–$90°C$, the COP_c approaches its maximal value as the isobars have a plateau that corresponds to the formation of calcium chloride dihydrate (Fig. 5.8a). Evidently, so high parametric sensitivity is a consequence of the steep isobars of SWS-1L.

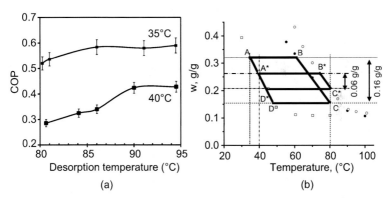

Figure 16.5 (a) Experimental cooling COP at different desorption and condensation temperatures ($T_L = 10°C$) (re-plotted from the data of Ref. [20]; (b) AC cycles plotted over the experimental SWS-1L adsorption/desorption isobars (□ — 12.3 mbar, • — 56.3 mbar, ○ — 74 mbar). Figure (b) reprinted from Ref. [20], Copyright 2004, with permission from Elsevier.

The COP_c gradually increases with a rise in desorption temperature up to 0.6 at 85–90°C, so that SWS-1L can be used for AC driven by low-temperature heat (heat wastes, solar energy, coolant from various engines, etc.). Although the design of this AC unit was not optimized, the COP_c is high enough mainly because of the prominent properties of SWS-1L, first of all, due to a large amount of water exchanged during the cycle (Δw = 0.16 g/g at T_M = 35°C).

The specific cooling power generated in the evaporator ranged between 35 and 60 W/kg, which was due to a quite low ratio S/m = <heat transfer surface>/<adsorbent mass> = (0.4 m^2)/(1.1 kg) = 0.36 m^2/kg, and large SWS-1L grains (0.8–1.6 mm). Indeed, the SCP significantly increases (up to 200–300 W/kg) in a finned flat-tube HEx with grains of 0.25–0.50 mm (not presented).

Another very important fact revealed in Ref. [20] is the stability of SWS-1L after 60 complete cycles: No sorbent destruction and dust formation were found, and no traces of calcium chloride were detected in the desorbed water. This indicates an encouraging hydrothermal stability of SWS-1L and its feasibility of practical application in AC technologies, although more cycling tests are welcome.

As the next step, both the steady-state and dynamic performances of SWS-1L were numerically studied [22] in a two-bed AC unit similar to a commercial chiller tested in Ref. [23]. Details of the modeling can be found elsewhere [22]. An ideal AC cycle of the studied system is plotted from the equilibrium sorption data of SWS-1L (A–B–C–D) and compared with that for a common RD silica gel (a–b–c–d) used in the commercial chiller (Fig. 16.6a). This figure clearly demonstrates that SWS-1L exchanges much larger mass of water in the cycle than the silica gel (0.21 g/g instead of 0.11 g/g). There are two direct consequences of this fact [22]:

1. The cycle time increases from 420 to 630 s.
2. The COP is 20–25% larger for the SWS-1L-based unit (Fig. 16.6b).

It was concluded in Ref. [22] that "the COP of SWS-1L + water system is higher than that of RD-water system... This much larger COP shows significant advantage of the new working pair as compared with the conventional unit. Thus, from the present simulation results, it is found that the newly SWS-1L-based adsorption chiller provides a promising unit for cooling applications."

A four-bed adsorption heat pump with the composite adsorbent SWS-1L and water as working pair was numerically studied in Ref. [24]. The authors found that using this composite to substitute a regular-density silica gel in standard AC units allows the COP value to be increased by 51%.

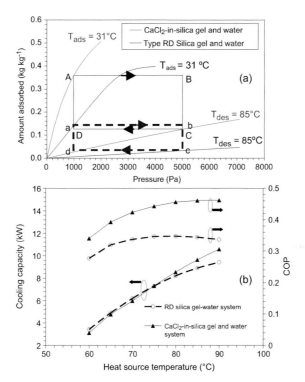

Figure 16.6 (a) *P–T–x* diagram of an adsorption cooling cycle as the adsorption uptake against the pressure (see text above); (b) Influence of driving heat source temperature on cyclic average cooling capacity and COP. Reprinted from Ref. [22], Copyright 2009, with permission from Elsevier.

All these basic and applied studies convincingly demonstrated that adsorption chillers based on SWS-1L may have better performance than the existing commercial chillers. The best SWS-1L performance corresponds to the driving temperature of 80–90°C. If the temperature available for regeneration is lower than 70°C, it is not sufficient to remove water from SWS-1L, and a composite that bounds water less strongly is required. As pointed out in Chapter 5, a variation in the salt's nature is the most efficient way to harmonize an adsorbent with the AC cycle. Since lithium nitrate is less hygroscopic than calcium chloride, it was suggested for AC driven by heat with $T_H < 70°C$ [25]. Water is absorbed due to hydration reaction (5.6).

Two new $LiNO_3$-based composites were synthesized at BIC with silica gel KSK, SWS-9L [26], and expanded vermiculite, SWS-9V [27], as porous hosts. Details of the salt and choice of matrices can be found in the original papers [26, 27].

The former composite, indeed, can be regenerated at quite a low temperature of 65°C, which was confirmed by testing in a lab-scale AC unit with a flat-plate finned-tube adsorber (Fig. 16.7a) at ITAE-CNR, Italy [26]. The amount of water exchanged at T_L = 10°C, T_c = 35°C, T_a = 30°C, and T_H = 65°C reaches 0.19 g/g (Fig. 16.7b). The initial performance of the composite is quite good [26], but its hydrothermal stability appears to be poor. The problem is caused by a significant swelling of the solid phase due to reaction (5.6).

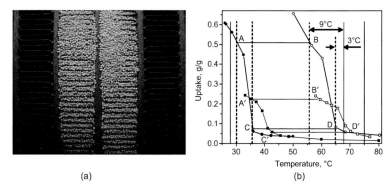

Figure 16.7 (a) View of the finned flat-tube heat exchanger [26]; (b) equilibrium isobars of water adsorption at 12.6 mbar (•) and desorption at 56.2 mbar (○) and appropriate AC cycle for the composites SWS-9V (A–B–D–C) and SWS-9L (A′–B′–D′–C′). Figure (a) reprinted from Ref. [26] and Figure (b) reprinted from Ref. [27], Copyright 2012, with permission from Elsevier.

To ensure a space for the swelled salt, LiNO$_3$ was inserted into the macropores of the expanded vermiculite. The specific pore volume of this host material reaches 3.0 cm^3/g, which allows a large mass of the salt to be housed inside the pores (63 mass.% for SWS-9V). As a result, this composite exchanges 0.4 g H$_2$O/g at the boundary conditions of Fig. 16.7b. Since crystals of the confined salt are quite large, the sorption isobars are very steep, and water is exchanged in the exceptionally narrow temperature ranges of 33–36°C (for adsorption at 12.6 mbar) and 62–65°C (for desorption at 56.2 mbar). At T_M = 35°C, T_L = 10°C, and high driving temperature (T_H = 90°C), the AC prototype provides the best performance with COP = 0.66 and SCP = 230 W/kg. Quite good performance (COP = 0.59 and

SCP = 95 W/kg) is obtained at a driving temperature of only 68°C [27].

In sum, SWS-9V is able to exchange a large amount of water (up to 0.4 g H_2O/g) under operating conditions typical for AC units driven by a thermal source with T_H < 70°C. Further considerations show that the new sorbent could represent a promising solution for heat storage as well. It demonstrated a remarkable heat storage capacity (1.15 MJ/kg) [27], which is far superior to common materials suggested for the storage of heat with a temperature potential of 60–80°C.

The considered composites exemplify how the intelligent choice of salt and porous matrix allows nanotailoring of new efficient solid sorbents adapted to particular AC cycles. Several other CSPMs with $Ca(NO_3)_2$ [28], LiCl [29], and LiBr [30, 31] have also been suggested and studied for the transformation of low-temperature heat with water as an adsorbate.

16.2.2 Composite Sorbents of Methanol

Reviews on methanol adsorbents for TD heating and cooling applications are given in Refs. [32, 33]. They are mostly various activated carbons, including activated carbon fibers (ACFs) [34]. These adsorbents commonly need regeneration at T_H > 90–100°C, and adsorption properties of various carbons differ not so much [35]. As a result, different carbons demonstrate similar COP in air-conditioning (0.3–0.4) and ice-making cycles (0.17–0.28) [36]. Therefore, new methanol adsorbents better adapted to AHT cycles are highly welcome.

Sixteen new CSPMs for sorption of methanol were synthesized and systematically evaluated for AHT in Ref. [37]. Various halides (LiCl, LiBr, $CaCl_2$, $CaBr_2$, $NiBr_2$, $CoCl_2$, etc.) were inserted into the pores of three commercial silica gels (with an average pore size of 6–15 nm). Here we only briefly introduce the most important materials and findings of this and other studies (see also Ref. [5]).

The composite sorbent of *methanol* "LiCl/silica gel" with an average pore size d_{av} = 15 nm (see also Chapter 7) was intently prepared and tested for adsorption air-conditioning cycle in Ref. [38]. This composite exchanges 0.6 g/g at T_L = 7°C, T_c = 35°C, T_a = 30°C, T_H = 85–90°C, and allows the theoretical COP much larger than for a common carbon AC-35 (Fig. 16.8a) [38]. This extraordinary sorption is due to the chemical reaction (7.2) in the course of which 1 g LiCl absorbs 2.26 g methanol!

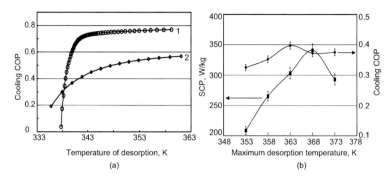

Figure 16.8 (a) Cooling COP for composite LiCl (20%)/SiO$_2$ (1) and commercial activated carbon AC-35 (2), calculated at T_L = 10°C and T_c = T_a = T_M = 35°C; (b) experimental specific cooling power and cooling COP for LiCl (20%)/SiO$_2$ versus desorption temperature (T_L = 10°C and T_c = T_a = T_M = 30°C). Reprinted with permission from Ref. [38], Copyright 2007, American Chemical Society.

Testing this composite in a lab-scale chiller in the ITAE-CNR results in somewhat lower, but still quite good, output parameters, namely, COP = 0.4 and SCP = 300–350 W/kg at the desorption temperature 85–95°C (Fig. 16.8b) [38]. The total sorbent loading is 0.62 kg. A portion of the tested sorbent is extracted from the chiller after 100 cycles and compared with the fresh sorbent. It is demonstrated that the porous structure, phase composition, and sorption properties of the composite remained stable after the cycling, which allows its good durability upon further cycling to be expected. Hence, this composite can be efficiently used for compact adsorption chiller driven by low-grade heat.

Due to the extremely steep sorption isobars and large methanol uptake, this composite was found to be promising for a special version of AC cycle driven by heat from a cooling jacket of cars [39].

A similar composite LiCl/silica, but with smaller pores of 8 nm, was tested in a finned-tube adsorption chiller with heat and mass recovery [40]. The total adsorbent loading was 40 kg. Quite a similar COP of 0.41 and SCP of 250 W/kg were obtained at T_H = 85°C, T_L = 15°C, and T_M = 30°C. The temperature sufficient for the composite regeneration (80–85°C) is lower than for the silica gel–water chiller

(85–90°C). This gives one more confirmation that the LiCl/silica composite is very attractive for AC driven by low-temperature heat.

16.2.3 Composite Sorbents of Ammonia

It sounds incredible, but the first demonstration of solid sorption refrigeration was made as early as in 1848 by Faraday [41], who presented a cooling unit based on ammonia sorption by AgCl. Nowadays, carbons are the most popular ammonia adsorbents, which are commercially available in granular, extruded, and consolidated forms as well as fibers, felts, and cloths. A comprehensive comparison of 26 activated carbon–ammonia pairs was performed in Ref. [42] for three types of AC cycles (single bed, two-bed, and infinite number of beds) at typical conditions for ice making, air conditioning, and heat pumping applications. That analysis revealed that both cooling and heating COPs are not very sensitive to the nature of carbon, being quite close, especially for single- and two-bed configurations. Thus, although a variety of commercial carbons are available, the variation in their adsorption properties is not sufficient to satisfy the demands of an ample variety of AC cycles. Additionally, the driving temperature between 100°C and 200°C is commonly required; therefore, low-temperature heat sources cannot be applied.

Application of composites "salt/matrix" was suggested to solve or at least smoothen these problems: Spinner et al. [43] proposed to mix calcium chloride with expanded graphite and then make a monolith block with advanced heat conductivity. Vasiliev et al. [44, 45] studied mixtures of metallic chlorides impregnated into active carbon fibers. New composite sorbents of *ammonia* composed of various salts ($CaCl_2$, $BaCl_2$, etc.) confined to porous inorganic oxides were synthesized in BIC (Chapter 8).

A composite based on $BaCl_2$ impregnated into expanded vermiculite was specifically developed for AC applications. This salt adsorbs ammonia in one stage with the formation of the ammonia complex compound $BaCl_2 \cdot 8NH_3$ (reaction 8.9). This results in a large mass of ammonia to be sorbed, which is important for adsorptive chilling. The thermodynamic cycle of an AC based on the "ammonia–$BaCl_2$/vermiculite" working pair is plotted on the basis of equilibrium data obtained in Refs. [46, 47] (Fig. 16.9a). It is worthy to remind that the vermiculite does not affect the equilibrium between ammonia and the salt (Fig. 8.3). The experimental equilibrium pressure over the composite is in good agreement with the pressure over a bulk BaCl$_2$ calculated according to the data presented in Ref. [48]. The theoretical estimation of COP for this cycle showed a quite high value of 0.6 [49], which encouraged us to take the next step toward evaluating the real performance of this composite.

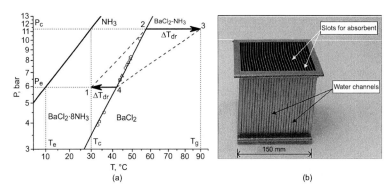

Figure 16.9 (a) The thermodynamic cycle of an adsorption chiller based on the ammonia–BaCl$_2$/vermiculite working pair plotted as ln$P$ versus 1/T on the basis of equilibrium data from Ref. [47]; (b) plate heat exchanger generator designed at the University of Warwick (4 mm of the sorbent layer). Reprinted from Ref. [49], Copyright 2010, with permission from Elsevier.

This composite was tested at the University of Warwick (UK) in a lab-scale prototype (Fig. 16.9b) under typical AC conditions [49]. The total mass of the composite loaded inside the generator is 570 g, and the grain size is 1–2 mm. It is demonstrated that this sorbent can provide effective operation of the chiller using a low potential heat source (80–90°C), giving the COP as high as 0.54 and SCP = 300–680 W/kg, which is quite interesting for practice.

The composite sorbent of ammonia CaCl$_2$/expanded graphite [50] was selected for adsorption ice maker to operate on fishing boats [51]. The unit provided the specific cooling power of 770 W/kg and a COP of 0.39, when the evaporation temperature was as low as −20°C.

Many other composite sorbents of ammonia were suggested and studied: compressed blocks of expanded graphite with various salts [43, 52], composites based on activated carbon fiber Busofit with CaCl$_2$ [44, 45], NaBr/expanded graphite [53], LiCl/expanded graphite [54], CoCl$_2$/ACF [55], BaCl$_2$/expanded graphite [56], CaCl$_2$/

expanded graphite [57], and others (see also Ref. [5]). Sorption properties of these composites significantly differ depending on the salt and matrix, which allows their fitting to particular AC cycles.

16.3 Desiccant Cooling

Desiccant cooling (DC) systems are basically open-cycle systems that use water vapor as a refrigerant. The thermally driven cooling cycle is a combination of evaporative cooling with air dehumidification by a solid desiccant. The first patent on a rotary desiccant air-conditioning cycle was introduced in 1955 [58].

The common DC technology uses rotary desiccant wheels. The main components of a solid DC system are shown in Fig. 16.10. It provides a conditioned air as follows. A warm and humid air enters the slowly rotating desiccant wheel and is dehumidified by adsorption of water (stage 1–2). Since the air is heated up by the adsorption heat, a heat recovery wheel is passed (2–3), resulting in a significant pre-cooling of the supply air stream. Subsequently, the air is humidified and thus further cooled by a controlled humidifier (3–4) according to the set values of supply air temperature and humidity. The exhaust air stream of the rooms is humidified (6–7) close to the saturation point to exploit the full cooling potential in order to allow an effective heat recovery (7–8). Finally, the adsorption wheel has to be regenerated (9–10) by applying heat in a comparatively low-temperature range from 50 to 75°C and to allow a continuous operation of the dehumidification process.

Typical adsorbents employed in open-cycle desiccant cooling systems are microporous silica gels, molecular sieves, and alumina [59–61]. Nowadays, CSPMs are used in desiccant cooling systems along with the conventional adsorbents. Jia et al. [62] developed a novel composite desiccant material based on silica gel and LiCl for a honeycombed rotary wheel. Experimental comparison of this new desiccant wheel and a conventional one using silica gel showed that the moisture removal capacity of the former wheel was larger by 20–50% [63]. Thus, a highly hygroscopic salt LiCl embedded in the silica pores significantly improves both the moisture sorption capacity and the DC performance. The new desiccant wheel can be driven by a lower regeneration temperature for acquiring the same amount of moisture removal.

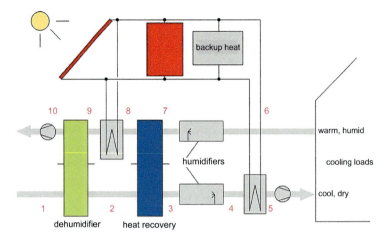

Figure 16.10 Schematic of a desiccant cooling system (see explanation in the text).

16.4 Adsorption Heat Storage

Adsorption heat storage (AHS) can improve the efficiency of energy units by ensuring a temporal coherence between heat production and consumption. The AHS is promising for heat storage due to its higher heat storage capacities when compared to sensible or latent heat storage materials [64–67]. The AHS can operate as closed (evacuated) [13] and open (coupled to the ambient) [66, 68] cycles. Although the thermodynamics is similar, an open AHS cycle is considered on a psychrometric chart of humid air (not presented) as described in Ref. [66]. Practical implementations of the AHS open and closed systems are quite different as well.

The density A of AHS is considered to lie between 100 and 500 kWh/m^3 or (0.36–1.8) GJ/m^3, being larger than that of sensible (c.a. 50 kWh/m^3) and latent (c.a. 100 kWh/m^3) heat storage [65, 69]. At the annual heating demand of 25 kWh/m^2, typical for low-energy buildings in Europe, a 100 m^2 house consumes for heating about 9 GJ per year. These heat needs can be, in principle, covered by solar energy; for instance, in the UK, the amount of solar radiation incident on a correctly orientated roof of a typical house exceeds its energy consumption during a year [70]. Accordingly, efficient interseasonal heat storage is necessary in this case. The volume of a heat

Adsorption Heat Storage | **333**

storage material is reasonable (2–10 m³) only if the storage density is above 1 GJ/m³ (270 kWh/m³). In actual practice, the volume of the heat storage unit is larger than the material volume itself by a factor of 2–3 due to the volumes of HExs, evaporator/condenser, pumps/fans, pipes, etc.

For AHS, the amount of heat needed for complete water desorption or the maximal heat storage density A_{max} is

$$A_{max}^{mas} = \Delta H_{des} w_{max} \text{ (per 1 kg dry adsorbent)} \qquad (16.8)$$

or

$$A_{max}^{vol} = \Delta H_{des} w_{max} \rho \text{ (per 1 m}^3 \text{ adsorbent)} \qquad (16.9)$$

where ΔH_{des} is the average heat of water desorption (in kJ/kg), w_{max} is the maximal mass of water adsorbed per 1 kg adsorbent, and ρ is the adsorbent apparent density. These densities are displayed in Table 16.4 [71] for several adsorbents tested for AHS (common zeolites, CSPMs, FAMs, MOFs). The CSPMs based on $CaCl_2$ and $LiNO_3$ confined to the expanded vermiculite are far superior to other adsorbents and allow a A_{max}^{mas} value of 1.7–4.2 MJ/kg to be reached.

Table 16.4 Maximal mass of water adsorbed w_{max}, average desorption heat ΔH_{des}, maximal heat storage densities A_{max}^{mas} and A_{max}^{vol} (calculated by Eqs. (16.8) and (16.9))

Adsorbent	w_{max} (g/g)	ΔH_{des} (MJ/kg)	A_{max}^{mas} (MJ/kg)	A_{max}^{vol} (GJ/m³)	Reference
Silica gel Fuji RD	0.4	2.40	0.96	0.77	[72]
Zeolite 13X	0.34	3.8	1.29	0.83	[73]
Zeolite 4A	0.22	3.05	0.67	0.49	[74]
Zeolite MgX	0.45	2.68	1.21	—	[75]
SWS-1L	0.65	2.65	1.72	1.55	[76]
SWS-9V	1.80	2.30	4.15	1.16	[77]
SWS-1V	1.80	2.35	4.2	1.25	[78]
AQSOA-Z02	0.33	3.25	1.07	0.55	[79]
AlPO-Tric	0.31	3.17	0.98	—	[80]
MIL-101	1.40	1.83	2.57	—	[81]
MIL-125NH$_2$	0.47	2.85	1.33	0.39	[82]
MOF-841	0.48	3.05	1.47	—	[83]

Source: Ref. [71].

Under real conditions, the storage density is lower than the maximal one calculated by Eqs. (16.8) and (16.9) as only a part of the sorbed water is involved in the storage process. To realize how much water can be involved, we consider a model 3T cycle (Fig. 3.3a) with $T_L = 10°C$, $T_M = 35°C$, and $T_H = 90°C$, and evaluate the mass of water exchanged along the cycle Δw and appropriate cycle heat storage density A_{max}^{vol}, which appears to be lower than the maximal one displayed in Table 16.4 by a factor of 2–12 (Table 16.5). The significant reduction in the density is especially typical for zeolites, which need a regeneration temperature much higher than the set temperature $T_H = 90°C$. It confirms that a high affinity of the adsorbent and the adsorbate for each other or/and a large total sorption capacity are not important of themselves. The heat storage density is defined by the amount of adsorbate exchanged within the particular AHS cycle, which means under certain conditions of heat storage process [7]. As before, the CSPMs are among the best materials for adsorptive storage of low-temperature heat.

Table 16.5 Mass of water exchanged Δw and appropriate heat storage densities A_{max} for a model AHS cycle ($T_L = 10°C$, $T_M = 35°C$, and $T_H = 90°C$)

Adsorbent	Δw (g/g)	A_{max}^{mas}	A_{max}^{vol}	Reference
		(MJ/kg)	(GJ/m^3)	
Silica gel Fuji RD	0.12	0.29	0.23	[84]
Zeolite 13X	0.03	0.11	0.07	[73]
SWS-1L	0.17	0.45	0.32	[76]
SWS-9V*	0.43	1.15	0.45	[77]
SWS-1V	0.33	0.76	0.21	[78]
AQSOA-Z02	0.19	0.62	0.31	[85]
AQSOA-Z02**	0.26	0.84	0.42	[86]
MIL-125NH$_2$	0.21	0.64	0.19	[83]

*$T_L = 10°C$, $T_M = 30°C$, and $T_H = 70°C$
**$T_L = 15°C$, $T_M = 28°C$, and $T_H = 90°C$
Source: Ref. [71].

Systematic experimental study of the CSPMs for AHS was performed in Refs. [87, 88]. The authors tested various hygroscopic salts ($CaCl_2$, $MgSO_4$, $Ca(NO_3)_2$, $LiNO_3$, and $LiBr$) and host matrices (silica gel, zeolite, activated carbon, and mineral vermiculite) to obtain their combinations promising for open AHS systems and studied them by a variety of physicochemical methods (BET, BJH, TG, DSC, SEM, EDX, etc). It is found that the sorption capacity of all composite materials is significantly higher (up to 1.9 g/g) compared to their raw host matrices alone, suggesting that the addition of the salt is beneficial for moisture and, hence, heat storage. It is demonstrated that the pore structure may get damaged for non-vermiculite matrices; however, no damage was observed in the vermiculite as a host. Vermiculite with either lithium bromide or calcium chloride appears to have significantly larger AHS potential when compared to both the raw matrices and the other composites [87, 88].

Liu et al. [89] used another cheap mineral material as a host porous matrix—Wakkanai siliceous shale (WSS). WSS is a type of siliceous mudstone that is widely distributed in Wakkanai, the northern part of Hokkaido (Japan). It is a natural mesoporous material composed of silicon dioxide (SiO_2) with pores of 5–40 nm size. A new AHS composite was formed by inserting LiCl into the WSS pores. To overcome the problems of low heat and mass transfer and high hydrodynamic resistance in an open AHS system, WSS was formed by adding a binder into a honeycomb ceramic element of 10 cm × 10 cm × 20 cm with 36 cells/cm^2 and a cell wall thickness of 0.28 mm (Fig. 16.11a). The composite containing 9.6 wt.% LiCl showed a good potential for storing heat with a volumetric heat storage density of 0.180 GJ/m^3 and could supply air heated up to 53°C. No material degradation was observed during at least 250 cycles at a quite low desorption temperature of 60°C.

An open-type AHS setup equipped with 40 kg of composite sorbent $CaCl_2$/(mesoporous silica gel) was tested as a batch adsorber in Refs. [90, 91]. The concentration of the $CaCl_2$ impregnating solution was found to be an important factor affecting the storage properties of the composites. The composite prepared by impregnating silica gel with a 30 wt.% $CaCl_2$ solution showed a high storage capacity of 1.02 MJ/kg (0.81 GJ/m^3), which remained stable after 500 consequent sorption/desorption cycles. The heat discharging temperature varied from 47°C to 30°C and the sorbent can be charged at temperature 90°C or below. Mathematical modeling of the AHS process predicted that the specific heat storage capacity can be increased up to 1.35 MJ/kg at a regeneration temperature of 100°C [92].

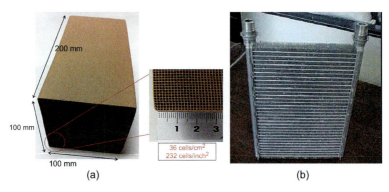

Figure 16.11 Various configurations of "adsorber–heat exchanger" (Ad–HEx) unit: open systems—a honeycomb ceramic element made of WSS [89] (a); closed systems—overall view of the Ad–HEx unit (b) [26]. Figure (a) reprinted from Ref. [89], Copyright 2015, and Figure (b) reprinted from Ref. [26], Copyright 2012, with permission from Elsevier.

SWS-type materials were successfully used also in closed AHS systems [77]. For instance, lithium nitrate was introduced into silica gel [26] and vermiculite [77] matrices and tested in the typical AHS cycle (Fig. 16.7b). LiNO$_3$ is a salt that binds water less strongly than calcium chloride, lithium bromide, or calcium nitrate. Therefore, this composite sorbent was specifically developed to operate at low regeneration temperature (<65–70°C). The sorbent particles were embedded inside a compact heat exchanger of a finned flat-tube type (Fig. 16.11b). The composite LiNO$_3$/vermiculite (SWS-9V) exchanges 0.43 g H$_2$O per gram of sorbent, which corresponds to a heat storage capacity of 1.15 MJ/kg or 0.45 GJ/m^3 related to the *total volume* of the Ad–HEx unit.

This brief overview of the current practice in AHS shows that the CSPMs have a great potential to obtain a high AHS density and to reduce an AHS system size as compared to the other heat storage materials. The main advantages of these composites are as follows:

- Low charging temperature (solar energy absorbed by flat receivers can be stored).
- Large storage capacity.
- Possibility to controllably modify their sorption properties to match conditions of particular heat storage cycles (e.g., variation of the salt and matrix nature, etc.).

The main shortcoming is a relatively low-temperature lift. It can be enhanced by applying salts with higher affinity for water vapor at the expense of an appropriate increase in charging temperature.

The use of CSPMs for regeneration of heat and moisture in ventilation systems will be separately considered in Chapter 18.

References

1. Meunier, F. (2001) Adsorptive cooling: A clean technology, *Clean Prod. Processes*, **3**, pp. 8–20.

2. Yonezawa, Y., Matsushita, M., Oku, K., Nakano, H., Okumura, S., Yoshihara, M., Sakai, A., and Morikawa, A. (1989) Adsorption Refrigeration System, US Patent No. 4881376.

3. Mycom. Technical brochure; http://mayekawa.elitetopdown.net/Descargas/The%20 World%20of%20MYCOM.pdf.

4. Jakob, U. and Kohlenbach, P. (2010) Recent developments of sorption chillers in Europe, *IIR Bulletin*, pp. 34–40.

5. Gordeeva, L. G. and Aristov, Yu. I. (2012) Composites "salt inside porous matrix" for adsorption heat transformation: A current state of the art and new trends, *Int. J. Low Carbon Tech.*, **7**, pp. 288–302.

6. Aristov, Yu. I. (2007) Novel materials for adsorptive heat pumping and storage: Screening and nanotailoring of sorption properties (review), *J. Chem. Eng. Jpn.*, **40**, pp. 1139–1153.

7. Aristov, Yu. I. (2014) Concept of adsorbent optimal for adsorptive cooling/ heating, *Appl. Therm. Engn.*, **72**, pp. 166–175.

8. Dubinin, M. M. (1960) Theory of physical adsorption of gases and vapour and adsorption properties of adsorbents of various natures and porous structures, *Bull. Division Chem. Soc.*, **9**, pp. 1072–1078.

9. Aristov, Yu. I., Sharonov, V. E., and Tokarev, M. M. (2008) Universal relation between the boundary temperatures of a basic cycle of sorption heat machines, *Chem. Eng. Sci.*, **63**, pp. 2907–2912.

10. Pons, M., Meunier, F., Cacciola, G., Critoph, R., Groll, M., Puigjaner, L., Spinner, B., and Ziegler, F. (1999) Thermodynamic based comparison of sorption systems for cooling and heat pumping, *Int. J. Refrig.*, **22**, pp. 5–17.

11. Meunier, F., Poyelle, F., and LeVan, M. D. (1997) Second-law analysis of adsorptive refrigeration cycles: The role of thermal coupling entropy production, *Appl. Therm. Eng.*, **17**, pp. 43–55.

12. Alefeld, G. and Radermacher, R. (1994) *Heat Conversion Systems* (CRC Press, Boca Raton).

13. Raldow, W. M. and Wentworth, W. E. (1979) Chemical heat pumps: A basic thermodynamic analysis, *Sol. Energy*, **23**, pp. 75–79.

14. Sharonov, V. E. and Aristov, Yu. I. (2008) Chemical and adsorption heat pumps: Comments on the second law efficiency, *Chem. Eng. J.*, **136**, pp. 419–424.

15. Meunier, F. (1985) Second law analysis of a solid adsorption heat pump operating on reversible cascade cycles, *Heat Recovery Syst. CHP*, **5**, pp. 133–141.

16. Shelton, S. V. (1986) Solid adsorbent heat pump system, US Patent No. 4610148.

17. Aristov, Yu. I. (2009) Optimum adsorbent for adsorptive heat transformers: Dynamic considerations, *Int. J. Refrig.*, **32**, pp. 675–686.

18. Aristov, Yu. I. (2013) Experimental and numerical study of adsorptive chiller dynamics: A loose grains configuration, *Appl. Therm. Eng.*, **61**, pp. 841–847.

19. Aristov, Yu. I., Restuccia, G., Cacciola, G., and Parmon, V. N. (2002) A family of new working materials for solid sorption air conditioning systems, *Appl. Therm. Eng.*, **22**, pp. 191–204.

20. Restuccia, G., Freni, A., Vasta, S., and Aristov, Yu. I. (2004) Selective water sorbents for solid sorption chiller: Experimental results and modelling, *Int. J. Refrig.*, **27**, pp. 284–293.

21. Aristov, Yu. I., Tokarev, M. M., Cacciola, G., and Restuccia, G. (1996) Selective water sorbents for multiple applications: 1. $CaCl_2$ confined in mesopores of the silica gel: Sorption properties, *React. Kinet. Cat. Lett.*, **59**, pp. 325–334.

22. Saha, B. B., Chakraborty, A., Koyama, S., and Aristov, Yu. I. (2009) A new generation cooling device employing $CaCl_2$-in-silica gel-water system, *Int. J. Heat Mass Transfer*, **52**, pp. 516–524.

23. Boelman, E. C., Saha, B. B., and Kashiwagi, T. (1995) Experimental investigation of a silica gel–water adsorption refrigeration cycle: The influence of operating conditions on cooling output and COP, *ASHRAE Trans. Res.*, **101**(2), pp. 358–366.

24. San, J. Y. and Hsu, H. C. (2009) Performance of a multi-bed adsorption heat pump using SWS-1L composite adsorbent and water as the working pair, *Appl. Therm. Eng.*, **29**, pp. 1606–1613.

25. Simonova, I. A. and Aristov, Yu. I. (2008) Composite sorbents "lithium nitrate in silica" for transformation of low temperature heat, *Altern. Energy Ecol.*, **11**, pp. 95–99.

26. Aristov, Yu. I., Sapienza, A., Freni, A., Ovoschnikov, D. S., and Restuccia, G. (2012) Reallocation of adsorption and desorption times for optimization of cooling cycles, *Int. J. Refrig.*, **35**, pp. 525–531.

27. Sapienza, A., Glaznev, I. S., Santamaria, S., Freni, A., and Aristov, Yu. I. (2012) Adsorption chilling driven by low temperature heat: New adsorbent and cycle optimization, *Appl. Therm. Eng.*, **32**, pp. 141–146.

28. Freni, A., Sapienza, A., Glaznev, I. S., Aristov, Yu. I., and Restuccia, G. (2012) Testing of a compact adsorbent bed based on the selective water sorbent "silica modified by calcium nitrate," *Int. J. Refrig.*, **35**, pp. 518–524.

29. Gong, L. X., Wang, R. Z., Xia, Z. Z., and Chen, C. J. (2011) Design and performance prediction of a new generation adsorption chiller using composite adsorbent, *Energy Convers. Manag.*, **52**, pp. 2345–2350.

30. Gordeeva, L. G., Resticcia, G., Cacciola, G., and Aristov, Yu. I. (1998) Selective water sorbents for multiple applications: 5. LiBr confined in mesopores of silica gel: Sorption properties, *React. Kinet. Cat. Lett.*, **63**, pp. 81–88.

31. Dawoud, B. (2007) A hybrid solar-assisted adsorption cooling unit for vaccine storage, *Renew. Energy*, **32**, pp. 947–964.

32. Aristov, Yu. I. (2013) Challenging offers of material science for adsorption heat transformation: A review, *Appl. Therm. Eng.*, **50**, pp. 1610–1618.

33. Wang, L. W., Wang, R. Z., and Oliveira, R. G. (2009). A review on adsorption working pairs for refrigeration, *Renew. Sustain. Energy Rev.*, **13**(3), pp. 518–534.

34. Attan, D., Alghoul, M. A., Saha, B. B., Assadeq, J., and Sopian, K. (2011) The role of activated carbon fiber in adsorption cooling cycles, *Renew. Sustain. Energy Rev.*, **15**, pp. 1708–1721.

35. Henninger, S. K., Schicktanz, M., Hugenell, P. P. C., Sievers, H., and Henning, H.-M. (2012) Evaluation of methanol adsorption on activated carbons for thermally driven chillers part I: Thermophysical characterization, *Int. J. Refrig.*, **35**, pp. 543–553.

36. Schicktanz, M., Hugenell, P. P. C., and Henninger, S. K. (2012) Evaluation of methanol adsorption on activated carbons for thermally driven chillers, part II: The energy balance model, *Int. J. Refrig.*, **35**, pp. 554–561.

37. Gordeeva, L., Freni, A., Restuccia, G., and Aristov, Yu. I. (2007) Methanol sorbents "salt inside mesoporous silica": The screening of salts for adsorptive air conditioning driven by low temperature heat, *Ind. Eng. Chem. Res.*, **46**, pp. 2747–2752.

38. Gordeeva, L. G., Freni, A., Restuccia, G., and Aristov, Yu. I. (2007) Influence of characteristics of methanol sorbents "salt in mesoporous silica" on the performance of adsorptive air conditioning cycle, *Ind. Eng. Chem. Res.*, 46, pp. 2747–2752.

39. Gordeeva, L. G., Freni, A., Restuccia, G., and Aristov, Yu. I. (2007) Adsorptive air conditioning systems driven by low temperature heat: Choice of the working pairs, *J. Chem. Eng. Jpn.*, **40**, pp. 1287–1291.

40. Gong, L. X., Wang, R. Z., Xia, Z. Z., and Lu, Z. S. (2012) Experimental study on an adsorption chiller employing lithium chloride in silica gel and methanol, *Int. J. Refrig.*, **35**, pp. 1950–1957.

41. Critoph, R. E. and Zhong, Y. (2005) Review of trends in solid sorption refrigeration and heat pumping technology, *Proc. Instit. Mech. Eng. E*, **219**, pp. 285–300.

42. Tamainot-Telto, Z., Metcalf, S. J., Critoph, R. E., Zhong, Y., and Thorpe, R. (2009) Carbon–ammonia pairs for adsorption refrigeration applications: Ice making, air conditioning and heat pumping, *Int. J. Refrig.*, **32**, pp. 1212–1229.

43. Spinner, B. (1993) Ammonia-based thermochemical transformers, *Heat Recov. Syst. CHP*, **13**, pp. 301–307.

44. Vasiliev, L. L., Kanonchik, L., Antujh, A. A., and Kulakov, A. G. (1996) NaX, carbon fibre and $CaCl_2$ ammonia reactors for heat pumps and refrigerators, *Adsorption, 2*, pp. 311–316.

45. Vasiliev, L., Nikanpour, D., Antukh, A., Snelson, K., Vasil'ev, L., and Lebru, A. (1999) Multisalt-carbon chemical cooler for space applications, *J. Eng. Phys. Thermophys.*, **72**, pp. 595–600.

46. Sharonov, V. E., Veselovskaya, J. V., and Aristov, Yu. I. (2006) Ammonia sorption on composites "$CaCl_2$ in inorganic host matrix": Isosteric chart and its performance, *Int. J. Low Carbon Tech.*, **1**(3), pp. 191–200.

47. Veselovskaya, J. V., Tokarev, M. M., and Aristov, Yu. I. (2010) Novel ammonia sorbents "porous matrix modified by active salt" for adsorptive heat transformation: 1. Barium chloride in various matrices, *Appl. Therm. Eng.*, **30**, pp. 584–589.

48. Touzain, Ph. (1999) Thermodynamic values of ammonia-salts reactions for chemical sorption heat pumps, *Proc. Int. Sorption Heat Pump Conf.*, Munich, Germany, March 24–26, 1999, pp. 225–238.

49. Veselovskaya, J. V., Critoph, R. E., Thorpe, R. N., Metcalf, S., Tokarev, M. M., and Aristov, Yu. I. (2010) Novel ammonia sorbents "porous matrix modified by active salt" for adsorptive heat transformation: 3. Testing of "BaCl$_2$/vermiculite" composite in the lab-scale adsorption chiller, *Appl. Therm. Eng.*, **30**, pp. 1188–1192.

50. Lu, Z. S., Wang, R. Z., Wu, J. Y., and Chen, C. J. (2006) Performance analysis of an adsorption refrigerator using activated carbon in a compound adsorbent, *Carbon*, 44, pp. 747–752.

51. Wang, R. Z. and Oliveira, R. G. (2006) Adsorption refrigeration: An efficient way to make good use of waste heat and solar energy, *Prog. Energy Combust. Sci.*, **32**, pp. 424–458.

52. Coste, C., Mauran, S., and Crozat, G. (1986) Gaseous–solid reaction. US Patent No. 4,595,774.

53. Oliveira, R. G., Wang, R. Z., Kiplagat, J. K., and Wang, C. Y. (2009) Novel composite sorbent for resorption systems and for chemisorption air conditioners driven by low generation temperature, *Renew. Energy*, **34**, pp. 2757–2764.

54. Kiplagat, J. K., Wang, R. Z., Oliveira, R. G., and Li, T. X. (2010) Lithium chloride–expanded graphite composite sorbent for solar powered ice maker, *Solar Energy*, **84**, pp. 1587–1594.

55. Aidoun, Z. and Ternan, M. (2002) Salt impregnated carbon fibres as the reactive medium in a chemical heat pump: The NH$_3$–CoCl$_2$ system, *Appl. Therm. Eng.*, **22**, pp. 1163–1173.

56. Li, T. X., Wang, R. Z., Wang, L. W., and Kiplagat, J. K. (2009) Study on the heat transfer and sorption characteristics of a consolidated composite sorbent for solar-powered thermochemical cooling systems, *Solar Energy*, **83**, pp. 1742–1755.

57. Wang, K., Wu, J. Y., Wang, R. Z., and Wang, L. W. (2006) Effective thermal conductivity of expanded graphite–CaCl$_2$ composite adsorbent for chemical adsorption chillers, *Energy Convers. Manag.*, **47**, pp. 1902–1912.

58. Pennington, N. A. (1955) Humidity changer for air conditioning, USA, Patent No. 2,700,537.

59. Halliday, S. P., Beggs, C. B., and Sleigh, P. A. (2002) The use of solar desiccant cooling in the UK: A feasibility study, *Appl. Therm. Eng.*, **22**, pp. 1327–1338.

60. Sumathy, K., Yeung, K. H., and Yong L. (2003) Technology development in the solar adsorption refrigeration systems, *Prog. Energy Combust. Sci.*, **29**, pp. 301–327.

61. Liang, C. H., Zhang, L. Z., and Pei, L. X. (2010) Performance analysis of a direct expansion air dehumidification system combined with membrane-based total heat recovery, *Energy*, **35**, pp. 3891–3901.

62. Jia, C. X., Dai, Y. J., Wu, J. Y., and Wang, R. Z. (2007) Use of compound desiccant to develop high performance desiccant cooling system, *Int. J. Refrig.*, **30**, pp. 345–353.

63. Jia, C. X., Dai, Y. J., Wu, J. Y., and Wang, R. Z. (2006) Experimental comparison of two honeycombed desiccant wheels fabricated with silica gel and composite desiccant material, *Energy Convers. Manag.*, **47**, pp. 2523–2534.

64. Yu, N., Wang, R. Z., and Wang, L. W. (2013) Sorption thermal storage for solar energy, *Prog. Energy Combust. Sci.*, **39**, pp. 489–514.

65. N'Tsoukpoe, K. E., Liu, H., Le Pierrès, N., and Luo, L. (2009) A review on long-term sorption solar energy storage, *Renew. Sustain. Energy Rev.*, **13**, pp. 2385–2396.

66. Hauer, A. (2007) Evaluation of adsorbent materials for heat pump and thermal energy storage applications in open systems, *Adsorption*, **13**, pp. 399–405.

67. Aydin, D., Casey, S. P., and Riffat, S. (2015) The latest advancements on thermochemical heat storage systems, *Renew. Sustain. Energy Rev.*, **41**, pp. 356–367.

68. Close, D. and Dunkle, R. (1977) Adsorbent beds for energy-storage in drying of heating systems, *Solar Energy*, **19**, pp. 233–238.

69. Aristov, Yu. I., Restuccia, G., Tokarev, M. M., and Cacciola, G. (2000) Selective water sorbents for multiple applications: 10. Energy storage ability, *React. Kinet. Cat. Lett.*, **69**, pp. 345–353.

70. CIBSE, Guide A - Environmental Design, 7th Edition, Chartered Institute of Building Service Engineers, London, 2006.

71. Aristov, Yu. I. (2015) Progress in adsorption technologies for low-energy buildings (review), *Future Cities Environ.*, **1**, doi:10.1186/s40984-015-0011-x.

72. Ng, K. S., Chua, H. T., Chung, C. Y., Loke, C. H., Kashiwaga, T., Akisawa, A., and Saha, B. B. (2001) Experimental investigation of the silica gel–water adsorption isotherm characteristics, *Appl. Therm. Eng.*, **21**, pp. 1631–1642.

73. Dubinin, M. M. (1966) Adsorption equilibrium of water on NaX zeolite, *Coll. Czech Chem. Commun.*, **31**, pp. 406–414.

74. Gorbach, A., Stegmaier, M., and Eigenberger, G. (2004) Measurement and modeling of water vapor adsorption on zeolite 4A: Equilibria and kinetics, *Adsorption*, **10**, pp. 29–46.

75. Stach, H., Mugele, J., Jaenchen, J., and Weiler, H. (2005) Influence of cycle temperatures on the thermochemical heat storage densities in the systems water/microporous and water/mesoporous adsorbents, *Adsorption*, **11**, pp. 393–404.

76. Aristov, Yu. I., Tokarev, M. M., Cacciola, G., and Restuccia, G. (1996) Selective water sorbents for multiple applications: 1. $CaCl_2$ confined in mesopores of the silica gel: Sorption properties, *React. Kinet. Cat. Lett.*, **59**, pp. 325–334.

77. Sapienza, A., Glaznev, I. S., Santamaria, S., Freni, A., and Aristov, Yu. I. (2012) Adsorption chilling driven by low temperature heat: New adsorbent and cycle optimization, *Appl. Therm. Eng.*, **32**, pp. 141–146.

78. Aristov, Yu. I., Restuccia, G., Tokarev, M. M., Buerger, H.-D., and Freni, A. (2000) Selective water sorbents for multiple applications. 11. $CaCl_2$ confined to expanded vermiculite, *React. Kinet. Cat. Lett.*, **71**, pp. 377–384.

79. Goldsworthy, M. J. (2014) Measurements of water vapour sorption isotherms for RD silica gel, AQSOA-Z01, AQSOA-Z02, AQSOA-Z05 and CECA zeolite 3A, *Micropor. Mesopor. Mater.*, **196**, pp. 59–67.

80. Ristic, A., Zabukovec Logar, N., Henninger, S. K., and Kaucic, V. (2012) The performance of small-pore micro-porous aluminophosphates in low-temperature solar energy storage: The structure–property relationship, *Adv. Funct. Mater.*, **22**, pp. 1952–1957.

81. Akiyama, G., Matsuda, R., Sato, H., Hori, A., Takata, M., and Kitagawa, S. (2012) Effect of functional groups in MIL-101 on water sorption behaviour, *Micropor. Mesopor. Mater.*, **157**, pp. 89–93.

82. Gordeeva, L. G., Solovyeva, M. V., and Aristov, Yu. I. (2015) NH_2-MIL-125 as a promising material for adsorptive heat transformation and storage, *Energy*, **100**, pp. 18–24.

83. Furukawa, H., Gándara, F., Zhang, Y.-B., Jiang, J., Queen, W. L., Hudson, M. R., and Yaghi, O. M. (2014) Water adsorption in porous metal-organic frameworks and related materials, *J. Am. Chem. Soc.*, **136**, pp. 4369–4381.

84. Glaznev, I. S. and Aristov, Yu. I. (2010) The effect of cycle boundary conditions and adsorbent grain size on the water sorption dynamics in adsorption chillers, *Int. J. Heat Mass Transfer*, **53**, pp. 1893–1898.

85. Sapienza, A., Santamaria, S., Frazzica, A., Freni, A., and Aristov, Yu. I. (2014) Dynamic study of adsorbers by a new gravimetric version of the large temperature jump method, *Appl. Energy*, **113**, pp. 1244–1251.

86. Frazzica, A., Sapienza, A., and Freni, A. (2014) Novel experimental methodology for the characterization of thermodynamic performance of advanced working pairs for adsorptive heat transformers, *Appl. Therm. Eng.*, **72**, pp. 229–236.

87. Casey, S. P., Elvins, J., Riffat, S. B., and Robinson, A. (2014) Salt impregnated desiccant matrices for 'open' thermochemical energy storage: Selection, synthesis and characterisation of candidate materials, *Energy Buildings*, **84**, pp. 412–425.

88. Casey, S. P., Aydin, D., Riffat, S. B., and Elvins, J. (2015) Salt impregnated desiccant matrices for 'open' thermochemical energy storage: Hygrothermal cyclic behaviour and energetic analysis by physical experimentation, *Energy Buildings*, **92**, pp. 128–139.

89. Liu, H., Nagano, K., and Togawa, J. (2015) A composite material made of mesoporous siliceous shale impregnated with lithium chloride for an open sorption thermal energy storage system, *Solar Energy*, **111**, pp. 186–200.

90. Zhu, D., Wu, H., and Wang, S. (2006) Experimental study on composite silica gel supported $CaCl_2$ sorbent for low grade heat storage, *Int. J. Therm. Sci.*, **45**, pp. 804–813.

91. Wu, H., Wang, S., and Zhu, D. (2007) Effects of impregnating variables on dynamic sorption characteristics and storage properties of composite sorbent for solar heat storage, *Solar Energy*, **81**, pp. 864–871.

92. Wu, H., Wang, S., and Zhu, D. (2009) Numerical analysis and evaluation of an open-type thermal storage system using composite sorbents, *Int. J. Heat Mass Transfer*, **52**, pp. 5262–5265.

Chapter 17

Adsorptive Transformation of Heat: Pressure-Driven Cycles

The *pressure-driven* (PD) adsorptive heat transformation (AHT) cycles are much less common than the temperature-driven (TD) cycles and suggested mainly for the temperature upgrading mode (3 in Fig. 3.2) [1, 2]. This chapter carries out a preliminary examination of PD AHT because

- They can essentially differ from the TD cycles [3, 4];
- A new PD AHT cycle "Heat from Cold" (HeCol) has recently been suggested for upgrading the ambient heat in cold countries [2]. Composites "salt in porous matrix" (CSPMs) can be very promising for the practical realization of the new cycle [5–7].

Here we consider the basics of PD cycles and the first realization of the HeCol cycle by using "salt/matrix" composites. This study is supported by a current grant of the Russian Science Foundation (N 16-19-10259).

17.1 Pressure-Driven Adsorptive Cycles

Here a brief description of the isothermal and non-isothermal versions of PD AHT cycles is given. Their thermodynamic analysis is performed based on the first and second laws.

Nanocomposite Sorbents for Multiple Applications
Yu. I. Aristov
Copyright © 2020 Jenny Stanford Publishing Pte. Ltd.
ISBN 978-981-4267-50-2 (Hardcover), 978-981-4303-15-6 (eBook)
www.jennystanford.com

17.1.1 Thermodynamic Charts of a PD Cycle

17.1.1.1 Isothermal PD cycle

A common (isothermal) PD cycle consists of two isosters and two isotherms (Fig. 17.1a). The adsorbent is regenerated at constant temperature T_M by dropping the vapor pressure over the adsorbent (stage 4–1). It is somewhat similar to the pressure swing adsorption process used for gas drying and separation [8].

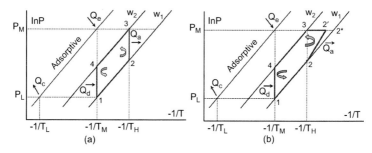

Figure 17.1 *P–T* diagram of the isothermal (a) and non-isothermal (b) 3*T* PD adsorptive cycles for heat upgrading. Reprinted from Ref. [2], Copyright 2017, with permission from Elsevier.

The initial adsorbent state (point 1 in Fig. 17.1a) corresponds to temperature T_M and the pressure of the adsorptive vapor, $P_L = P_0(T_L)$, where $P_0(T_L)$ is the saturation vapor pressure at temperature T_L. The weak isoster is characterized by the equilibrium adsorbate content $w_1 = w(T_M, P_L)$. Then the adsorbent is heated up to temperature T_H (stage 1–2) at constant uptake w_1. At point 2, the adsorber is connected with the evaporator maintained at T_M, which is the temperature of the external heat source whose temperature level is to be upgraded. The evaporator generates the constant pressure $P_M = P_0(T_M)$ of adsorptive. This pressure jump $P_2 \rightarrow P_M$ causes the vapor adsorption that leads to an increase in the equilibrium adsorbate content up to $w_2 = w(T_H, P_M)$ (point 3). The heat Q_e of adsobate evaporation is absorbed in the evaporator at $T = T_M$, and the useful adsorption heat Q_a is released at constant temperature T_H (isotherm 2–3) in the heating circuit of a consumer. Then the adsorber is disconnected from the evaporator and cooled down to the intermediate temperature T_M (isoster 3–4) at a constant uptake w_2. At point 4, the adsorber is connected with the condenser maintained at temperature T_L and pressure $P_L = P_0(T_L)$. This pressure drop $P_4 \rightarrow P_L$ results in the desorption of the adsorbate to restore the initial uptake $w_1 = w(T_M, P_L)$ (point 1). The heat Q_d needed for adsobate desorption is supplied to the adsorbent at temperature T_M from the external heat source (isotherm 4–1). The excess of adsorbate is collected in the condenser, releasing the heat Q_c to the ambience and the cycle is closed. Work is required to pump the liquid adsorbate from the low-pressure level in the condenser to the high pressure level in the evaporator. However, this work is small as compared with heats Q_a and Q_d.

This cycle was theoretically analyzed in Refs. [1, 9] for upgrading industrial waste heat. It has recently been suggested for increasing the temperature potential of the ambient heat to the level sufficient for heating dwellings (see Ref. [2] and Sec. 17.2).

17.1.1.2 Non-isothermal PD cycle

Heat can also be upgraded under non-isothermal mode as demonstrated in Fig. 17.1b: At point 2, the adsorber is linked with the evaporator maintained at T_M; the heat of adsorption is released and the adsorbent temperature increases up to $T(2^*) > T_H$, if all thermal masses are equal to zero. If the thermal masses are nonzero, the adsorbent temperature increases up to $T(2') < T(2^*)$, because a part of adsorption heat is spent for heating the adsorbent and heat exchanger (HEx). After that, the rest of the adsorption process $(2^*–3)$ or $(2'–3)$ is isobaric (at $P = P_M = $ const.) and proceeds until the equilibrium is set at point 3 corresponding to (T_{II}, P_M). This non-isothermal cycle allows supplying a consumer with a somewhat lower quantity of heat, but with a higher quality, namely, at the temperature between $T(2^*)$ or $T(2')$ and T_H (Fig. 17.1b), which is higher than in the isothermal cycle (Fig. 17.1a).

It is worthy to mention that both PD cycles presented in Fig. 17.1 degenerate into a common thermochemical cycle typical of monovariant gas–solid chemical reactions, if isosters 1–2 and 3–4 coincide as shown in Fig. 16.3a.

17.1.2 First Law Analysis

If we neglect all thermal masses, the energy balance of the PD cycle is $Q_d + Q_e = Q_a + Q_c$, and the *first law efficiency* can be formally determined as $\eta_{FLT} = Q_a/(Q_d + Q_e) \approx 0.5$. This value, however, has no practical meaning because the heats Q_d and Q_e are taken for free from an inexhaustible natural heat reservoir at low temperature T_M. On the contrary, the useful heat Q_a may have a commercial value if used for heating.

The useful heat can be evaluated as follows: $Q_a = \Delta H_a \cdot \Delta w$, where ΔH_a is the average heat of adsorption [J/g] and Δw is the specific mass of adsorbed working fluid [g/(g adsorbent)]. Therefore, from the first law of thermodynamics, the adsorbent *optimal for the HeCol cycle* has to have a high adsorption heat ΔH_a and exchange a large mass of working fluid between isosters 1–2 and 3–4.

Among the common adsorbates for AHT, water has the largest adsorption heat, but it cannot be used in the HeCol cycle because of its high freezing temperature (0°C) and very low saturation vapor pressure at the cycle temperatures (0.5–10 mbar). Therefore, the best adsorbates for the HeCol cycle are methanol, ammonia, and freons.

The latter demand can be quantitatively formulated in terms of the Dubinin adsorption potential $\Delta F(P,T) = -RT \ln[P/P_0(T)]$. Taking into account the Polanyi principle of temperature invariance [10] and considering the density of the adsorbed phase constant under the cycle operating conditions, the adsorbed mass $\Delta w = w_2 - w_1 = w(\Delta F_2) - w(\Delta F_1)$, where $\Delta F_1 = -RT_M \ln(P_L/P_M)$ corresponds to the weak isoster 1–2 and $\Delta F_2 = -RT_H \ln(P_M/P_H)$ corresponds to the rich isoster 3–4 (Fig. 17.1). For methanol as a working fluid, the Δw value has to be larger than 0.2–0.4 g/g [5].

17.1.3   Second Law Analysis

For an ideal 3T PD AHT system, the entropy balance (the second law) is described by Eq. (16.4). If all the processes are completely reversible, the entropy generation is equal to zero ($\Delta S = 0$), and the Carnot COP can be calculated for the heat amplification mode [11]:

$$\text{COP}^{\text{Carnot}} = (1/T_L - 1/T_M)/(1/T_L - 1/T_H) \qquad (17.1)$$

The function $\text{COP}^{\text{Carnot}}(T_L)$ is calculated at various temperatures T_M and the constant temperature lift $T_H - T_M = 30$ K (Fig. 17.2). The desorption/evaporation temperature T_M is selected quite low, between 0 and 30°C, keeping in mind conditions of the new HeCol cycle considered below (see Sec. 17.2). The estimated efficiency is quite encouraging, which demonstrates that such a PD AHT cycle can, in principle, be of practical interest.

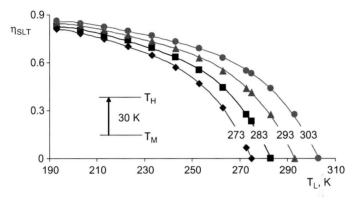

Figure 17.2 The Carnot efficiency for 3T PD AHT cycles as a function of the condenser temperature T_L at various evaporation/desorption temperatures T_M indicated near appropriate curves (in K) and $T_H - T_M$ = 30 K. Reprinted from Ref. [2], Copyright 2017, with permission from Elsevier.

It is worthy to mention that any AHT cycle driven by a finite pressure drop $\Delta P > 0$ is irreversible, and the entropy is inevitably generated at stage 4–1 due to the vapor expansion from P_4 to P_1. This generation is, however, much smaller than the thermal entropy production caused by the external thermal coupling in the TD AHT cycles [12]. This is because the mass of the gaseous adsorptive expanded during pressure equalization is very low as compared to the difference in the adsorbate mass at points 4 and 1 of the PD cycle [4]. Therefore, a PD AHT cycle is *more in equilibrium* than a TD AHT cycle working under the same boundary temperatures T_L, T_M, and T_H.

17.1.4 Adsorbent Optimal from Dynamic Point of View

The general methodology applied here to analyze which adsorbent is optimal for the PD AHT cycles is similar to that already applied for the TD AHT cycles (see Sec. 16.1.2.3 and Refs. [13–15]), because it is based on the fundamental principles of adsorption dynamics.

The desorption (regeneration) process starts at point 4 and is initiated by the pressure drop from the initial desorption pressure

P_4 to the final pressure $P_1 = P_o(T_L)$ at constant temperature T_M. At $t = 0$, right after this pressure drop, the driving force for heat transfer *equals zero* because the system is isothermal and there is no temperature gradient. At the same time, the driving force for mass transfer, the mechanism of which is vapor (adsorptive) diffusion, *is maximal*. Indeed, according to Fick's first law [16–18], the vapor diffusional flux is proportional to the concentration gradient dC/dr in the adsorbent grain. At constant temperature T, $C = P/RT$ and the flux is proportional to the pressure gradient dP/dr or, in the first approximation, to the current difference $\Delta P(t) = [P_{gr}(t) - P_L]$ between the vapor pressure inside and outside the grain. This difference is maximal at $t = 0$ and equals $\Delta P(0) = P_4 - P_1$. This pressure gradient causes diffusion of the adsorptive, occupying the pores, out of the grain. The pressure P inside the grain reduces, which, in turn, initiates desorption of the adsorbed vapor (adsorbate) from the grain surface. This process consumes heat and cools the grain, which creates a temperature gradient between the grain and the external heat source at T_M. As a result, the driving force for heat transfer appears. It clearly demonstrates that the heat and mass transfers in this system are strongly coupled, and the intensive vapor flux causes appropriate heat flux [3]. This coupling was also revealed by a numerical modeling of sorption dynamics caused by a large pressure jump [19].

If the diffusion rate (flux) is proportional to the current pressure difference $\Delta P(t) = [P_{gr}(t) - P_L]$, it can be maximized by maintaining the P_{gr} value at the highest possible level during the whole desorption process. This is possible if the isotherm of vapor desorption at $T = T_M$ is *step-like and the step is positioned at P = P_4* (Fig. 17.3). In this case, desorption starts at the maximal pressure difference $(P_4 - P_1)$ and this difference is nearly maintained until the whole adsorbate is removed. Therefore, such a step-like isotherm is optimal to ensure the fastest dynamics of desorption stage 4–1.

The same considerations applied to the adsorption stage 2–3 inevitably lead to a similar conclusion that to fasten this stage, the vapor adsorption isotherm at $T = T_H$ has to be *step-like with the step at P = P_2* (Fig. 17.3). In this case, adsorption starts right after the pressure jump, at the maximal pressure difference $(P_3 - P_2)$, which is nearly maintained constant until the adsorption is completed. Thus, the above analysis gives strict recommendation on how to accelerate *each isothermal stage* of a PD AHT cycle by using an optimal adsorbent with a properly positioned step-like isotherm.

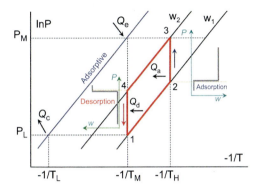

Figure 17.3 PD AHT cycle at the Clausius–Clapeiron diagram and schematics of adsorption isotherms of imaginary adsorbents dynamically optimal for desorption stage 4–1 (left $w(P)$ insert) and adsorption stage 2–3 (right $w(P)$ insert). Reprinted from Ref. [6], Copyright 2017, with permission from Elsevier.

Unfortunately, the best adsorbent for the desorption stage 4–1 appears to be the worst for the adsorption stage 2–3 and vice versa: Let us consider the adsorbent that gives up the whole adsorbate immediately at point 4 (P_4, T_M), which is optimal for desorption. At $T = T_H$, its isotherm of vapor adsorption is step-like with the step at point 3, which is at pressure P_M. Hence, during the adsorption stage 2–3, this material can start to adsorb the vapor only when the pressure inside the grain has reached P_M, so that the driving force for vapor diffusion becomes zero. This results in a negligible adsorption rate. To find a compromise between reasonable rates for both adsorption and desorption stages, the step should be at an intermediate point between the rich and weak isosters, e.g., between points 4 and 1 for the desorption stage (Fig. 17.4).

The position of this point provides an efficient tool to manage the rates of the isothermal stages of any PD AHT cycle. For instance, to make desorption faster, the step should be shifted close to point 4. On the contrary, the shift toward point 1 slows down the desorption stage and accelerates the adsorption one.

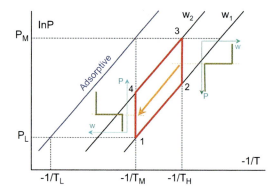

Figure 17.4 PD AHT cycle at the Clausius–Clapeiron diagram and schematics of adsorption isotherms, which provide compromise between good desorption and adsorption dynamics. Reprinted from Ref. [6], Copyright 2017, with permission from Elsevier.

17.2 New Cycle "Heat from Cold"

A new PD AHT cycle has been recently suggested for increasing the temperature potential of ambient heat to the level sufficient for heating dwellings [2]. The main feature of this cycle is that the adsorbent is regenerated by dropping the vapor pressure, and this drop is ensured by low ambient air temperature (e.g., T_L = −20 to −60°C). In this case, the regeneration of adsorbent does not need any supply of commercial energy, and it is easier at colder ambience. Since the useful heat that gains commercial value is obtained by means of a low ambient temperature, the new approach is called "Heat from Cold" (HeCol). It can be interesting for countries with cold climate (the northern parts of Russia, Europe, the USA, and Canada), and especially for the Arctic zone. Its schematics is displayed in Fig. 17.5, where the HeCol cycle is conditionally presented as two reversible Carnot cycles of heat engine (left) and heat pump (right).

Thus, the HeCol cycle uses the temperature difference between two natural thermostats, both being at low temperature (near 0°C and below). They can be, for instance, the ambient air with low temperature level (T_L) and a natural non-freezing water basin, such as sea, river, lake, underground water, etc., at medium temperature

(T_M), which is somewhat above 0°C (Fig. 17.5). Heat or cold from both the natural water basins is available for free. During winter in cold countries, the temperature difference between them can reach 30°C and more. This difference can be used to upgrade the temperature of the heat taken from the water reservoir up to a high level T_H sufficient for heating, thus gaining commercial value.

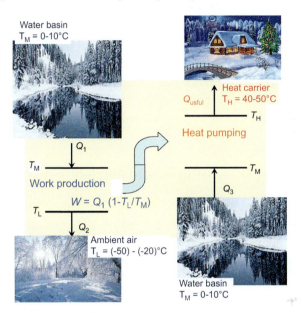

Figure 17.5 Schematics of the HeCol cycle.

17.2.1 Evaluation of Useful Heat

The useful heat $Q_{us} = Q_a = \Delta H_a \cdot \Delta w$ released during the isothermal adsorption stage 2–3 is evaluated for two sorbents of methanol: a commercial activated carbon ACM-35.4 [4] and composite "LiCl/silica gel." This composite was described in Sec. 7.3.2. The heat depends on the properties of the working "adsorbent–adsorbate" pair as well as on the cycle temperatures T_L, T_M, and T_H.

At fixed temperature of the water basin (T_M) and decreasing temperature of the ambient air (T_L), the useful heat Q_a gradually increases and reaches a limiting value. There is a strict conformity of the useful heat $Q_{us}(T_L)$ and the specific mass of methanol exchanged $\Delta w(T_L)$ [4] (not presented). This mass can reach 0.20–0.25 and 0.6–0.7 g/g, at $T_3 < (-50 \div -40°C)$, so that the useful heat $Q_{us} =$ 200–300 and 750–1000 kJ/kg for the tested carbon and composite, respectively. Both the values for the composite are much larger than those for the carbon. The advantage of the composite is due to the very steep isotherm and the large limiting value of methanol sorption (Figs. 7.12a and 7.13). On the contrary, the "methanol–ACM-35.4" pair exhibits a smooth adsorption isotherm that restricts the mass of methanol exchanged in the HeCol cycle. A decisive enhancement of Δw and Q_{us} can also be reached by using other CSPMs with a step-like isotherm between the borders of the HeCol cycle, like those based on $CaCl_2$, $CaBr_2$, and $CaCl_xBr_{(2-x)}$ confined to the silica gel mesopores [5].

17.2.2 Evaluation of Threshold Ambient Temperature

The calculations made in Ref. [4] clearly demonstrate that the performance of the HeCol cycle is better at colder outdoors: the lower the temperature T_L of ambient air, the more heat is transferred to the consumer at fixed temperatures T_M and T_H. It is important that for each combination of these two temperatures, there is a threshold ambient temperature above which the useful heat cannot be generated at all. This temperature can be calculated taking into account that the working pair under consideration follows the Polanyi principle of temperature invariance (Fig. 17.6a) [4]:

$$RT_M \ln\left(P_0(T_M)/P_0(T_L)\right) = RT_H \ln\left(P_0(T_H)/P_0(T_M)\right) \qquad (17.2)$$

The threshold temperature does not depend on the adsorbent's nature (Fig. 17.6a) as can be evaluated from Trouton's rule [20].

So the proposed cycle can be used to increase the temperature potential of natural water source from T_M to T_H only if the ambient temperature T_L is low enough. Therefore, this cycle can be of interest to countries with cold climate (e.g., the northern parts of Russia, Europe, the USA, and Canada), which occupy huge territories on the Earth (Fig. 17.7).

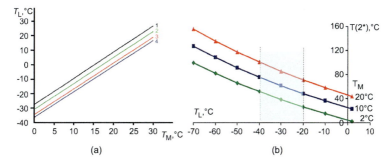

Figure 17.6 (a) The threshold temperature T_L as a function of the water basin temperature T_M at various T_H: 1 — 25, 2 — 30, 3 — 35, and 4 — 40°C. (b) The maximal heating temperature versus the condenser temperature T_L at various T_M indicated on the graph. Figure (a) reprinted by permission from Springer Nature from Ref. [4], Copyright 2018, and Figure (b) reprinted from Ref. [2], Copyright 2017, with permission from Elsevier.

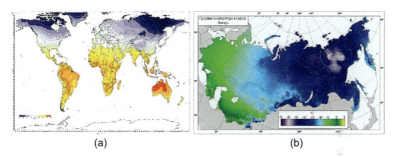

Figure 17.7 The average air temperature in January in the World (a) and the former USSR (b). Figure (a) reprinted from Ref. [2], Copyright 2017, with permission from Elsevier, and Figure (b) reprinted by permission from Springer Nature from Ref. [21], Copyright 2018.

17.2.3 Evaluation of Maximal Heating Temperature

The non-isothermal HeCol cycle allows a higher heating temperature to be obtained (Fig. 17.1b). The maximal heating temperature $T(2^*)$ can be determined by an intersection of the isoster 1–2 and the isobar $P = P_M$, if all thermal masses are neglected. It can be estimated by the empirical Trouton's rule, which provides a simple link between the boundary temperatures of a 3T adsorptive cycle as [20, 22] $T(2^*) =$

T_M^2/T_L (see Sec. 2.2.2). This empiric rule was shown to be valid for many adsorbate–adsorbent pairs promising for AHT, including the CSPMs [20].

Even at the most severe case of $T_M = 2°C$, the maximal heating temperature can reach 26–51°C at $T_L = -20$ to $-40°C$ (see the shaded area in Fig. 17.6b). The latter temperatures T_L are typical in winter in cold countries (Fig. 17.7). If a water basin is available at $T_M = 10$ or 20°C (e.g., underground water), the heating temperature can reach much higher levels of 43–71°C and 66–95°C, respectively (Fig. 17.6b).

17.2.4 First Testing of HeCol Prototype

The feasibility of the HeCol cycle was studied under the non-isothermal mode (Fig. 17.1b) in Refs. [5, 7]. A first lab-scale prototype of HeCol unit was designed, built, and tested. Four CSPMs were selected as methanol sorbents after analyzing their suitability for particular HeCol cycles [5, 7]. The amount of methanol exchanged was evaluated for several selected HeCol cycles with different operating conditions to assess the potential of the tested sorbents (see Refs. [5, 7]). Finally, two of the studied composites ($CaClBr/SiO_2$ and $LiCl/SiO_2$) were tested in the first lab-scale HeCol prototype.

17.2.4.1 Design of the first HeCol prototype

The adsorber and condenser/evaporator are the key parts of the first HeCol prototype (Figs. 17.8a,b). Two commercial finned flat-tube HExs (Fig. 17.8c) loaded with 500 and 700 g of the LiCl/silica and CaClBr/silica composites, respectively, as loose grains of 0.2–0.5 mm size were used as adsorber. The heat exchange surface of each HEx was $S = 1.24$ m^2. Vacuum valve V_0 served for connecting the adsorber and the condenser/evaporator.

Three thermocryostats were used to imitate natural thermostats with low and intermediate temperatures T_L and T_M, as well as the heat sink at T_H of the HeCol cycle. The thermocryostats were integrated into the prototype by pipelines, valves V_1–V_8, and pumps M_1–M_3 (Fig. 17.8a). Thus, at the adsorption mode, the adsorber and the evaporator were maintained at temperatures T_H and T_M, respectively; while at the regeneration mode, these temperatures were switched to T_M and T_L, respectively, as Fig. 17.9 schematically demonstrates. The temperature of the heat transfer fluid (HTF) at the inlet T_{in} and outlet T_{out} of the adsorber and evaporator/condenser was measured by K-type thermocouples. The flow rate f was measured by mechanical flow meters F_1–F_3. More experimental details can be found elsewhere [5, 7].

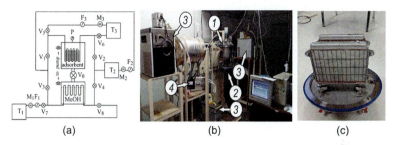

Figure 17.8 Schematics of the lab-scale HeCol prototype (a), its overall view (b), the adsorber composed of two HExs (c) [5, 7]. 1 — adsorber, 2 — condenser/evaporator, 3 — thermocryostats, 4 — vacuum pump, F1–F3 — flow meters, V_0 — vacuum valve, V_1–V_8 — valves, P — pressure gauge. Reprinted from Ref. [5], Copyright 2018, with permission from Elsevier.

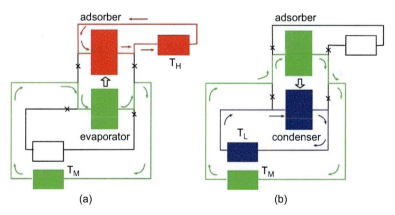

Figure 17.9 Thermal liquid flow diagram at the HeCol adsorption (a) and regeneration (b) modes. Reprinted from Ref. [5], Copyright 2018, with permission from Elsevier.

The heating power W and the heat Q consumed/released in the adsorber and evaporator/condenser were calculated as

$$W = [T_{out}(t) - T_{in}(t)]C_p f \rho \tag{17.3}$$

$$W_\Sigma = \int_{t=0}^{t=\infty} W dt \qquad (17.4)$$

where T_{out} and T_{in} are the outlet and inlet temperatures, C_p is the HTF specific heat, ρ is the HTF density, and f is the HTF flow rate.

17.2.4.2 Testing HeCol unit with the selected "salt/matrix" composites

Typical evolution of the adsorber inlet (T_{in}) and outlet (T_{out}) temperatures of the HTF during the heat release stage (2-2'-3) of the non-isothermal cycle (Fig. 17.1b) with operating conditions $T_L/T_M/T_H = -20/20/26°C$ is presented in Fig. 17.10a for the LiCl/silica composite. At $t = 300$ s, the sorbent starts to sorb methanol and the outlet temperature T_{out} increases from $T_H = 26°C$ to the maximal heating temperature $T(2')$ and then decreases. The inlet temperature T_{in} slightly increases in the course of adsorption, which indicates that the power of the thermostat at T_H is not sufficient to compensate the heat released during adsorption. After disconnecting the evaporator and adsorber (at $t = 740$ s), the adsorption stops and the outlet temperature gradually approaches the inlet one. Time evolution of the heating power follows the same trend as T_{out} (Fig. 17.10b).

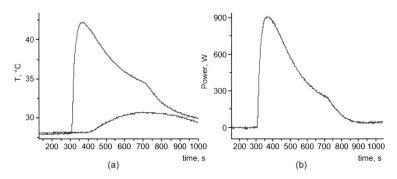

Figure 17.10 Time evolution of the adsorber inlet T_{in} (green) and outlet (red) T_{out} temperature (a) and the heating power (b) during the methanol sorption by the LiCl/silica composite at $T_L/T_M/T_H = -20/20/26°C$ and $f = 1.2$ L/min. Reprinted from Ref. [7], Copyright 2018, with permission from Elsevier.

The maximum outlet temperature T_{out} depends on the HTF flow rate and reach 45 and 41°C for the LiCl- and CaClBr-based composites (Fig. 17.11). At larger flow, the maximum temperature decreases and the cycle approaches to the isothermal one shown in Fig. 17.1a. On the contrary, the maximum heating power rises at the increase in the flow rate (Fig. 17.11b), because more heat is transferred to the consumer, however, at lower temperature. Thus, the HTF flow rate is an efficient tool to manage the temperature level of the released heat and the heating power to meet the requirements of a consumer. The maximal power approaches 3 kW or 6.0 kW/kg composite.

Comparison of the two tested sorbents, namely, LiCl/silica and CaClBr/silica demonstrates that

- At T_M = 20–25°C, the LiCl/silica composite allows to get a higher heating temperature of 40–46°C and a larger specific power of 4.5–6.0 kW/kg than those for CaClBr/silica (37–42°C and 2.5–3.2 kW/kg);
- The total release of the useful heat is larger for the LiCl/silica composite by a factor of 1.3–1.6;
- As opposed to the LiCl/silica, the CaClBr/silica, which reacts with methanol by the formation of complex CaClBr·4CH$_3$OH at lower $\Delta F_r \approx$ 4.0 kJ/mol, can be regenerated at higher temperature T_L and applied also for the cycles with $T_L/T_M/T_H$ = (−15 ÷ −10)/20/26°C.

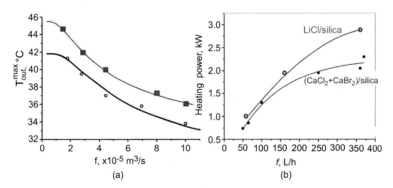

Figure 17.11 The maximum outlet temperature (a) and heating power (b) versus the HTF flow rate f. $T_L/T_M/T_H$ = −20/20/26°C. Reprinted from Ref. [7], Copyright 2018, with permission from Elsevier.

Thus, the composite sorbents with different affinity for methanol vapor can be used for realizing different HeCol cycles. The sorbent with higher affinity provides higher temperature lift and heating power but requires a lower temperature T_L for regeneration. The adsorbent with lower affinity can be regenerated at "milder" conditions (higher T_L) but ensures smaller temperature lift.

The testing of the first lab-scale HeCol unit with the two "salt/silica" composite sorbents demonstrates the possibility of the cycle realization with the maximum specific heating power of up to 6.0 kW/kg sorbent, and the temperature lift of 20–25°C. Varying the heat transfer liquid flow allows managing the process in order to get either a higher power or temperature. These first tests are quite encouraging and have definitely demonstrated the feasibility of the new HeCol cycle by using the intently synthesized composite sorbents of methanol "salt in porous matrix." The comprehensive study of a full scale HeCol prototype with the composite $LiCl/SiO_2$ completely confirmed this conclusion [23].

References

1. Chandra, I. and Patwardhan, V. S. (1990) Theoretical studies on adsorption heat transformer using zeolite-water vapour pair, *Heat Recov. Syst. CHP*, **10**, pp. 527–537.

2. Aristov, Yu. I. (2017) Adsorptive transformation of ambient heat: A new cycle, *Appl. Therm. Eng.*, **124**, pp. 521–524.

3. Aristov, Yu. I. (2017) Adsorptive transformation and storage of renewable heat: Review of current trends in adsorption dynamics, *Renew. Energy*, **110**, pp. 105–114.

4. Okunev, B. N., Voskresensky, N. M., Girnik, I. S., and Aristov, Yu. I. (2018) Thermodynamic analysis of a new adsorption cycle for ambient heat upgrading: Ideal heat transfer, *J. Eng. Thermophysics*, **27**, pp. 524–530.

5. Tokarev, M. M., Grekova, A. D., Gordeeva, L. G., and Aristov, Yu. I. (2018) A new cycle "Heat from Cold" for upgrading the ambient heat: The testing a lab-scale prototype with the composite sorbent CaClBr/silica, *Appl. Energy*, **211**, pp. 136–145.

6. Aristov, Yu. I. (2017) "Heat from cold": A new cycle for upgrading the ambient heat: Adsorbent optimal from the dynamic point of view, *Appl. Therm. Eng.*, **124**, pp. 1189–1193.

7. Gordeeva, L. G., Tokarev, M. M., Shkatulov A. I., and Aristov, Yu. I. (2018) Testing the lab-scale "Heat from Cold" prototype with the "LiCl/silica–methanol" working pair, *Energy Conv. Manag.*, **159**, pp. 213–220.

8. Ruthven, D. M., Farooq, S., and Knaebel, K. S. (1994) *Pressure Swing Adsorption* (Wiley), p. 376.

9. Frazzica, A., Dawoud, B., and Critoph, R. E. (2016) Theoretical analysis of several working pairs for adsorption heat transformer application, *Proc. HPC Conf.*, Nottingham, June 2016.

10. Polanyi, M. (1932) Theories of the adsorption of gas: A general survey and some additional remarks, *Trans. Faraday Soc.*, **28**, pp. 316–333.

11. Raldow, W. M. and Wentworth, W. E. (1979) Chemical heat pumps: A basic thermodynamic analysis, *Solar Energy*, **23**, pp. 75–79.

12. Meunier, F., Poyelle, F., and LeVan, M. D. (1997) Second-law analysis of adsorptive refrigeration cycles: The role of thermal coupling entropy production, *Appl. Therm. Eng.*, **17**, pp. 43–55.

13. Glaznev, I. S., Ovoshchnikov, D. S., and Aristov, Yu. I. (2009) Kinetics of water adsorption/desorption under isobaric stages of adsorption heat transformers: The effect of isobar shape, *Int. J. Heat Mass Transfer*, **52**, pp. 1774–1777.

14. Okunev, B. N., Gromov, A. P., Heifets, L. I., and Aristov, Yu. I. (2008) A new methodology of studying the dynamics of water sorption/desorption under real operating conditions of adsorption heat pumps: Modelling of coupled heat and mass transfer, *Int. J. Heat Mass Transfer*, **51**, pp. 246–252.

15. Okunev, B. N., Gromov, A. P., and Aristov, Yu. I. (2013) Modelling of isobaric stages of adsorption cooling cycle: An optimal shape of adsorption isobar, *Appl. Therm. Eng.*, **53**, pp. 89–95.

16. Fick, A. E. (1855) Ueber diffusion (on diffusion), *Ann. Phys.*, **94**, pp. 59–86.

17. Crank, J. (1975) *Mathematics of Diffusion* (Oxford University Press, London).

18. Kaerger, J., Ruthven, D. M., and Theodorou, D. N. (2012) *Diffusion in Nanoporous Materials*, vol. 1 (Wiley-VCH, Weinheim), 872p.

19. Okunev, B. N., Gromov, A. P., Heifets, L. I., and Aristov, Yu. I. (2008) Dynamics of water sorption on a single adsorbent grain caused by a large pressure jump: Modelling of coupled heat and mass transfer, *Int. J. Heat Mass Transfer*, **51**, pp. 5872–5876.

20. Aristov, Yu. I., Tokarev, M. M., and Sharonov, V. E. (2008) Universal relation between the boundary temperatures of a basic cycle of sorption heat machines, *Chem. Eng. Sci.*, **63**, pp. 2907–2912.

21. Gordeeva, L. G., Tokarev, M. M., Aristov, Yu. I. (2018) New adsorption cycle for upgrading the ambient heat. *Theor. Found. Chem. Eng.*, **52**, pp. 195–205.

22. Critoph, R. E. (1988) Performance limitation of adsorption cycles for solar cooling, *Solar Energy*, **41**, pp. 21–31.

23. Tokarev M. M. (2019) A double-bed adsorptive heat transformer for upgrading ambient heat: Design and first tests, *Energies*, **12**, 4037. doi:10.3390/en12214037.

Chapter 18

Regeneration of Heat and Moisture in Ventilation Systems

Air dehumidification is an important adsorption technology widely applied in countries with humid climate [1]. Quite the contrary, air humidification is required for countries where outdoor air is drier than it is necessary for comfortable conditions indoor. For instance, in countries with a cold climate, the absolute outdoor humidity can drop in winter to $0.1–1.0$ g/m^3, while comfortable humidity should always be higher than $5–7$ g/m^3. Therefore, the dry outdoor air has to be moisturized before being supplied inside dwellings. In this chapter, we present a new adsorptive approach for outdoor air moisturizing in ventilation systems in cold climates (the so-called VENTIREG [2, 3]). In addition, this method allows heat storage and heat exchange between warm outgoing air and cold incoming air, which prevents large heat losses in ventilation systems.

No requirements for an optimal adsorbent have been formulated for the VENTIREG process so far. Therefore, only intuitive considerations could help in choosing a proper adsorbent. These considerations together with a preliminary scanning of several available desiccants have allowed a composite $CaCl_2$/alumina (SWS-1A) to be selected as a promising water buffer for VENTIREG units [4]. This composite was first applied for gas drying [5] and has already been considered in Sec. 6.3.2. The material was tested in several VENTIREG prototypes under climatic conditions of Western

---

Nanocomposite Sorbents for Multiple Applications
Yu. I. Aristov
Copyright © 2020 Jenny Stanford Publishing Pte. Ltd.
ISBN 978-981-4267-50-2 (Hardcover), 978-981-4303-15-6 (eBook)
www.jennystanford.com

Siberia (Novosibirsk city, 55°02′N, 83°00′E, Russia) and appeared to be superior to common desiccants, such as pure silica gel and alumina. Physicochemical, technical, economic, and social aspects of the VENTIREG method are discussed in this chapter.

18.1 Climatic Features of Cold Countries

The current global changes in climate due to greenhouse gas emissions [6] have brought about an increased necessity of rational use of energy in dwellings. To reduce consumption of energy, there is a trend toward improving heat insulation and air-tightness. For the purpose of tightening a building's thermal envelope, providing adequate ventilation is required [7]. The main function of residential ventilation systems is to exchange stale, contaminated room air with fresh outdoor air. Particular features of residential ventilation systems strongly depend on standards of indoor air quality and outdoor climate conditions [8]. For relatively cold climates, mechanical ventilation with heat recovery plays a vital role in securing optimum air quality, thermal comfort, and saving thermal energy [7]. Estimates show that as much as 70% of the energy lost through mechanical ventilation can be recovered by the use of ventilation heat recovery systems [9]. For this purpose, many heat exchangers (counter-current units or enthalpy wheels) are used to transfer heat from the exhaust air to the supply air [9–13]. In Ref. [7], the low-grade heat recovered from the exhaust air is upgraded by a heat pump and used for heating the fresh supply of air.

For winter season in severe climates (typical for Russia, Canada, Northern Europe, and the USA), the difference ΔT between indoor and outdoor temperatures can reach 60°C or even more, which leads to enormous heat losses and freezing of moisture at the system exit. As a result, common heat recovery units integrated into ventilation systems may not be capable of working at these conditions. Moreover, such systems are not able to manage the indoor humidity, which dramatically reduces in the winter and greatly unbalances the indoor heat comfort [8]. Thus, to fill these three main gaps in the current technique, the following actions have to be performed:

- Efficient exchange of heat between the exhaust and supply air fluxes to reduce heat losses.
- Reasonable drying of the exhaust air to avoid ice formation at the system's exit.
- Moisturizing the supplied air to provide indoor conditions of human thermal comfort.

18.2 The VENTIREG Approach: Description and Testing

Here we present a recent approach (the so-called VENTIREG) for regenerating heat and moisture in ventilation systems in cold climates, which resolves the obstacles mentioned earlier. We discuss some qualitative issues that would help in selecting sorbents to be used in VENTIREG systems. Finally, this selection was performed after preliminary scanning of available desiccants.

18.2.1 Description of the Approach

To exchange the sensible heat between the inlet (fresh) and outlet (exhausted) air fluxes, a granulated layer of heat-storing material (HSM) is placed closer to the unit exit (1 in Fig. 18.1). Before this layer (closer to the room side), a layer of water-adsorbing material (2) is located. It serves as a water buffer. The unit operates in two intermittent modes (Fig. 18.1):

1. Outflow mode: A warm and humid indoor air is blown by an extract fan through the relatively dry adsorbent, which captures and retains the indoor moisture [3]. Dried and warm air enters layer 1 and heats it up. After that, the air flux switches.

2. Inflow mode: A dry and cold outdoor air is blown by a supply fan through warm layer 1 and is heated up to the temperature T_{in} close to that in the room, thus recovering the stored heat. Passing through the layer of the humid adsorbent, warm and dry air causes the retained water to be desorbed and come back to the room [3], thus maintaining the indoor moisture balance. Because of the finite heat capacity of layers 1 and 2,

temperature T of the incoming air slowly decreases, and the air flux switches when $T_{in} - T$ reaches a predetermined value ΔT_0, and so on.

It is expected that after a transient period, a steady-state regime is established with the time interval $\Delta\tau$ (a half-cycle time) between the flux switches. For continuous operation, two similar units should work in the opposite modes (Fig. 18.1).

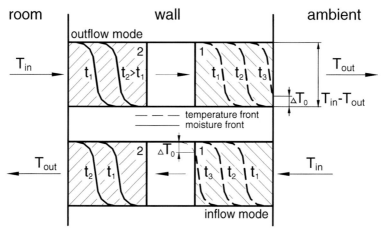

Figure 18.1 Scheme of the regeneration process: the temperature and moisture profiles at various times $t_1 < t_2 < t_3$. Reprinted from Ref. [3], Copyright 2008, with permission from Elsevier.

18.2.2 Adsorbent Selection: Intuition

No requirements for an optimal adsorbent for the VENTIREG units have been formulated. Here we briefly discuss some intuitive issues, which could help in choosing a proper adsorbent. In the next section, more direct selection will be performed on the basis of restricted scanning of several commercial desiccants.

The moisture regeneration process in a VENTIREG unit proceeds at a temperature close to the room temperature and, in whole, resembles the process of pressure swing adsorption (PSA) [14]. It consists of the following two stages:

1. The adsorption (or air-drying) phase: the moisture from the wet outgoing air is absorbed by the adsorbent and retained inside its pores.

2. The desorption (or return) phase: the absorbed moisture is blown away from the adsorbent by the dry incoming air and returned to the room.

An optimal adsorbent is deemed to be quite intelligent. On the one hand, it has to efficiently remove moisture from the wet air and retain it. On the other hand, it should readily give the sorbed water off, being subjected to the flux of the dry incoming air. Therefore, the adsorbent's affinity for water vapor should not be too strong or too weak. Further experimental and theoretical study will show how subtle this balance has to be.

18.2.3 Adsorbent Selection: Experiment

For making a preliminary evaluation, we experimentally investigated [4] the moisture exchange between a stationary adsorbent layer and the air passed through it, which can be considered a model of the moisture exchange in a VENTIREG system. The moisture exchange was realized in a periodic regime, as in the case of the PSA process: In the first (and any odd) half-cycle, the wet air passing through the adsorbent was completely or partially dried, and in the second (and each even) half-cycle, the relatively dry air supplied to the other end of the adsorbent layer was humidified, contacting with the wet adsorbent. We measured the relative and absolute humidity of the air and its temperature at the input and output of the adsorbent layer in the stationary regime. Then we calculated the degree of moisture exchange at the dehumidification and humidification stages as a function of the adsorbent nature, the size of its grains, the volume rate of the inlet air flow, and the duration of the process. In this way, we compared three commercial desiccant materials, namely, two standard dehumidifiers—a mesoporous alumina (Al_2O_3) of A1-type (Angarsk, Russia) and a microporous silica gel of KSM type (Mendeleevsk, Russia)—and a IK-011-1 sorbent (SWS-1A), which is the aluminum oxide A1 modified with calcium chloride (10–12 wt.%). Alumina oxide and SWS-1A grains of 1.8 and 4.5 mm in diameter and 6–8 mm in length, respectively, and spherical silica gel granules of 2–6 mm size were used.

The experiments were carried out in the following way: the adsorbent was placed into a cylindrical adsorber of 210 mm in diameter and 160 mm in length. Before loading, the adsorbent was held in air for 48 h at T = 20°C and a relative humidity (RH) of 30–35%. At the adsorption (humidification) stage, an air saturated with water vapor to RH = 29.5% ± 1.5% was supplied at room temperature to the adsorber inlet. The temperature and humidity of the air at the input and output of the adsorbent layer were measured by RH–T sensors of an IVA-6B type with an accuracy of ±1.0% and ±0.1°C, respectively. After a certain time, a reverse air flow with RH = 2.3 ± 1.2% was supplied at room temperature to the other end of the layer (desorption stage). The volume rate of the air flow was changed from 5.0 to 31.3 m³/h and measured by a flow meter. The portion of water absorbed at the end of the adsorption stage

$$\alpha = \int (d_{in} - d_{out})dt \Big/ \int d_{in} dt = S(BCDE)/S(ACDF)$$

and the portion of water released from the adsorbent at the end of the desorption stage

$$\beta = \int (d_{out} - d_{in})dt \Big/ \int d_{out} dt = S(KLMN)/S(JLMO)$$

can be calculated from the temporal evolution of the air absolute humidity $d(t)$ at the input and output of the adsorbent layer at the stationary regime (Fig. 18.2a,b). Here we considered the inlet and outlet air fluxes to be equal.

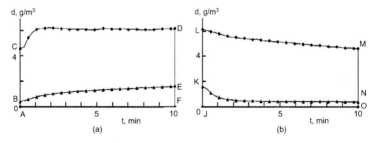

Figure 18.2 Evolution of the absolute humidity of air at the wet (●) and dry (▲) ends of the adsorber filled with SWS-1A (4.5/8 mm) during the half-cycle of adsorption (a) and desorption (b) in the stationary regime. The inlet air flow rate is 5.0 m³/h. Reprinted by permission from Springer Nature from Ref. [4], Copyright 2018.

The time of setting the stationary regime depends on the initial state of the adsorbent and the volume rate of the air flow. Under the conditions of our experiments, this time was approximately 30 min. Under the stationary regime, the amount of water absorbed (determined by area BCDE in Fig. 18.2a) was equal to the amount of water released at the desorption stage (proportional to the area KLMN in Fig. 18.2b) within ±(2–4)%. That means the process of moisture regeneration is reversible.

At the adsorption stage, the absolute air humidity at the adsorber inlet (at the "wet end of the adsorber") was maintained constant (d_{in} = 5.9 ± 0.2 g/m^3). The air passed through the adsorbent layer became dry. The air humidity at the adsorber outlet (at the "dry end of the adsorber") and the degree of drying α depended on the adsorbent's nature, the size of its grain (Table 18.1), and the volume rate of the inlet wet air. The outlet air humidity was slowly increasing with time (Fig. 18.2a) and corresponded to a dew point between c.a. −30 and −10°C (Table 18.1).

The values of d and α essentially depend on the adsorbent used. The drying degree increases in the series SWS-1A > (silica KSM) > Al_2O_3 (Table 18.1), which correlates with their static moisture capacity. Comparison of pure and modified alumina oxides shows that the introduction of the salt into the pores significantly increases the degree α of drying. For example, this degree for SWS-1A (1.8/6 mm) reaches 0.98 at the inlet flow rate of 5.0 m^3/h, i.e., almost all moisture delivered to the SWS layer is absorbed there. Even at the largest inlet flow rate of 31.3 m^3/h (contact time $\tau \approx 0.5$ s), the layer absorbs 74% of the incoming moisture (not presented). For the pure Al_2O_3 (1.8/6 mm), these values are lower (0.75 and 0.62, respectively). The microporous silica ensures the α values of 0.87 and 0.71. For larger grains, the degree of drying appropriately reduces (Table 18.1) likely due to intragrain diffusion limitations.

Table 18.1 The drying degree α and the dew point of the outlet air, T_{dew}, for the adsorbents tested at the inlet flow rate of wet air of 5.0 m^3/h (adsorption stage)

Sample	Size 4.5/8 mm		Size 1.8/6 mm	
	α	T_{dew} (°C)	α	T_{dew} (°C)
Silica gel*	—	—	0.87	−23 ÷ −20
Alumina	0.65	−23 ÷ −20	0.75	−23 ÷ −20
SWS-1A	0.80	−23 ÷ −20	0.98	−23 ÷ −20

*grain size 2–6 mm

At the subsequent desorption stage, the absolute air humidity at the adsorber inlet (the "dry end of the adsorber") was maintained quite low (d_{in} = 0.4 ± 0.1 g/m^3). This dry air, when passing through the wet adsorbent bed, was humidifying, so that the moisture was returned back to the air flux. The air outlet humidity was slowly decreasing with time from 5.9 to 4.9 g/m^3 (Fig. 18.2b), which was close to the inlet humidity at the previous adsorption stage (Fig. 18.2a).

The maximal degree of moisture return, β, again increases in the series SWS-1A > (silica KSM) > Al$_2$O$_3$. Thus, under the same conditions, the composite sorbent CaCl$_2$/alumina is capable of sorbing and desorbing a larger amount of water than the commercial microporous silica and alumina. A possible reason is the interaction between calcium chloride and water vapor [15]. Hence, an essential minimization of the amount of adsorbent and the unit size becomes possible. This reduction in adsorbent mass leads to lower hydrodynamic resistance of its layer. This could allow the use of cheap blade-type fans instead of centrifugal ones and reduce electricity consumption. At the moment, the VENTIREG unit III (see below) consumes 20–40 W of electric power for blowing air and gives a heating load of about 600–1400 W, which corresponds to a coefficient of performance (COP) of 25–35.

Another advantage of the composite sorbent is a higher absolute humidity of the return air at the desorption stage (Table 18.2). It seems that its affinity for water is quite suitable for the VENTIREG process to be efficiently realized. Because of its advanced properties, SWS-1A was selected for further testing in several VENTIREG prototypes described as follows.

Table 18.2 The return degree β and the absolute humidity of the returned air, d_{out}, for the adsorbents tested at the inlet flow rate of dry air of 5.0 m^3/h (desorption or return stage)

Sample	Size 4.5/8 mm		Size 1.8/6 mm	
	β	d_{out} (g/m^3)	β	d_{out} (g/m^3)
Silica gel*	—	—	0.90	5.2 → 4.6
Alumina	0.87	5.3 → 4.8	0.91	5.5 → 4.9
IK-011-1	0.90	5.6 → 4.9	0.98	5.8 → 4.9

*grain size 2–6 mm

18.3 Experimental Testing of VENTIREG Prototypes

To study and optimize the process of heat and moisture recovery, four experimental units with air flux up to 25 (I), 40 (II), 135 (III), and 220 (IV) m^3/h were built and tested (Fig. 18.3). The heat storage container was filled with gravel of irregular shape and 4–7 mm in size (except for unit I).

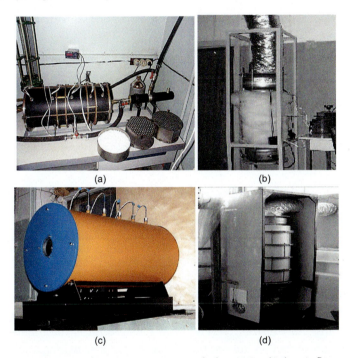

Figure 18.3 View of VENTIREG units tested: the maximal inlet air flow rates are 25 m^3/h (a), 40 m^3/h (b), 135 m^3/h (c), and 220 m^3/h (d). Figure (a) reprinted by permission from Springer Nature from Ref. [2], Copyright 2001, and Figure (c) reprinted from Ref. [3], Copyright 2008, with permission from Elsevier.

As air fluxes during the inflow and outflow modes are equal, the evolution of temperature T and humidity d during a cycling operation of the VENTIREG unit gives the efficiencies of heat regeneration ($\theta = S_{\text{ACDE}}/S_{\text{ABDE}}$) and moisture regeneration ($\beta = S_{\text{KMOP}}/S_{\text{KLOP}}$), where S is the area of appropriate figures in Fig. 18.4. As the outdoor temperature T_{out} and, hence, the difference ΔT_{MAX}

= $T_{in} - T_{out}$ changed during these experiments, it was convenient to present θ as a function of a dimensionless temperature difference $\Delta \tilde{T}$ = $\Delta T_o/\Delta T_{MAX}$ (Fig. 18.5a).

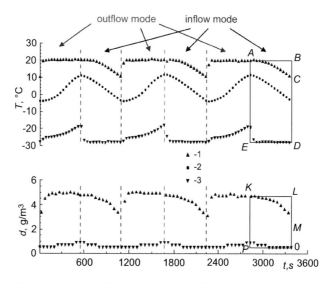

Figure 18.4 Evolution of air temperature T and absolute humidity d as a function of time t during a cycling operation of the VENTIREG unit III (1 — indoor side, 2 — at the middle of the unit, 3 — outdoor side). T_{in} = 20°C, T_{out} = −28°C, ΔT_o = 10°C. Air flux 123 m³/h. Reprinted from Ref. [3], Copyright 2008, with permission from Elsevier.

Similarly, the efficiency β of moisture regeneration can be presented as a function of a dimensionless difference in the absolute humidity $\Delta \tilde{d} = \Delta d / \Delta d_{MAX}$ (Fig. 18.5b), where Δd is the difference in the absolute humidity at the moment of flux switching, Δd_{MAX} is the maximal difference in the absolute humidity during a particular experiment. Experimental data obtained showed that θ > 0.9 at $\Delta \tilde{T}$ = 0–0.25 and β > 0.7 at $\Delta \tilde{d}$ = 0–0.7, so that efficient regeneration of both heat and moisture took place. Both these efficiencies can be easily and purposefully varied by managing the half-cycle time $\Delta \tau$, which is a linear function of the Reynolds number [3]. The range of its variation is quite wide (3–70 min).

It appears that for unit III, θ and β can be described by a unique equation

$$\theta(\beta) = 1 - 0.38 \cdot \Delta \tilde{T}(\Delta \tilde{d}) - 0.12 \cdot \Delta \tilde{T}(\Delta \tilde{d})^2 \qquad (18.1)$$

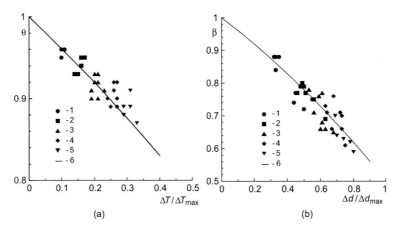

Figure 18.5 (a) Efficiency of heat regeneration θ (1–5) as a function of relative temperature difference $\Delta T_o/\Delta T_{MAX}$ for unit III at ΔT_o = 5°C (1), 7.5°C (2), 10°C (3), 12.5°C (4), and 15°C (5); 6 — calculated according to Eq. (18.1). (b) Efficiency of moisture regeneration β (1–5) as a function of relative difference in the absolute humidity $\Delta d/\Delta d_{MAX}$ for unit III at ΔT_o = 5°C (1), 7.5°C (2), 10°C (3), 12.5°C (4), and 15°C (5); 6 — calculated according to Eq. (18.1). Lengths of the adsorbent and gravel layers were 100 and 300 mm, respectively. Reprinted from Ref. [3], Copyright 2008, with permission from Elsevier.

The fact that both the efficiencies θ and β are described by the same equation can indicate a close link between the processes of heat and mass transfer when a humid air passes through an adsorbent layer [16].

At the same ΔT_o, the efficiency of heat regeneration increases with the rise in ΔT_{MAX}, which means the VENTIREG unit operates better at colder climates.

It is interesting to note that for unit II, θ and β can also be described by a similar equation with the coefficients close to those of Eq. (18.1):

$$\theta(\beta) = 1 - 0.30 \cdot \Delta \tilde{T}(\Delta \tilde{d}) - 0.12 \cdot \Delta \tilde{T}(\Delta \tilde{d})^2 \qquad (18.2)$$

18.4 Other Aspects

Here we consider important issues on further development of this approach, such as scaling-up, economic, hygienic, and social aspects.

Magnifying the VENTIREG unit from a laboratory to a larger size may decrease the process efficiency with a rise in unit diameter. The scale-up theory shows that this effect results from the radial non-uniformity of the velocity distribution in the layer (column) [17]. Application of appropriate mathematical models will permit the analysis, optimization, and design of VENTIREG units for heat and moisture recovery in ventilation systems for a single room, family house, and large residential buildings. The final aim is to meet the indoor air quality and energy standards in such dwellings. Harmonization of the adsorbent with the VENTIREG process is another important goal of future mathematical modeling. We have already mentioned above the moderate affinity of the adsorbent for water vapor as a desirable factor for the efficient realization of both drying and return phases. More efforts are still necessary to realize which adsorbent is optimal for the VENTIREG process. The first attempts to formulate quantitative requirements for the adsorbent optimal for the VENTIREG have been recently made in Ref. [18].

The economic impact of this unit is due to the expectation that the VENTIREG reliably supplies fresh air at a cost lower than common systems. The estimations based on the current performance give for a standard two-room apartment (47 m^2) in Novosibirsk, Russia, the unit capital cost of about 90–110 Euros, annual savings in operating costs of about 40–45 Euros, and a payback period of 2.0–2.6 years [19]. The energy saving is 43–45%. Larger saving can be expected for larger apartments under conditions of Northern Europe, where the cost of electricity is much higher. Further improvements in the system will increase the economic impact.

The social impact arises mainly from the fact that VENTIREG units provide a comfortable and healthy environment in dwellings, which tightly satisfy a specified range of human comfort. The final result is the improvement in human living standards (Fig. 18.6).

Hygiene is another crucial issue. Indeed, the exhausted indoor air may contain various contaminants, including pathogenic bacteria and microbes. Before testing the unit, we did not perform any special precautions against these health hazards. After the operation run (winter 2005–06), we found no organic contaminations or bacteria growth inside the unit. A possible reason is that liquid water does not form inside the device during its cyclic operation. Indeed, water caught by the sorbent is stabilized in the form of salt crystalline hydrates and solution. These hydrates are solid substances; the salt concentration in the solution is higher than 25 wt.%. So a large concentration avoids or limits multiplication of bacteria. This is an essential advantage of this device over standard ventilation and air-conditioning units, water reservoirs, and other water-containing systems.

Figure 18.6 Artistic representation of a VENTIREG (Image courtesy: A. S. Shorin).

Thus, the suggested VENTIREG unit exchanges stale, contaminated room air with fresh outdoor air, recovering up to 95% of energy and 70–90% of moisture from the exhaust air and prevents the formation of ice at the unit's exit. For countries with a cold climate, this makes it possible to bring in more conditioned fresh air at a lower cost. The prototypes III and IV supply fresh air at 135–220 m^3/h with effective regeneration of heat and moisture. This provides 0.5–1.0 air changes per hour for a typical one-family detached house. The unit requires very little maintenance, has a low capital cost, and is compact and energy efficient.

In this chapter, we demonstrated that the VENTIREG approach can be efficiently applied by using the composite CaCl$_2$/alumina (SWS-1A) as an adsorbent. The VENTIREG prototypes presented earlier were considered in Ref. [20] as one of the most efficient energy exchangers that have been suggested and tested for countries with a cold climate.

References

1. Wang, R. Z. and Oliveira, R. G. (2006) Adsorption refrigeration: An efficient way to make good use of waste heat and solar energy, *Prog. Energy Combust. Sci.,* **32**, pp. 424–458.

2. Aristov, Yu. I., Mezentsev, I. V., and Mukhin, V. A. (2006) A new approach to heat and moisture regeneration in the ventilation system of rooms. I. Laboratory prototype of the regenerator, *J. Eng. Phys. Thermophys.,* **79**(3), 569–576.

3. Aristov, Yu. I., Mezentsev, I. V., and Mukhin, V. A. (2008) A new approach to regenerating heat and moisture in ventilation systems, *Energy Buildings*, **40**, pp. 204–208.

4. Aristov, Yu. I., Mezentsev, I. V., and Mukhin, V. A. (2005) Investigation of the moisture exchange in a stationary adsorbent layer through which air is passed, *J. Eng. Phys. Thermophys.,* **78**, pp. 248–255.

5. Aristov, Yu. I. (2004) Selective water sorbents for air drying: From the lab to the industry, *Cat. Industry*, **6**, pp. 36–41.

6. The Kioto Protocol, https://unfccc.int/resource/docs/convkp/kpeng.pdf (last opened on March 14, 2019).

7. Riffat, S. B. and Gillott, M. C. (2002) Performance of a novel mechanical ventilation heat recovery heat pump system, *Appl. Therm. Eng.,* **22**, pp. 839–845.

8. ASHRAE (American Society of Heating, Refrigeration and Air-conditioning Engineers). (1995) *ASHRAE Handbook—HVAC Applications* (ASHRAE, Atlanta, GA, USA).

9. Steimle, F. and Roben, J. (1992) Ventilation requirements in modern buildings, in: *Proc. 13th AIVC Conf.*, Nice, France, 1992, pp. 414–422.

10. Sphaier, L. A. and Worek, W. M. (2004) Analysis of heat and mass transfer in porous sorbents used in rotary regenerators, *Int. J. Heat Mass Transfer*, **47**, pp. 1415–1430.

11. Shah, R. K. and Skiepko, T. (1999) Influence of leakage distribution on the thermal performance of a rotary regenerator, *Appl. Therm. Eng.,* **19**, pp. 685–705.

12. Buyukalaca, O. and Yilmaz, T. (2002) Influence of rotational speed on effectiveness of rotary type heat exchanger, *Heat Mass Transfer*, **38**, pp. 441–447.

13. Wu, Z., Melnik, R. V. N., and Borup, F. (2006) Model-based analysis and simulation of regenerative heat wheel, *Energy Buildings*, **38**, pp. 502–514.

14. Ruthven, D. M., Farooq, S., and Knaebel, K. S. (1994) *Pressure Swing Adsorption* (VCN Publishers, New York).

15. Gordeeva, L. G. and Aristov, Yu. I. (2012) Composites "salt inside porous matrix" for adsorption heat transformation: A current state of the art and new trends, *Int. J. Low Carbon Tech.*, **7**, pp. 288–302.

16. Close, D. J. and Banks, P. J. (1972) Coupled equilibrium heat and single adsorbate transfer in fluid flow through a porous medium. I. Predictions for a silica gel air drier using characteristic charts, *Chem. Eng. Sci.*, **27**, pp. 1143–1155.

17. Boyadijev, Ch. B. (2006) Diffusion models and scale-up, *Int. J. Heat Mass Transfer*, **49**, pp. 796–799.

18. Shkatulov, A. I., Gordeeva, L. G., Huinink, H., Girnik, I. S., and Aristov, Yu. I. (2020) Adsorption method for moisture and heat recuperation in ventilation: composites "LiCl/matrix" tailored for cold climate, *Energy*, in press.

19. Aristov, Yu. I., Mezentsev, I. S., and Mukhin, V. A. (2006) New approach to regenerating heat and moisture in ventilation systems. 2. Prototype of real unit, *J. Eng. Phys. Thermophys.*, **79**, pp. 151–157.

20. Alonso, M. J., Liu, P., Mathisen, H. M., Ge, G., and Simonson, C. (2015) Review of heat/energy recovery exchangers for use in ZEBs in cold climate countries, *Building Environ.*, **84**, pp. 228–237.

Chapter 19

Maintaining Relative Humidity

Nowadays, an increasing attention is being paid to the preservation of cultural heritage because of our civil responsibility. Specific parameters, including air temperature T and especially relative humidity (RH), are of primary importance because they strongly affect the degradation rate of artworks [1].

In Sec. 3.2.3, we considered requirements of an optimal adsorbent for maintaining RH in museums, libraries, and archives. Its isotherm of water adsorption should be stepwise, and the step of sorption should occur at the required RH (see Fig. 3.4b). The optimal RH for the conservation of various exhibits varies from 15 to 75%; RH = 45–60% is recommended for a majority of artworks [1].

In this chapter, we consider composites "salt in porous matrix" (CSPMs) specifically developed for adsorptive systems for maintaining RH. First, we present a novel composite buffer of water that can maintain RH at 45–60% [2]. It was tested both under laboratory conditions and in a real situation in the Museum of the History and Culture of Siberian Nations and the State Scientific Library (both in Novosibirsk, Russia), hereinafter referred to as the Museum and the Library, respectively. We have demonstrated its feasibility for efficient smoothing of daily and seasonal RH variations. The buffering ability of this CSPM appeared to be superior to common conditioned silica gels [1] and the buffer Art-sorb® [3].

Nanocomposite Sorbents for Multiple Applications
Yu. I. Aristov
Copyright © 2020 Jenny Stanford Publishing Pte. Ltd.
ISBN 978-981-4267-50-2 (Hardcover), 978-981-4303-15-6 (eBook)
www.jennystanford.com

380 | *Maintaining Relative Humidity*

Finally, we consider a family of new CSPM buffers intently synthesized to cover various RH ranges (10–20%, 20–30%, and 60–75%) [4]. These composites have been tested in the laboratory with controlled air exchange and demonstrated quite promising properties. These buffers can be applied for passive humidity control in exhibit cases and picture frames in museums, conservation bookcases in libraries, and filing cabinets in archives.

19.1 Selection of Salt and Matrix

The considered approach for maintaining RH [2] is based on the use of a gas–solid chemical reaction

$$S + nH_2O \Leftrightarrow S{\cdot}nH_2O \qquad (19.1)$$

between an inorganic salt S and water vapor with the formation of the salt crystalline hydrate $S{\cdot}nH_2O$. A specific thermodynamic feature of this reaction is that, according to Gibbs phase rule, the "salt–water vapor" system is mono-variant (see Sec. 2.5). Hence, at a fixed temperature T, the transition of the salt to the crystalline hydrate proceeds at a constant partial pressure of water vapor, P_{H_2O}. As a result, the constant value RH = $P_{H_2O}/P_o(T)$ is maintained in the "salt–vapor" system, where $P_o(T)$ is the saturated vapor pressure at temperature T. Therefore, such a reaction can be efficiently used for maintaining RH [2].

The requirements of an optimal sorbent (see Sec. 3.2.3) are based solely on thermodynamic considerations. However, several practical shortages should also be taken into account in the case of hydration–dehydration reactions of a bulk salt (Chapter 1):

- Salt swelling/expansion in the course of chemical reaction (19.1).
- Hysteresis between hydrate decomposition and synthesis reactions.
- Slow hydration kinetics due to the formation of a hydrate layer on the external surface of salt.

To overcome these problems, embedding a salt inside the pores of a host matrix was suggested. Such a composite consists of an active salt and a porous matrix, which is a host for the salt. In the latter sections, we will consider the effects of both the components on the maintained RH value.

19.1.1 Requirements of Optimal Adsorbent

The equilibrium RH value for reaction (19.1) depends, first of all, on the thermodynamic parameters of this reaction, hence on the nature of salt S. A great number of various salts and their crystalline hydrates provide an actual chance to select the proper reaction that keeps the RH value required for safe conservation in a particular case. Some of the promising salt and hydration transitions are listed in Table 19.1 (e.g., transition $1 \Rightarrow 7$ means the formation of heptahydrate from monohydrate of the salt).

Table 19.1 The RH of transition between different crystalline hydrates of various bulk salts and amount Δx of water exchanged in reaction (19.1) related to 1 kg of the starting solid

Salt	Transition	RH (%)	Δx (g/kg)
Na_2SO_4	$1 \Rightarrow 7$	80 (50–60)	788
	$7 \Rightarrow 10$	91	196
Na_2SO_3	$0 \Rightarrow 7$	67–75	1000
Na_2HPO_4	$7 \Rightarrow 12$	72–83	336
	$2 \Rightarrow 7$	52–57	506
	$0 \Rightarrow 2$	43–47 (40–45)	254
$MgSO_4$	$7 \Rightarrow 12$	72–77 (64–68)	325
	$4 \Rightarrow 6$	30–37 (25–30)	188
	$2 \Rightarrow 4$	15–17	231
$MgHPO_4$	$1 \Rightarrow 3$	23–28	261
$LiNO_3$	$0 \Rightarrow 3$	22–23 (14–17)	783
$Ca(NO_3)_2$	$0 \Rightarrow 2$	15–17 (10–13)	220
$CaCl_2$	$2 \Rightarrow 4$	16–18 (11–13)	245
	$0 \Rightarrow 2$	2–3	324

Note: In parentheses, the hydration RH is given for several salts in the confined state (the average pore size is 15 nm) [4].
Source: Reprinted with permission from Ref. [2], Copyright 2009, Taylor & Francis Informa UK Ltd.

Thus, a variation in the salt's nature allows fixing the RH in passive hydrostats at a desirable level ranging between 0 and 100%. The specific water capacity for pure salts can be significantly larger than that for other known water buffers and reaches 500–1000 g/kg (Table 19.1). Of course, in a CSPM the capacity gets lower due to the presence of a host matrix, which is an inert component of the composite. However, if the salt content in the composite is large enough, this reduction can be fully acceptable (see below).

First, we focus on nanotailoring an efficient buffer for the RH range between 45 and 60% as this humidity is required for a majority of artworks, especially those of organic origin. Sodium sulfate was selected as an active salt as its bulk monohydrate reacts with vapor:

$$Na_2SO_4 \cdot H_2O + 6H_2O \Leftrightarrow Na_2SO_4 \cdot 7H_2O \qquad (19.2)$$

at RH = 80%, which allows as many as six molecules of water (or 788 g/kg $Na_2SO_4 \cdot H_2O$) to be bound (Table 19.1). This considerably exceeds the amount exchanged using the salt solutions, conditioned silica gels, and the Art-sorb. The equilibrium RH value looks too high; however, one should take into account that it reduces down to the required values of 50–60% [5] when the salt is placed inside the pores of the silica gel Grace Davison with d_{av} = 15 nm (Fig. 19.1).

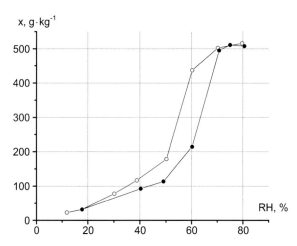

Figure 19.1 The isotherms of water adsorption (●) and desorption (o) for the composite (28 wt.% Na_2SO_4)/(silica gel Grace Davison) measured at T = 23°C [2].

The composite (28 wt.% Na_2SO_4)/(silica gel Grace Davison) has quite steep sorption isotherms with essential hysteresis. A steep rise in uptake from 0.22 to 0.50 is observed when the RH increases from 60 to 70%. A steep release drop from 0.44 g/g to 0.18 g/g is observed when the RH decreases from 60 to 50%. Thus, this material can efficiently maintain the RH level between 50 and 60% if the ambient humidity is lower than 50%. Indeed, in this case, only the desorption branch is important. The desorption curve concerns Russia, Canada, Northern Europe, and the USA, where humidification is the only necessary action, especially during winter. Additional fine-tuning of the RH value can be made by varying the size of silica pores (see the following sections).

19.1.2 Effect of the Matrix

The main function of the matrix is scattering the active salt into a large number of tiny pieces. Each salt nanoparticle interacts with water much faster than a lump of salt, and for the former, the synthesis–decomposition hysteresis can be reduced or even avoided. Besides, a host matrix provides a network for efficient transfer of heat and vapor, which greatly enhances the hydration process. As it appears, a host matrix should have a large volume of pores to enhance the content of an active salt in the composite. Moreover, matrices with large pore size and volume can better accommodate the swelling of salt due to the chemical reaction between the salt and water [6].

Moreover, as clearly stated earlier, the confinement of salt to the pores can cause a reduction in its hydration temperature, which appears to depend on the pore size of the host matrix. Indeed, this temperature is higher (accordingly, the equilibrium RH value is lower) if the salt is located in smaller pores (see Fig. 19.2). A typical increase in this temperature is 10–20°C (Fig. 19.2b) or 10–20% in terms of the RH (at the reaction temperature close to 20°C). Possible reasons for this effect are discussed in Chapter 6.

We synthesized and tested three composites A1, A2, and A3, based on sodium sulfate confined to three silica gels whose characteristics are displayed in Table 19.2. The salt content was 26 ± 2 wt.%. The water sorption properties of the composites were measured by a thermogravimetric method [2].

The water desorption curves for these water buffers demonstrate a region where the release of water is sharper than for the commercial buffer Art-sorb (Fig. 19.2). The location of this region depends, first of all, on the pore size of the host silica gel (Table 19.2): In smaller pores, desorption (dehydration) occurs at lower RH. This finding is in line with our previous observations discussed in Chapter 6.

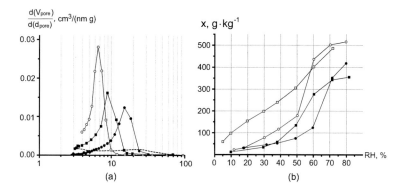

Figure 19.2 Composite water buffers Na$_2$SO$_4$/silica gel: (a) pore size distribution, (b) isotherms of water desorption measured at T = 23°C. A1 (•), A2 (■), A3 = ARTIC-1 (○), and Art-sorb (□).

Table 19.2 Texture of the host silica gel and the salt content for composite buffers A1–A3

Composite	Silica pore size (nm)	Silica pore volume (cm^3/g)	Salt content (wt.%)
A1	16	1.1	27.3
A2	9	1.0	29.2
A3	6	0.9	25.8

Composite A1 desorbs the maximal amount M of water in the range of RH between 60 and 70%, which is equal to M = 220 g/kg. The maximum M for the A2 and A3 composites shifts toward lower RH of 50–60% and equals 140 and 250 g/kg, respectively. It is worth noting that the buffering capacity of Art-sorb was 60, 90, and 80 g/kg in the RH range of 40–50, 50–60, and 60–70%, respectively [3]. Thus, the novel composite sorbents exceed Art-sorb regarding the buffering capacity at all RH ranges that are of practical interest. As described earlier, the RH range between 45 and 60% is of primary importance for keeping organic artworks. Composite A3 exchanges as much as 300 g/kg in this RH range, which is 2.3 times that of Art-sorb (130 g/kg). At the RH range of 50–60%, the buffering capacity of composite A3 is 250 g/kg, which is 2.6 times that of the buffering capacity of Art-sorb. Further below, we will call composite A3 as ARTIC-1 (**ART** + **I**nstitute of **C**atalysis) and study it in more detail as described in the next section. For all the composites, the sorption at RH > 70% remains almost constant, indicating that only solid crystalline hydrates form inside the pores, but not the salt solution. As there is no liquid phase in the system, this prevents the salt from leaking out of the pores and chemically affecting the exhibits.

Taking into account the thermodynamic mono-variance of the "salt–water" system, one could expect that the transformation between different hydrates should occur in the RH range much narrower than the experimentally observed one (Figs. 19.1 and 19.2). This broadening can be caused by the fact that the hydration/dehydration temperature (and, consequently, RH) depends on the size of the salt crystals. As the salt forms the crystallites of various dimensions inside the silica pores, they transform to hydrate at different RH, resulting in the extension of the transition RH range. For mono-sized pores, one can expect smaller broadening.

19.2 ARTIC-1 Testing

The new composite water buffer ARTIC-1 was tested further both in a laboratory prototype of demonstration showcase and under real conditions of the Museum and the Library.

19.2.1 Laboratory Tests

The demonstration showcase for testing ARTIC-1 and its comparison with Art-sorb was made of polymethyl methacrylate (Fig. 19.3). It had dimensions of $60 \times 35 \times 25$ cm^3 and volume $V = 45$ dm^3. The buffer (ARTIC-1 or Art-sorb) was placed in a textile bag and centered on the bottom of the box. The sorbent volume was equal to $V_{ad} = 150$ cm^3 so that the ratio V/V_{ad} was 300. The tests were performed at ambient temperature under flow and static operation modes. At the former mode, dry air was saturated with water vapor in an evaporator up to RH = 30% and then passed through the box at the rate $v = 1.9$ cm^3/s. This corresponds to the air exchange Ω inside the showcase of 3.5 1/day. The air exchange typical of common well-designed showcases is much smaller, $\Omega = 0.3$–1.0 1/day [1, 7, 8]. The lower air exchange rate promotes better buffering ability of the showcases. Hence, the adsorbent was tested under conditions harsher than those in the actual practice and a better performance of the sorbent can be expected in common showcases with lower air exchange rate. The RH of the inlet air was controlled by a hygrometer IVA-6B, whose sensors were located in two different positions, namely, near the air inlet and over the adsorbent bag (Fig. 19.3), and revealed the difference in readings less than 1%.

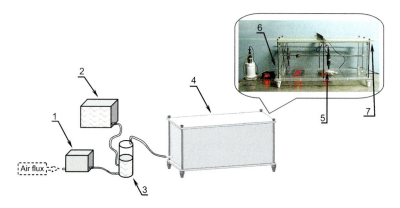

Figure 19.3 Scheme of the test rig unit: 1 — air flux regulator; 2 — thermostat for maintaining constant temperature in evaporator 3; 4 — showcase with sorbent bag 5; 6 — air inlet; 7 — air outlet [2].

The initial RHs were 63 and 58% over ARTIC-1 and Art-sorb, respectively (Fig. 19.4). After the dry air entered the showcase, the humidity gradually decreased as the sorbent dried progressively. The daily RH drift in the presence of ARTIC-1 was 0.47% so that the humidity inside the showcase was maintained within the necessary range of 50–60% for more than 20 days. Then the sorbent had to be reconditioned again. For the same volume of Art-sorb (V/V_{ad} = 300), the daily drift was 1.07%, which maintained the RH between 50 and 60% for less than 10 days. Thus, its efficiency is smaller by a factor of 2.3, which is in good agreement with the difference in static water capacities of these sorbents in the RH range of 50–60% (Fig. 19.2).

It is evident that the decrease in the air exchange Ω between the showcase and the environment results in better performance of the sorbent, whose "working time" is inversely proportional to the air exchange. For instance, at the air exchange of $\Omega = 0.3$ 1/day typical for a well-designed showcase, ARTIC-1 would maintain the necessary RH for longer than half a year without sorbent reconditioning. It is quite sufficient to go through the winter when room heating greatly reduces the indoor RH. Thus, for a showcase of volume $V = 1$ m^3 with $\Omega = 0.3$ 1/day, it is possible to manage the required RH for half a year with $V_{ad} = 3.33$ dm^3 or just 3 kg of ARTIC-1.

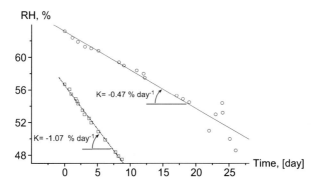

Figure 19.4 The dynamic RH inside the showcase with ARTIC-1 (○) and Artsorb (□) [2].

In the static tests, the showcase was completely sealed (air exchange $\Omega = 0$) and cyclically heated and cooled with an amplitude of 10°C/day. This regime imitated the worst case of daily temperature cycles in museums not equipped with heating, ventilation, and air conditioning (HVAC) systems. The absolute humidity inside the airtight showcase without any humidity buffering remained constant. As a result, the RH was enslaved to the temperature variation and underwent large fluctuations caused by the temperature change (Fig. 19.5). Inside the showcase equipped with ARTIC-1, the excursion of RH was smoother. At increasing/decreasing temperature, the adsorbent started to desorb/adsorb water vapor and regularized RH changes caused by the temperature drift. The amplitude of RH variation was 4–5 times lesser than without any adsorbent and 2.5–3 times lesser than that outside the box (Fig. 19.5). The inbox average RH noticeably exceeded the external one and fluctuated near the required value of 55%.

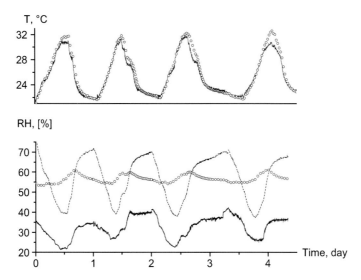

Figure 19.5 Damping down the RH alterations (the bottom graph) caused by the daily variation in the temperature (the top graph): (○) — temperature and RH inside the showcase with ARTIC-1; solid line — the external temperature and RH; dotted line — the RH alterations inside the showcase without any adsorbent [2].

19.2.2 Tests in the Museum and the Library

19.2.2.1 Tests in the museum

These tests were performed in showcases with limited air exchange, in which organic artifacts were displayed in the presence of ARTIC-1. The showcase used for the experiments was 27 × 68 × 38 cm³ in dimensions (with volume 70 dm³) and well sealed to minimize air exchange with the environment. The transparent material of the showcase allowed visual control over the state of the exhibit. The exhibit was the head of a female mummy found in mound No. 2 (Upper Caljin-2, Ukok Plateau, Altai Mountains) [9] and having partially preserved mummified tissues. The artifact was placed in the showcase together with the ARTIC-1 composite sorbent saturated with water to the required initial level of moisture content (initial weight ≈ 100 g).

The thermohygrometer readings inside the airtight showcase were compared with those in the exhibition hall as well as inside the demonstration container with the female mummy over a period from July 26 to October 1, 2007. In two days after the start of the experiment, the humidity dropped from 63 to 55% (Fig. 19.6a). The next two months brought a decrease of another 7%, after which humidity was stabilized at 48%. The initial drop in RH probably resulted from the fact that the head was initially dry and gained moisture from the showcase's internal space. The sorbent partially offset the change until the equilibrium was restored at RH = 48%. Indeed, weighing the sorbent revealed a 15 g loss of water (the buffer-specific water content decreased from 0.21 to 0.06 g/g).

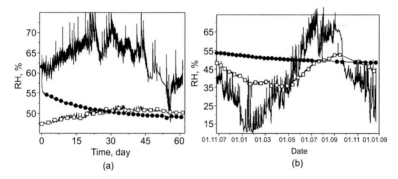

Figure 19.6 Relative humidity evolution in the Museum exhibition hall (solid line), inside the half-air-tight sarcophagus with a female mummy (□) and inside the airtight showcase equipped with ARTIC-1 (●). The first (a) and second (b) experimental steps (see the text) [2].

The second stage of the experiment started after the weight loss was compensated to a moisture content of 0.28 g/g. Within the first 3–4 days, the humidity in the showcase slightly increased, evidently because of the moisture exchange with the displayed fragment (head). As a result, the moisture content in the latter increased to the normal one and the equilibrium value of RH inside the showcase was settled at 54% (Fig. 19.6b). Within the next 2 months, the humidity was almost constant. The testing was continued in 2008, showing a very slow humidity drift in the whole year despite wide seasonal variation outside the showcase (Fig. 19.6b). That is, ARTIC-1 worked as an ideal humidity buffer under the conditions of a completely sealed showcase.

Then we continued this investigation in the actual showcase where the male mummy was exposed in the same hall of the Museum. For this sarcophagus, the intensity of air exchange was measured to be 1/25 per day. Taking into account this value, the required amount of ARTIC-1 (115 g) was estimated and placed inside the showcase with the male mummy (Fig. 19.7a). The adsorbent was conditioned to the water content of 0.29 g/g, which is the equilibrium water content at RH = 55%.

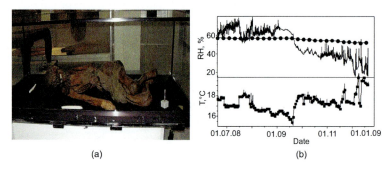

Figure 19.7 (a) Sarcophagus with the male mummy, RH sensor and ARTIC-1 encased in three textile saccules (38, 38 and 39 g) [2]; (b) the RH (solid line) and temperature (■) evolution in the Museum exhibition hall and inside the sarcophagus with the male mummy (●) containing ARTIC-1 (see text) [2].

The test was started on June 9, 2008. The RH was stabilized at 57% and maintained at this value for the first 3 months (Fig. 19.7b). The adaptation period, observed in the testing of ARTIC-1 in the small showcase with the mummy's head, was absent in that experiment, probably because the male mummy had been kept under proper humidity conditions (at RH ≈ 55%) before the testing was started. In October 2008, the central heating was initiated and the level of RH in the exhibition hall decreased, which caused appropriate reduction in the RH inside the sarcophagus down to 53% (Fig. 19.7b).

19.2.2.2 Tests in the library

Tests in the Library were aimed at safely conserving ancient books and were performed in a standard box made of acid-free cardboard. To reduce the air exchange with the environment, the box was placed in a plastic bag with a tight-fitting cap (Fig. 19.8a). The RH sensor was placed in the cardboard box. ARTIC-1 saturated with moisture up to $x = 0.40$ g/g and an old book were placed in the plastic bag.

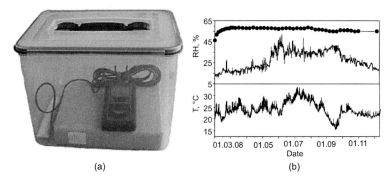

Figure 19.8 (a) Conservation bag for books (inside the carton box) equipped with ARTIC-1 (the white bag) [2]; (b) RH evolution in the exhibition hall of the Library (solid line) and in the bag equipped with ARTIC-1 (•) [2].

After an equilibration period of 10 days, the RH was fixed at 59% (Fig. 19.8b). During the next 10 months, the humidity was reduced to 55%, while the average RH in the depositary was as low as 25–30%. Hence, under those conditions, ARTIC-1 could smooth out the daily fluctuations and maintain the recommended RH in this box for c.a. 1 year without regeneration. The experiments performed in the Library completely confirmed that ARTIC-1 is an effective humidity buffer when the hygrostat is airtight.

Thus, the new adsorption buffer for maintaining RH between 45 and 60%, ARTIC-1, was developed and tested under the storage and exhibition conditions of the Museum and the Library. Its buffering capacity within the mentioned RH range reaches 0.30 g H_2O per gram of sorbent, which significantly exceeds the capacity of known conventional analogues. The testing of ARTIC-1 in a lab-scale prototype as well as in actual showcases for the presentation of unique organic artifacts in the Museum and the containers for conserving rare manuscripts in the Library demonstrated the high efficiency of ARTIC-1. These trials confirmed that ARTIC-1 is an effective water buffer for use in real, sufficiently airtight showcases and containers.

19.3 New CSPM Buffers for RH < 45%

Several new CSPMs were synthesized and studied for relatively low RH. According to the data of Table 19.1, we selected lithium and calcium nitrates as active salts for this RH range. In bulk, these salts can maintain RHs of 22–23 and 15–17%, respectively. These salts were inserted in two matrices with quite different pore sizes. A silica gel KSK has mesopores with an average size of 15 nm and a specific volume of 1 cm^3/g. A vermiculite has wide pore size distribution between 100 and 10,000 nm; its pore volume is 3.5 cm^3/g, which allows insertion of quite a large mass of salt inside the pores, up to 63–72 mass.%.

We found that confining the salts in the pores of both matrices shifted the step corresponding to hydration reaction to smaller RH values (Fig. 19.9). For the bulk salts, the hydration isotherms were very steep due to the mono-variant character of the equilibrium. Similar step-like isotherms were observed in the vermiculite pores, but the equilibrium RH was a bit smaller than in bulk (Fig. 19.9, curve 3). The composites based on Ca(NO$_3$)$_2$ and LiNO$_3$ were found to exchange quite a large quantity of water (approximately 200 and 400 g/kg) and are expected to be used for maintaining RH in the ranges of 13–14 and 18–22%, respectively.

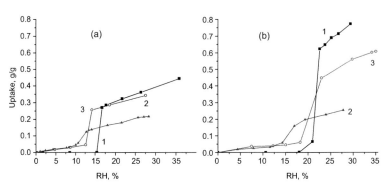

Figure 19.9 Water sorption as a function of the relative humidity for Ca(NO$_3$)$_2$ (a) and LiNO$_3$ (b): bulk (1) and confined to silica gel KSK (2) and vermiculite (3).

For CSPMs based on silica gel, hydration occurred at lower RH than in the vermiculite pores (Fig. 19.9, curve 2), which was due to smaller pores of this matrix. The transition region was smoother, and the mass of water exchanged was much smaller (approximately 100 g/kg). Thus, these composites are hardly promising for application in passive hydrostats.

The vermiculite-based composites were tested in a plastic container with an air exchange of 1.0 volume per day. We found that both composites can, indeed, stabilize the RH value near a certain level (see, e.g., Fig. 19.10 for LiNO$_3$/vermiculite), which is somewhat higher than that predicted from the equilibrium curves in Fig. 19.9. This can be due to the dynamic character of the water exchange process and will be a subject of further detailed analysis.

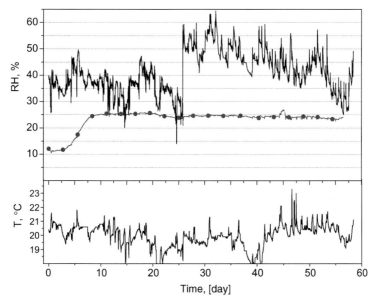

Figure 19.10 Evolution of RH and temperature in the container with LiNO$_3$/vermiculite (•) (symbols) and in the ambient (black lines). Reprinted by permission from Springer Nature from Ref. [10], Copyright 2012.

To sum up, a family of new CSPM buffers of water has been developed for various ranges of RH. These are nanocomposites "salt in porous matrix" based on metal sulfates and nitrates. These composites have been tested in laboratory prototypes of showcase with controlled air exchange and demonstrated quite promising properties. These buffers are also intended to be applied to passive humidity control in exhibit cases and picture frames in museums, as well as in containers transporting artworks and books.

Further testing of these novel materials is required to comprehensively investigate their mechanical, chemical, and hydrothermal stability and apply them in the everyday practice of libraries and archives. This asks for common efforts of an interdisciplinary team of professionals from materials science and conservation sectors within research projects at national and international scales.

References

1. Tomphson, G. (1986) *The Museum Environment* (London-Boston), p. 270.

2. Glaznev, I. S., Salnikova, I. V., Alekseev, V. N., Gordeeva, L. G., Elepov, B. S., Kundo, L. P., Shilova, I. A., and Aristov, Yu. I. (2009) ARTIC-1: A new humidity buffer for showcases, *Studies Conserv.*, **54**, pp. 133–148.

3. http://www.artsorbonline.com/ (last opening on January 4, 2020).

4. Aristov, Yu. I., Glaznev, I. S., Salnikova, I. V., Alekseev, V. N., Gordeeva, L. G., Shilova, I. A., Borodikhin, A. Yu., Kundo, L. P., and Elepov, B. S. (2009) ARTIC: A new humidity buffer for showcases. In: *Research in Book and Paper Conservation in Europe: A State of the Art*, P. Engel (Ed.) (Impressum ©Verlag Berger, Horn/Wien), pp. 207–240.

5. Gordeeva, L. G., Savchenko, E. V., Glaznev, I. S., Malakhov, V. V., and Aristov, Yu. I. (2006) Impact of phase composition on water adsorption on inorganic hybrides "salt/silica," *J. Coll. Interface Sci.*, **301**, pp. 685–691.

6. Aristov, Yu. I., Restuccia, G., Tokarev, M. M., Buerger, H.-D., and Freni, A. (2000) Selective water sorbents for multiple applications. 11. $CaCl_2$ confined to expanded vermiculite, *React. Kinet. Cat. Lett.*, **71**, pp. 377–384.

7. Thickett, D., Fletcher, P., Calver, A., and Lambarth, S. (2007) The effect of air tightness on RH buffering and control. In: *Museum Microclimates,* T. Padfield and K. Borchersen (Eds.) (National Museum of Denmark, Copenhagen), pp. 245–251.

8. Weintraub, S. (2002) Demystifying silica gel. In: *Objects Specialty Group Postprints,* V. Greene and P. Griffin (Eds.) (American Institute for Conservation, Washington, D.C.).

9. Polosmak, N. V. (2001) *Ukok Riders* (Infolio Press, Novosibirsk) (in Russian).

10. Aristov, Yu. I., Vasiliev, L. L., Glaznev, I. S., Gordeeva, L. G., Zhuravlev, A. S., and Kovaleva, M. N. (2012) Physicochemical bases of autonomous maintenance of humidity and temperature in closed spaces, *J. Eng. Phys. Thermophys.*, **85**, 977–986.

Chapter 20

Shifting Chemical Equilibrium

It is well known that the conversion of reactants in many important catalytic processes is limited by the chemical equilibrium of the reversible reaction. To increase conversion, it is desirable to shift the equilibrium of such a process toward the formation of products. This can be performed by removing a product from the reaction mixture by its adsorption on appropriate material [1, 2].

As discussed in Sec. 3.2.4, the thermodynamic requirements of an optimal adsorbent for the examined applications boil down to a large change in the adsorption uptake either between the two adsorbent states characterized by the fixed sets (T_1, P_1) and (T_2, P_2) or at fixed T and P. These demands were formulated [3] for the particular case of methanol synthesis $CO + 2H_2 = CH_3OH$, which is a very important industrial process. Composite sorbents $Ca(NO_3)_2/SiO_2$, $LiBr/SiO_2$, and $CaCl_2/SiO_2$ were intently designed for shifting the equilibrium of this reaction. These composites were tested in a lab-scale tubular flow reactor. It was shown that one of these composites efficiently sorbed methanol from the gaseous mixture under the required reaction conditions and demonstrated good thermal stability.

20.1 Introduction to the Problem

Methanol synthesis from carbon monoxide and hydrogen is one of the most important processes in the chemical industry. Conversion

Nanocomposite Sorbents for Multiple Applications
Yu. I. Aristov
Copyright © 2020 Jenny Stanford Publishing Pte. Ltd.
ISBN 978-981-4267-50-2 (Hardcover), 978-981-4303-15-6 (eBook)
www.jennystanford.com

of the reactants in a conventional tubular fixed-bed reactor is limited by the reaction thermodynamics. Due to the unfavorable equilibrium, methanol synthesis is conducted over a catalyst at high pressure (5–10 MPa), but only a fraction of raw materials is converted. To utilize reactants at a high enough rate, they should be separated from the reaction products and recycled. Recycling of reactants unfavorably affects energy consumption and the cost of product.

Removing the product from a reaction gas mixture appears to be a promising way of increasing the conversion of reversible reactions [2, 4]. Integrating an adsorptive process with a chemical reaction in multifunctional reactors is an important means of process intensification having favorable economics [5–8]. In any approach, the characteristics of an adsorbent dramatically affect process performance. The most important parameters are (a) adsorbent selectivity with respect to methanol and (b) high sorption capacity at operating conditions that allow minimization of the adsorbent recycle ratio (the adsorbent weight circulated per 1 ton of methanol produced). Amorphous LA-25 low alumina cracking catalyst (Akzo, Amsterdam) was used as the adsorbent of methanol [5, 7]. It adsorbed methanol selectively, and its equilibrium uptake w was 0.08–0.13 g/g under operating conditions (temperature range of 473–533 K and methanol partial pressure $P_{\mathrm{MeOH}} \geq 0.7$–0.9 MPa) [5]. The dynamic content of methanol in the adsorbent under reaction conditions was even less and did not exceed 0.02 g/g [7]. Low adsorption capacity leads to a rather high adsorbent recycle ratio of 15–20 tons of the adsorbent per 1 ton of methanol. Hence, new methanol adsorbents with advanced capacity are welcome.

Composites "salt in a porous matrix" (CSPMs) proposed for methanol sorption possess an outstanding capacity up to 0.6–0.8 g/g (see Chapter 7 and Refs. [3, 9]). Furthermore, a two-component nature of these materials opens challenging opportunity of designing the composite, whose properties meet the requirements of a particular application, by varying the nature of salt and matrix, the pore structure of the matrix, salt content, and synthesis conditions [10].

The main goal of the study [3] was preparation of CSPMs specialized for shifting the equilibrium of methanol synthesis toward the product by means of methanol sorption. First, thermodynamic requirements for an ideal adsorbent optimal for this application are formulated on the basis of the Polanyi principle of temperature invariance. Then, real adsorbents, whose properties are close to the ideal one, are synthesized. The novel adsorbents are tested in a lab-scale reactor of methanol synthesis.

20.2 Theoretical Considerations

As usual, we first consider the requirements to an optimal adsorbent for methanol synthesis. Based on these requirements, we select several promising salts to be experimentally tested.

20.2.1 Formulating Requirements to Optimal Adsorbent

General requirements for shifting chemical equilibrium were considered in Sec. 3.2.4. In Ref. [3] they are adapted for typical conditions of methanol synthesis. It was previously shown that the sorption of both methanol and water on CSPMs follows the Polanyi invariance [10, 11]. Thus, for formulating the mentioned requirements, one has to evaluate the adsorption potential ΔF_{syn} under typical conditions of methanol synthesis (Table 20.1). We consider two operation modes. The first mode corresponds to common methanol synthesis: a stoichiometric initial mixture CO/ $H_2 = 1/2$, $P = 50$ bar, and $T = 180–240°C$ [12]. The second mode is related to the synthesis in a lab-scale tubular flow reactor described as follows: the initial mixture composition is $CO/CO_2/N_2/H_2 = 30.0/2.0/3.2/64.8$, the total pressure $P = 20$ bar, and temperature $T = 200–240°C$ [13]. For both modes, the equilibrium methanol content in the reaction mixture and, hence, the value of ΔF_{syn} can vary in a wide range depending on particular operating conditions. For the first mode, the adsorbent used for methanol fixation has to sorb it at more severe conditions, that is, at a higher value of ΔF_{syn} (4.0–15.1 kJ/mol) because of the lower total pressure in the reactor (Table 20.1).

One more important demand is imposed by the fact that the methanol adsorbed should be easily extracted from the adsorbent, for instance, by purging [7] or pressure swing desorption [14]. Hence, the bonding between the methanol molecules and the adsorbent should not be too strong in order to minimize purge gas and energy consumption. It is optimal that desorption proceeds at the free sorption energy ΔF_{des}, which is just a little bit higher

Shifting Chemical Equilibrium

than ΔF_{syn}. Hence, a compromise between large sorption at a high temperature and easy regeneration has to be reached. And finally, the optimal adsorbent should be chemically inert and thermally stable under synthesis conditions.

Table 20.1 The equilibrium methanol concentration C_{eq} (CH$_3$OH) and the threshold adsorption potential ΔF_{syn} for the two modes of methanol synthesis

T (°C)	C_{eq} (CH$_3$OH) (%)	ΔF_{syn} (kJ/mol)
$P = 20$ bar, CO/CO$_2$/N$_2$/H$_2$ = 30.0/2.0/3.2/64.8		
180	49.5	4.0
200	34.9	7.0
220	21.6	10.6
240	11.4	15.1
$P = 50$ bar, CO/H$_2$=1/2		
200	69	0.6
220	57	2.8
240	44	5.4

Source: Reprinted by permission from Springer Nature from Ref. [3], Copyright 2011.

20.2.2   Selection of Proper Salts

A great number of salts (S) form crystalline solvates due to reaction (7.1) with methanol. Literature data on "methanol–salt" equilibrium for bulk salts [15] (Table 7.2) as well as for CSPMs, based on LiBr, LiCl, and CaCl$_2$ confined to silica mesopores [9, 11], were analyzed in order to select salts appropriate for shifting synthesis equilibrium under the two mentioned modes. The majority of the salts considered react with methanol vapor at the adsorption potential $\Delta F \leq 5$ kJ/mol, which is lower than the required ΔF_{syn} value. Only LiBr, Mg(NO$_3$)$_2$, and CaCl$_2$ appear to be of interest, because they react with methanol at ΔF = 9.5, 6.9, 12.0 kJ/mol, respectively, thus meeting the formulated requirements. Despite the proper ΔF_{syn} value of Mg(NO$_3$)$_2$, this salt can hardly be used because of its thermal instability. Decomposition of Mg(NO$_3$)$_2$ can proceed at a temperature over 100°C due to hydrolysis [16]:

$$Mg(NO_3)_2 + H_2O = Mg(OH)NO_3 + HNO_3 \qquad (20.1)$$

$$Mg(OH)NO_3 + H_2O = Mg(OH)_2 + HNO_3 \qquad (20.2)$$

The adsorption potential of $Ca(NO_3)_2$ solvatation, $\Delta F = 4.5$ kJ/mol, is just a bit lower than the required value. However, one may expect that confinement of this salt to matrix pores will shift the solvatation toward higher adsorption potentials [9, 10]. Thus, LiBr, $CaCl_2$, and $Ca(NO_3)_2$ have been selected for synthesis of CSPMs, which are expected to be efficient for methanol reversible binding under typical conditions of methanol synthesis.

20.3 Experimental Study of Sorbent-Assisted Methanol Synthesis

20.3.1 Composites Preparation

Silica gels Davisil Grade 646 (specific area $S_{sp} = 300$ m^2/g, pore volume $V_p = 1.15$ cm^3/g, average pore diameter $d_n = 15$ nm, and particle diameter $D_p = 250$–500 μm) and Grace Davison SP18-8749.01 ($S_{sp} = 580$ m^2/g, $V_p = 0.85$ cm^3/g, $d_n = 6$ nm, $D_p = 250$–500 μm) were used as host matrices. The composites were synthesized by the dry impregnation method (see Chapter 4). The silica gels were dried at 433 K for 2 h, impregnated with the appropriate amount of aqueous salt solution, and dried again at 433 K until the sample weight remained constant. The composites LiBr (31%)/SiO_2 (15 nm), $CaCl_2$ (28%)/SiO_2 (6 nm), and $Ca(NO_3)_2$ (34%)/SiO_2 (15 nm) were prepared. In this chapter, the scheme of sample designation is "salt (salt content in mass.%)/SiO_2 (pore size in nm)" [3].

20.3.2 Testing Facilities

The new composites were tested by the continuous flow method at the total pressure $P = 20$ bar and temperature $T = 493$ K. The scheme of a lab-scale reactor/adsorber system is presented in Fig. 20.1. Methanol was synthesized in the reactor loaded with a Cu–Zn–Al catalyst of 8 g weight mixed with quartz particles. The composition of the feeding gas mixture was as follows: 30.0 vol.% CO, 2.0 vol.% CO_2, 3.2 vol.% N_2, and 64.8 vol.% H_2. The contact time τ was 20–25 s. The gas mixture at the reactor outlet was passed through an adsorber loaded with 9.4–12.6 g of the composites synthesized. The composition of the outlet mixture was analyzed by a chromatographic technique using flame-ionization and thermal conductivity detectors.

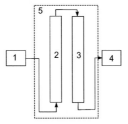

Figure 20.1 Scheme of the lab-scale device for methanol synthesis. 1 — unit of reactants dosing, 2 — reactor, 3 — adsorber, 4 — gas chromatograph, 5 — oven. Reprinted by permission from Springer Nature from Ref. [3], Copyright 2011.

20.3.3 Methanol Sorption under Reaction Conditions

The equilibrium uptake of methanol as a function of the adsorption potential under typical conditions of methanol synthesis is presented for the LiBr- and CaCl$_2$-based composites in Fig. 20.2. The data for commercial LA-25 adsorbent [17] are presented for comparison as well. Actually, these composites demonstrate much higher sorption capacity in the whole range of adsorption potentials, which gives a basis for their good performance in shifting the equilibrium of methanol synthesis.

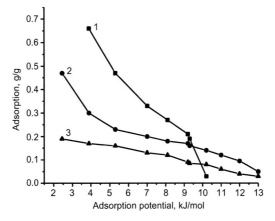

Figure 20.2 Methanol sorption as a function of the adsorption potential for the LiBr/SiO$_2$(15) (1) and CaCl$_2$/SiO$_2$(15) (2) composites, and commercial adsorbent LA-25 (3). Adapted from Ref. [3], Copyright 2011, Springer Nature.

Before the adsorption experiments, in a blank test methanol was synthesized without the adsorbent to measure the composition of gas mixture at the reactor outlet. The methanol concentration in the outlet mixture varied from 17 to 23 vol.%, which was close to the equilibrium methanol concentration $C_{eq}(CH_3OH)$ = 22 vol.% (Fig. 20.3). Methane, ethane, propane, and dimethyl ether were detected in negligible amounts.

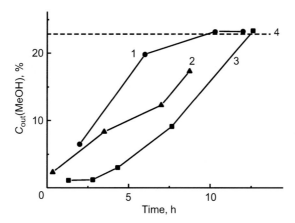

Figure 20.3 Methanol concentration at the outlet of the reactor and adsorber with LiBr/SiO$_2$(15) (1), CaCl$_2$/SiO$_2$(6) (2), Ca(NO$_3$)$_2$/SiO$_2$(15) (3) as well as the equilibrium methanol concentration (4). Adapted from Ref. [3], Copyright 2011, Springer Nature.

The starting methanol concentration at the adsorber outlet $C_{out}(CH_3OH)$ = 1.1–6.0 vol.% for all the composites was significantly lower than $C_{out}(CH_3OH) \approx C_{eq}(CH_3OH)$ in the blank experiment (Fig. 20.3). This demonstrated that all the composites adsorbed methanol under conditions of its synthesis reaction. In the course of reaction, the outlet methanol concentration was continuously increasing and approaching its equilibrium value (Fig. 20.3). It is important that the outlet concentration remained lower than the blank value for 8–12 h, showing the large potential of the sorbents tested. The LiBr/SiO$_2$ composite appeared to be less effective, and the methanol concentration reached the blank value in 8–9 h.

The CaCl$_2$/SiO$_2$ composite demonstrated better performance, giving $C_{out}(CH_3OH)$ < 10% for 5 h. This agrees well with the data on methanol sorption equilibrium on the composites (Fig. 20.2), which show a higher adsorption ability of CaCl$_2$-based composite at $\Delta F > 10$ kJ/mol. The Ca(NO$_3$)$_2$/SiO$_2$ composite demonstrated the best performance, providing C_{out}(CH$_3$OH) < 10% for 8 h. The total amounts of methanol sorbed during the tests were 0.07, 0.10, and 0.15 g/g for the LiBr/SiO$_2$, CaCl$_2$/SiO$_2$, and Ca(NO$_3$)$_2$/SiO$_2$ composites.

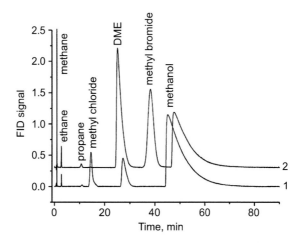

Figure 20.4 Gas chromatograms of the outlet gas mixture after 5 h on run using the CaCl$_2$-based sorbent (1) and after 1 h on run using the LiBr-based sorbent (2). Reprinted by permission from Springer Nature from Ref. [3], Copyright 2011.

It is worth noting that the composites were tested under rather rigorous experimental conditions ($P = 20$ bar, $T = 220°C$) at the value of adsorption potential $\Delta F = 10.6$ kJ/mol. Higher pressure typical for conventional methanol synthesis ($P = 50–100$ bar) would give larger equilibrium methanol pressure and, consequently, higher adsorption potential, which could favor larger sorption capacity of the composites. Thus, at high-pressure methanol synthesis, the tested sorbents are expected to demonstrate better performance.

20.3.4 Stability of Composites

Another parameter of practical importance is the thermal stability of the adsorbent used for shifting the equilibrium. When using the

Ca(NO$_3$)$_2$/SiO$_2$ composite, the outlet mixture contained the same components as in the blank experiment (CH$_3$OH, minor amounts of DME and light alkanes). In contrast, an additional peak foregoing the peak of methanol was detected on the chromatographic diagrams for the tests with CaCl$_2$/SiO$_2$ and LiBr/SiO$_2$ composites (Fig. 20.4). It might be related to methyl chloride or methyl bromide formation due to the following reactions:

$$CaCl_2 + CH_3OH = Ca(OH)Cl + CH_3Cl \qquad (20.3)$$

$$LiBr + CH_3OH = LiOH + CH_3Br \qquad (20.4)$$

Furthermore, when testing the LiBr/SiO$_2$ composite, the starting concentration of dimethyl ether in the outlet mixture, C_{out}(DME) = 1.38 vol.%, was essentially larger than that for the blank experiments (0.3–0.4 vol.%), which probably indicates dehydroxylation of methanol on the adsorbent. Then C_{out}(DME) decreased to the value of the blank experiment.

X-ray diffraction (XRD) analysis of the phase composition of fresh and tested composites was performed in order to detect if any transformation of phase composition occurred. The XRD patterns of the Ca(NO$_3$)$_2$/SiO$_2$ and LiBr/SiO$_2$ composites did not change during the testing. In contrast, additional reflections at 2θ = 30.2, 36.9, and 43.3° appeared at the XRD pattern of the CaCl$_2$/SiO$_2$ composite (not presented), showing the appearance of a new phase [3]. The reflections could be related to Ca(OH)Cl (PDF#73-1885) and probably micro-quantity of hydrogenation phases, which confirmed possible contribution of reaction (20.3).

Thus, the composite adsorbent Ca(NO$_3$)$_2$/SiO$_2$ has been intently designed for shifting the equilibrium of methanol synthesis by removing methanol from the reaction mixture. Testing this composite has shown that it efficiently adsorbs methanol from the gaseous mixture under lab-scale operating conditions of methanol synthesis and demonstrates good thermal stability. A detailed study of the Ca(NO$_3$)$_2$/SiO$_2$ adsorbent, which includes measurements of its methanol adsorption equilibrium, testing its performance in shifting the equilibrium of methanol synthesis, and its long-term stability under conditions of synthesis, is needed to be performed in the future for further progress in this field.

References

1. Kuczynski, M., Oyevaar, M. H., Pieters, R. T., and Westerterp, K. R. (1987) Methanol synthesis in a countercurrent gas–solid–solid trickle flow reactor: An experimental study, *Chem. Eng. Sci.*, **42**, pp. 1887–1898.

2. Roes, A. W. M. and Van Swaaij, W. P. M. (1979) Hydrodynamic behavior of a gas–solid counter-current packed column at trickle flow, *Chem. Eng. J.*, **17**, pp. 81–89.

3. Gordeeva, L. G., Khassin, A. A., Chermashentseva, G. K., and Krieger, T. A. (2012) New adsorbents of methanol for the intensification of methanol synthesis, *React. Kinet. Mech. Cat.*, **105**, pp. 391–400.

4. Carr, R. W. (1992) Continuous reaction chromatography. In: *Preparative and Production Scale Chromatography*, G. Ganetson and P. E. Barker (Eds.) (Marcel Dekker, New York), pp. 421–447.

5. Agar, D. W. (1999) Multifunctional reactors: Old preconception and new dimensions, *Chem. Eng. Sci.*, **54**, pp. 1299–1305.

6. Rawadieh S. and Gomes, V. G. (2007) Catalyst-adsorbent configuration in enhancing adsorptive reactor performance, *Int. J. Chem. Reactor Eng.*, **5**, article A108, http://www.bepress.com/ijcre/vol5/A108.

7. Westerterp, K. R. and Kuczynski, M. (1987) A model for a countercurrent gas–solid–solid trickle flow reactor for equilibrium reactions. The methanol synthesis, *Chem. Eng. Sci.*, **42**, pp. 1871–1885.

8. Kruglov, A. V. (1994) Methanol synthesis in a simulated countercurrent moving-bed adsorptive catalytic reactor, *Chem. Eng. Sci.*, **49**, pp. 4699–4716.

9. Gordeeva, L. G. and Aristov, Yu. I. (2009) Nanocomposite sorbents "salt inside porous matrix" for methanol sorption: Design of phase composition and sorption properties, practical applications. In: *Porous Materials: Structure, Properties and Applications*, T. Wills (Ed.) (Nova Science Publishers, Inc., NY), pp. 109–138.

10. Gordeeva, L. G., Freni, A., Restuccia, G., and Aristov, Yu. I. (2007) Influence of characteristics of methanol sorbents "salt in mesoporous silica" on the performance of adsorptive air conditioning cycle, *Ind. Eng. Chem. Res.*, **46**, pp. 2747–2752.

11. Aristov, Yu. I., Gordeeva, L. G., Pankratiev, Yu. D., Plyasova, T. M., Bikova, I. V., Freni, A., and Restuccia, G. (2007) Sorption equilibrium of methanol on new composite sorbents "CaCl$_2$/silica gel", *Adsorption*, **13**, pp. 121–127.

12. Haut, B., Halloin, V., and Ben Amor H. (2004) Development and analysis of multifunctional reactor for equilibrium reactions: Benzene hydrogenation and methanol synthesis, *Chem. Eng. Proc.*, **43**, pp. 979–986.

13. Minyukova, T. P., Shtertser, N. V., Khassin, A. A., Plyasova, L. M., Kustova, G. N., Zaikovskii, V. I., Shvedenkov, Yu. G., Baronskaya, N. A., van den Heuvel, J. C., Kuznetsova, A. V., Davydova, L. P., and Yurieva, T. M. (2008) Evolution of Cu–Zn–Si oxide catalysts in the course of reduction and reoxidation as studied by in situ X-ray diffraction analysis, transmission electron microscopy, and magnetic susceptibility methods, *Rus. Kinet. Catal.*, **49**, pp. 821–830.

14. US Patent 2344329 A.

15. Loid, E., Brown, C. B., Bonnel, D. G. R., and Jones, W. J. (1928) Equilibrium between alcohols and salts. Part II, *J. Chem. Soc. Part I*, **0**, pp. 658–666.

16. Jacobson, C. A. (Ed.) (1951) *Encyclopedia of Chemical Reactions VII.* (Reinold Publishing Corporation, New York).

17. Kuczynski, M. and Westerterp, K. R. (1986) Retrofit methanol plants with this converter system, *Hydrocarbon Process.*, **65**, pp. 80–83.

Chapter 21

Active Heat Insulation

In this chapter, we consider the application of composites "salt in a porous matrix" (CSPMs) in heat protection and fire extinguishing systems [1, 2]. As it has been shown earlier, CSPMs are capable of absorbing and retaining a large amount of water. Moreover, their pores may be completely filled with liquid (aqueous salt solution) even at RH = 30–50%. Thus, at elevated temperatures when the sorbed water evaporates, this can dramatically affect the heat transfer through this porous solid. As a result, a strong retardation of the heat front propagation inside the solid can be anticipated.

The evaluation of heat transfer in such multi-phase porous systems is a complex problem as the evaporation, capillary, and gravity effects have to be simultaneously taken into consideration [3, 4]. Several complex models have been developed so far to describe this effect [3]. Another approach to this problem comes from the theory of heat propagation in a medium with phase transition like melting/solidification. The first published discussion was of Stefan [5] although some general results were first obtained by Neumann (see Ref. [6]). All available exact solutions are comprehensively described in Refs. [3, 6].

In this chapter, we present experimental data on heat transfer in wet CSPM layer heated by a very intense heat flux. Thus, this study models the use of CSPMs as a sort of heat-protecting materials. It

---

Nanocomposite Sorbents for Multiple Applications
Yu. I. Aristov
Copyright © 2020 Jenny Stanford Publishing Pte. Ltd.
ISBN 978-981-4267-50-2 (Hardcover), 978-981-4303-15-6 (eBook)
www.jennystanford.com

has demonstrated that these composites could be very promising for heat protection and fire extinguishing.

21.1 Experimental Details

Preparation of the samples for this study included four stages:

1. Impregnation of a KSK silica gel powder (particles size < 100 μm, pore volume 0.9–1.0 cm^3/g, average pore size 15 nm) with water (SG-W sample), CaCl$_2$, or MgCl$_2$ aqueous solutions (SG-CaCl$_2$ or SG-MgCl$_2$);
2. The soaked samples were mold under a pressure of 200–300 bar as tablets of 24.5 mm in diameter, with the thickness L variable between 4 and 20 mm;
3. Heating at 200°C to evaporate water (for the salt-containing tablets). Densities of the dry tablets were 0.80 (SG-CaCl$_2$) and 0.73 g/cm^3 (SG-MgCl$_2$) as compared with 0.50 g/cm^3 for the pure silica gel;
4. Saturation with water vapor in a desiccator where a fixed relative humidity between 30 and 70% was maintained. The SG-CaCl$_2$ and SG-MgCl$_2$ samples were efficient sorbents of water that are capable of sorbing under mentioned conditions the equilibrium water amount $w = 0.3$–0.5 g/g.

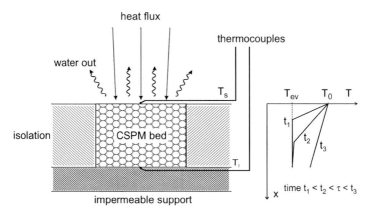

Figure 21.1 Scheme of the heat transfer experiments (left) and tentative temperature distribution in the sorbent bed at different times (right). Reprinted by permission from Springer Nature from Ref. [1], Copyright 2001.

The sample was placed on a horizontal support impermeable for water vapor and heated from the opposite permeable side by a very intense heat flux (Fig. 21.1). The flux is generated by concentrated light of a 10 kW Xenon lamp of a DKSHRb-10000 type. The average heat flux W_{in} was measured by a calorimetric method and was varied from 50 to 500 kW/m^2. The temperatures T_s of the upper (heated and permeable) and T_l of the lower (adjacent to the impermeable support) surfaces were measured using K-type thermocouples. Thermal insulation was put to avoid heat losses from the lateral brick surface.

21.2 Heat Front Propagation

At the incident heat flux W_{in} = 150–400 kW/m^2, the temperature T_s of the upper (heated) surface is found to be rapidly increasing up to the limiting values of 500–1200°C, which depend on W_{in} (not presented), while the temperature T_l of the protected surface remains much lower (Fig. 21.2). Moreover, for water-containing samples, the dependence $T_l(t)$ has a considerable plateau at T_p = 100–120°C, showing a significant heat front retardation. As the plateau gets larger at higher amounts of water inside the bed, the plateau seems to be essentially connected with the water evaporation rather than the salt presence. Indeed, a similar plateau is observed for SG-W samples that do not contain any salt. It is convenient to use the plateau duration τ to characterize the heat retardation effect and to study its dependence on sample properties (thickness, water content, salt nature), incident heat flux, etc.

As it has been mentioned earlier, the water content w strongly affects the heat propagation through the CSPM layer, so that the delay time τ is nearly proportional to the mass of water m_{con} to be evaporated. Therefore, a long retardation time for the CSPM layer is due to a large amount of water retained inside its pores. It is worth noting that the water-free samples do not demonstrate any retardation effect at all.

Sample thickness L is a very efficient tool to change the delay time, which is found to be about quadratic in L:

$$\tau \sim L^n \tag{21.1}$$

where $n = 2.0 \pm 0.2$ (Fig. 21.3). This equation is quite typical for the kinetics of heat front propagation in a medium with phase transition like melting [5–7]. The consequence of Eq. (21.1) is that the value τ/L^2 does not depend on l and may be chosen as a convenient parameter to compare different samples and to study the influence of the incident heat flux as well.

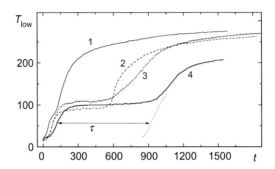

Figure 21.2 Experimental dependence of the temperature (°C) of the lower face of the SWS-coating on the heating time (s) at $W_{in} = 150 \pm 30$ kW/m²: (1) SG-W, $w = 0.05$ g/g, $L = 10$ mm; (2) the same sample, $w = 0.9$ g/g; (3) SG-MgCl$_2$, $w = 0.6$ g/g, $L = 10$ mm; (4) SG-CaCl$_2$, $w = 0.55$ g/g, $L = 13$ mm. Reprinted by permission from Springer Nature from Ref. [1], Copyright 2001.

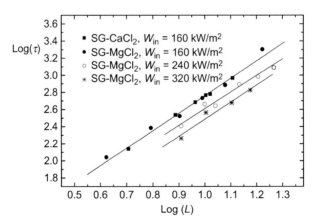

Figure 21.3 Experimental dependencies of the delay time on the sample thickness at different values of the incident heat flux. Reprinted by permission from Springer Nature from Ref. [1], Copyright 2001.

The surprising thing is that the intermediate *heat fluxes* (W_{in} = 100–400 kW/m²) have, rather, a weak effect on the delay time, whereas at lower fluxes (< 100 kW/m²), the τ value drastically increases (Fig. 21.4).

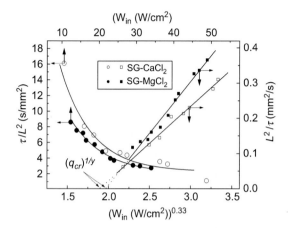

Figure 21.4 Influence of the heat flux W_{in} on the parameter τ/L^2 and determination of the critical heat flux W_{cr} (straight lines). Reprinted by permission from Springer Nature from Ref. [1], Copyright 2001.

This probably indicates that there is a threshold for water evaporation, that is, a minimum heat flux necessary for total water removal. This is in line with the theory of evaporation in porous media [3] stating that no vapor region is present inside the pores until heat flux exceeds some critical value W_{cr} depending on the impregnated liquid and host-matrix properties. For fluxes lower than the critical one, the gravity and capillarity prevent vaporization and increase evaporation time τ significantly.

Precise analytic description of heat transfer in the horizontal porous bed saturated with liquid, adjacent to lower impermeable surface and heated from the upper permeable side, asks for taking into account the evaporation, capillary, and gravity effects [3] simultaneously, and, as to our knowledge, has not been done yet in the literature. To analyze the data obtained, here we use a simple one-dimensional model based on the theory of heat propagation in a medium with phase transition like melting [5, 6]. We assume (a) the temperature T inside the layer to depend just on the distance x from the heated surface and time t, and (b) the upper surface temperature T_0 to be constant. Then the heat balance in the layer dx, where phase transition takes place, at time between t and $t + dt$, can be written as

Active Heat Insulation

$$\lambda(dT/dx)_x \, dt = \rho \, \Delta H_{ev} \, dx \qquad (21.2)$$

where λ is the heat conductivity coefficient of the dry composite layer, $(dT/dx)_x$ is a temperature gradient at distance x, ρ and c_p are the density and specific heat of the water-containing layer, and ΔH_{ev} is the heat of water evaporation. Approximating $(dT/dx)_x \approx (T_0 - T_{ev})/x$, one can easily integrate Eq. (21.2) with the boundary condition $T(x=0) = T_0 = $ const. to obtain the time for completing the phase transition (complete water evaporation) in the whole sample volume

$$\tau = \rho \, \Delta H_{ev} \, L^2/2\lambda \, (T_0 - T_{ev}) \qquad (21.3)$$

This equation allows describing the main features of the heat transfer in the system under study, because it represents well the experimentally observed dependence $\tau(L)$ and predicts a sharp increase in the protection time τ if the upper surface temperature T_0 tends to reach T_{ev}. Thus, it explains the threshold effect at incident fluxes close to the critical one reported above.

An analytic dependence $\tau(W)$ can be evidently obtained from Eq. (21.3) if the link between T_0 and W is known. If approximating the dependence $W(T_0)$ as $W \approx \alpha T_0^\gamma$, where the parameter γ summarizes the relative contribution of reflection, convective and radiation heat losses from the upper sample surface, at $T_0 \gg T_{ev}$ Eq.(21.3) may be rewritten as

$$\tau = \rho \, \Delta H_{ev} \, L^2/2\lambda \, \beta \, (W^{(1/\gamma)} - W_{cr}^{\,(1/\gamma)}) \qquad (21.4)$$

where $\beta \, W_{cr}^{\,(1/\gamma)} = T_{ev}$. Special measurements performed at $T_{up} = 700{-}1200°C \gg T_{ev}$ showed that for the systems investigated, $\gamma = 3.0 \pm 0.5$ (not presented) [1]. So high γ value demonstrates clearly that most of the incident energy dissipates from the sample surface due to radiation losses. It is reasonable since the T_0 value usually reaches 700–1200°C. Plotting (τ/L^2) versus $(W_{in})^{0.33}$ allows obtaining the critical heat flux $W_{cr} = (40 \pm 10)$ kW/m^2 (Fig. 21.4).

If Eq. (21.4) is compared with the typical time of the heat front propagation in a homogeneous and isotropic solid without any phase transition $\tau = \rho \, cLl^2/4\lambda$ (here c is the specific heat capacity of a solid) [6], one can introduce the apparent coefficient of heat conductivity for a porous layer saturated with liquid $\lambda_{app} = \lambda \, [c \, (T_0 - T_{ev})/\Delta H_{ev}] = (0.03{-}0.3)\lambda$. Hence, the evaporation of water may indeed retard the heat transfer and lead to apparent heat conductivities 3–30 times lower (for $(T_0 - T_{ev})$ = 1000 and 100°C, respectively) than the actual conductivity λ of the dry material. As for porous solids $\lambda \approx$ 0.1–0.2 W/mK, the λ_{app} value can be as low as 0.03–0.006 W/mK. For comparison, typical high-temperature insulating materials (laminated and corrugated asbestos, diatomaceous earth, etc.) possess λ > 0.05 W/mK at room temperature and λ > 0.1–0.3 W/mK at 1000°C [8]. Equation (21.3) directly predicts a rise in τ if impregnating liquids with a high evaporation heat and matrices with low heat conductivity are used.

21.3 Further Considerations

What should be an optimal adsorbent for adsorptive heat protection? These requirements have not been formulated so far, but some views can be expressed. The maximal retardation time is expected for an adsorbent that has a maximal mass of water adsorbed per its unit volume. Hence, the adsorbent should have a large pore volume to provide the necessary space for loading the sorbed water and this space has to be filled with the salt solution as completely as possible before heat exposure. The hygroscopic salt has to provide this complete pore filling, water retaining, and self-recovering the water loading after the exposure. The salt content should be selected accordingly. For the salt-containing layer, the heat-protecting effect can be completely reproduced as soon as the layer restores its saturated state by spontaneous moisture sorbing from the ambient air. Quite the contrary, the salt-free silica gel can spontaneously adsorb only 5–7 wt.% of water. Therefore, for increasing its heat protection properties, it should be directly filled with liquid water (see case 2 in Fig. 21.2). This oversaturated state is not in equilibrium with the ambient air, and the layer continuously dries.

Thus, the data obtained show that the consolidated layers of the CSPM spontaneously saturated with water have the ability of considerable heat front retardation and demonstrate abnormally low heat conductivity at T = 100–1200°C. The retardation time is strongly dependent on the incident heat flux, sample thickness, and water content and may change over a wide range. It is shown that these dependences can be described satisfactorily within the framework of a simple one-dimensional model that takes into account the evaporation of liquid inside a porous medium. The results obtained may serve as a basis for the practical development of new efficient and self-recovered heat- and fire-protecting porous coatings.

Figure 21.5 Heat protection with SWS, as viewed through the eyes of an artist (Image courtesy: A. S. Shorin).

References

1. Tanashev, Yu. Yu., Parmon, V. N., and Aristov, Yu. I. (2001) Retardation of a heat front in a porous medium containing an evaporating liquid. *J. Eng. Phys. Thermophys.*, **74**, pp. 1053–1058.
2. Parmon, V. N., Krivoruchko, O. P., and Aristov, Yu. I. (1999) Use of modern composite materials of the chemical heat accumulator type for fire protection and fire extinguishing. In: *Prevention of Hazardous Fires and Explosions*, V. E. Zarko, V. Weiser, N. Eisenreich, and A. A. Vasil'ev (Eds.) (Kluwer Academic Publishers, The Netherlands), pp. 34–48.
3. Kaviany, A. (1991) *Principles of Heat Transfer in Porous Media* (Springer-Verlag, New York).
4. Luikov, A. V. (1966) *Heat and Mass Transfer in Capillary Porous Bodies* (Pergamon Press, USA).

5. Stefan, J. (1891) Stefan problem, *Ann. Phys. Chem. (Wiedemann) N.F.*, **42**, pp. 269–286.

6. Carslaw, H. S. and Jaeger, J. C. (1988) *Conduction of Heat Transfer in Solids*, 2nd Ed. (Oxford Science Publications, UK).

7. Riesmann-Weber, H. (1912) Die Partiellen Differentialgleichungen der Mathematischen Physik, 5th ed., **2**, p. 121.

8. Rohsenow, W. M., Hartnett, J. P., and Ganic, E. N. (Eds.) (1985) *Handbook of Heat Transfer Fundamentals*, 2nd ed. (McGraw-Hill, New York).

Postface

The foregoing chapters mark the first journey through the field of nanoComposite sorbents "**S**alt in **P**orous **M**atrix" (CSPMs). Of all the approaches we have tried to convey on this journey, the most important is the *nanotailoring* concept. Only recently, the impressive progress in materials science has allowed the ambitious task of target-oriented synthesis of adsorbents with desired properties to be achieved. The idea of this approach is to have an *optimal adsorbent* for each adsorptive technology (process, cycle) and its particular conditions so that the properties of the adsorbent can allow a perfect performance of the process or cycle. Therefore, first, it is necessary to realize what an optimal adsorbent is. The second, practical step is the synthesis of a real material with sorption properties close or even identical to those of the optimal adsorbent.

For composites "salt + matrix," the tools for tailoring their properties are much wider than for single-component materials. The principle CSPM component is the salt, which mainly determines the composite sorption properties. An important finding presented and discussed in this book is the change in the salt sorption properties if it is confined to sufficiently small pores of the host matrix. The matrix serves just like a dispersing medium that provides the required salt dimensions. It also accommodates the salt swelling in the course of vapor sorption, transfers heat through its skeleton, and vapor through the pore space.

The nanotailoring approach to CSPM synthesis has appeared very fruitful. The first systematic studies on composite sorbents of water were performed at the end of the past century at the Boreskov Institute of Catalysis, Russia. After that, plenty of CSPMs was synthesized in many laboratories all over the world and tested as sorbents of water, methanol, ethanol, ammonia, and carbon dioxide. As a result of the booming interest, the number of research papers related to the study of CSPM-type materials have increased sharply (see Fig. P.1) and are approaching 1,000. As "nobody can embrace the unembraceable" [1], I have mostly used the results obtained in my laboratory as examples in this book. I hope colleagues working in this area will understand this.

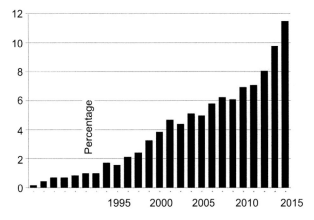

Figure P.1 Percentage of publications on CSPMs in different years.

The initial period of accumulating the basic knowledge about CSPMs, which took about 10 years, revealed outstanding properties of these materials. As a consequence, the list of potential applications of CSPMs extended from sole adsorptive heat transformation to many others, and promising applications can be expected in the future. Some of the applications considered in this book are gas drying, maintaining relative humidity, extraction of water from the atmosphere, regeneration of heat and moisture in ventilation systems, shifting chemical equilibrium, and active heat insulation.

This transformation of fundamental knowledge (truth) to practical benefit (use) makes me remember a wonderful thought by Dmitry Mendeleev, the Russian chemist who invented the Periodic table more than 150 years ago: "Truth matters on its own, without any references to immediate use. Usefulness is a response to stern human necessity, whereas searching for bits of truth is a matter of pure human curiosity. And, from my standpoint, all things that are progressive, important, and the most useful directly follow from this human craving. Benefits will come, they will be found without us looking for them if a truth is being discovered by itself and for itself. This is what science teaches us" [2]. Frans de Waal even believes that the thirst for knowledge among scientists fills the spiritual void that other people fill with religion [3].

This book is a reflection of the level of knowledge accumulated about CSPMs so far, but the journey through the CSPM world does not end here. Here it is appropriate to quote the words of Tomas Mann about the bottomless learning of the past [4]: "Again and further are the right words, for the unreachable plays a kind of mocking game with our research ardour; it offers apparent holds and goals, behind which, when we have gained them, new reaches of the past still open out—it happens to the coastwise voyager, who found no end to his journey." These words can be fully attributed to the composites described in this book, the study of which not just answered many questions but opened up new and alluring horizons.

Y. I. A.

Endnotes

1. Prutkov, K. (1884) *Thoughts and Aphorisms* (Moscow).
2. Mendeleev, D. I. (1912) *To the knowledge of Russia*, 7th edition, Suvorov Publishing House, Sankt-Petersburg.
3. de Waal, F. (2014) *The bonobo and the atheist: In search for humanism among the primates*, 1st edition, W. W. Norton & Company.
4. Mann, T. (1978) *Joseph and his brothers*, Penguin Books.

Index

absolute humidity 46, 367, 368, 370, 372, 387
AC *see* adsorptive chiller
AC cycle 322–326, 328, 329
ACF *see* activated carbon fiber
activated carbon 2, 4, 7, 100, 170, 172, 197, 205, 327, 329, 335
activated carbon fiber (ACF) 3, 21, 177, 185, 187, 327, 330
active salt 57–62, 65–67, 72, 84, 145, 152, 170, 203, 380, 382, 383, 392
adsorbate 16, 39–42, 214, 316, 319, 327, 334, 347, 348, 350, 351, 353, 356
adsorbent 1–7, 16, 17, 23, 24, 37, 38, 40, 41, 43–45, 48–50, 52, 53, 144, 291–294, 298, 299, 313–321, 333, 334, 346–349, 351, 352, 366–370, 373–375, 387, 388, 396, 397, 401–403
alumino-phosphate 2
ammonia 169, 329
commercial 124, 400
conventional 331
dry 214, 333, 365
humid 365
mesoporous 23
natural 2
non-microporous 20
novel 1, 5, 397
saturated 40
single-component 6
synthetic 2
traditional 8, 9, 84
two-component 6
wet 367

adsorbent bed 39, 223, 292, 294, 299, 370
adsorbent grain 26, 27, 30, 50–52, 266, 299, 350
adsorbent layer 49–51, 367–369, 373
adsorber 37, 39, 40, 42, 319, 346, 347, 356–358, 368–370, 399–401
batch 335
cylindrical 367
fixed bed 37
adsorption 15, 16, 19–21, 23, 24, 27, 36–38, 44, 45, 58–61, 97–99, 113–115, 144, 294, 296–299, 301, 303, 305, 314–316, 347, 348, 350, 351, 368, 397–400
heterogeneous 99
isobaric 41, 314, 319
multilayer 18, 23
adsorption chiller 4, 119, 223, 325, 328, 330
adsorption equilibrium 16–22, 24, 26, 28, 30, 48, 51, 121, 319
adsorption heat 16, 17, 36, 297, 331, 346–348
adsorption heat storage (AHS) 332, 333, 335, 336
adsorption heat transformer (AHT) 21, 39, 40, 50, 221, 291, 295, 296, 305, 313, 314, 319, 345, 348
adsorption ice maker 330
adsorption isobar 320–21
adsorption isotherms 16, 18, 23, 24, 36, 43, 45, 50, 128, 160, 351, 352, 354

adsorption kinetics 26, 27, 29, 30
adsorption process 6, 19, 27, 30,
49, 61, 264–266, 346, 347, 366
adsorption rate 26, 28, 50, 267,
299, 320, 351
adsorption stage 37, 291, 315,
317, 350, 351, 368–370
adsorption technology 1, 3, 4, 36,
37, 39, 78, 86, 116, 281, 363
adsorption uptake 19, 48, 50, 325,
395
adsorptive chiller (AC) 53, 109,
124, 127, 128, 313, 316, 317,
322, 323, 325, 329
adsorptive cooling 144, 169, 171,
322, 323, 325, 327, 329
adsorptive cycle 41, 345–347, 349,
351, 355
adsorptive heat pump (AHP) 4
adsorptive heat transformation
39, 186, 239, 264, 291, 313,
345
AHP *see* adsorptive heat pump
AHS *see* adsorption heat storage
AHT *see* adsorption heat
transformer
AHT cycle 40, 41, 221–223, 291,
292, 298, 299, 305, 313, 314,
327, 345, 349, 352
air conditioning 1, 144, 189,
314–316, 329, 387
air flux 198, 365, 370–372, 386
alumina 108, 109, 121, 170, 172,
174, 177, 178, 183, 185, 187,
203, 205, 220, 221, 274, 369,
370
   activated 2, 4
   mesoporous 367
   pure 186
alumina mesopores 177, 181, 183
ammonia sorption 117, 169–173,
175, 177–186, 188–191, 329
Avrami–Erofeev equation 30

BET equation *see* Brunauer–
Emmett–Teller equation
Bruker model BVT-3000 249
Brunauer–Emmett–Teller equation
(BET equation) 22
bulk hydrate 73, 96, 97, 99, 126,
132, 245, 247, 248, 252, 253,
256
bulk salt 8, 9, 72, 73, 84, 85,
87–92, 115, 117, 120, 122,
163, 177, 178, 182–184, 212,
260, 263, 264, 380, 381
bulk solution 98, 116, 122, 131,
135, 156, 230, 233, 234, 242
bulk system 70, 126, 155, 162, 232

Carnot efficiency 42, 319, 349
chemical heat accumulator 9
chemical heat transformer 39
coefficient of performance (COP)
41, 318, 320, 323–328, 330,
370
coherent scattering domain (CSD)
120, 127
composite 7–10, 63–64, 66, 67, 87,
88, 90, 91, 107, 145–148, 150,
152, 153, 158–165, 170–177,
187–191, 198–200, 209, 210,
329–331, 335, 336, 383–385,
392, 393, 395, 396, 399–402
   alumina-based 174
   anhydrous 213
   dry 222
   $LiNO_3$-based 325
   silica-based 219
   single-salt 152, 188
   vermiculite-based 393
COP *see* coefficient of performance
crystalline hydrate 70, 95, 97, 99,
100, 113, 217, 245, 249, 251,
260, 380, 381
crystallization 230, 231, 234, 235,
237, 242, 305

CSD *see* coherent scattering domain

CSPMs *see* nanocomposite sobents "salt in porous matrix"

Debye–Huckel model 136, 137
Debye law 211
Debye temperature 210
desiccant cooling systems 331
desorption 46, 47, 114, 161, 162, 294, 296–298, 301, 303, 305–307, 313, 314, 318, 319, 321, 326, 347, 349–352, 367, 368, 382, 383
  isobaric 40
  pressure swing 397
desorption rate 298, 305
desorption stage 48, 298, 314, 316, 350, 351, 368–370
desorption temperature 40, 42, 298, 319, 323, 328, 335
dew point 37–39, 100, 120, 369
differential scanning calorimetry (DSC) 68, 229, 237, 241, 335
diffusion 51, 61, 143, 264, 266, 273, 287, 350
  gas molecule 50
  intragrain 28, 283, 287, 289
  methanol 300
  molecular 26, 261
diffusional resistance 51, 52, 284, 285, 287
diffusion flux 270–272, 302
diffusion mechanism 26, 222, 258, 307
dispersion 61, 62, 64, 65, 68, 90, 131, 146, 148, 149, 175, 176, 187, 197, 200
DMTA *see* dynamic mechanical thermal analysis
Donnan model 137
DSC *see* differential scanning calorimetry

Dubinin adsorption potential 38, 314, 348
Dubinin-Radushkevich equation 19, 314
dynamic mechanical thermal analysis (DMTA) 231, 232

enthalpy 85, 90, 157, 163, 170, 179, 183, 184, 187, 191, 201, 231, 232
entropy 85, 90, 163, 170, 178, 179, 183, 184, 187, 191, 209, 211, 214, 319
equilibrium uptake 19, 25, 36, 37, 45, 59, 88, 97, 285, 314, 396, 400

Fickian diffusion model 16, 93, 269, 273
Fickian diffusivity 27
Fickian expression 26
flux 26, 27, 271, 272, 298, 350, 367, 372, 409, 411
  diffusional 307
  intensive vapor 350
  parallel vapor 276
fossil fuel 39
Freundlich equation 17

gas diffusion 16, 27, 50, 264
gas drying 4, 36, 37, 41, 42, 111, 209, 346, 363
gas transport 26–28, 50, 52, 222
Gibbs's phase rule 15, 24, 25, 153, 380
grain 28, 30, 50–52, 59–61, 241, 242, 264, 266, 267, 275, 281, 289, 295, 297, 298, 300, 302, 305, 307, 350, 351
  alumina 286
  irregular 322
  nonporous 28
  spherical 28, 29, 31

424 | *Index*

grain size 38, 52, 264, 266, 267, 281, 285–287, 290, 292, 293, 299, 300, 307, 369, 370
grain surface 28, 61, 234, 277, 300, 302, 350

Haber–Bosch process 169
Harkins–Jura equation 18
heat conductivity 209, 217, 220, 329, 412, 413
heat exchanger (HEx) 46, 307, 320, 322, 324, 326, 333, 336, 347, 356, 357, 364
heat from cold (HeCol) 345, 348, 352, 353, 355, 357, 359
heat pump 4, 40, 144, 352, 364
  four-bed adsorption 324
heat storage 9, 53, 73, 109, 111, 327, 332, 335, 363
heat-storing material 365
heat transfer 218, 220, 222, 296, 298, 303, 305, 308, 350, 407, 411, 412
heat transfer fluid (HTF) 357, 358
heat transformation 39, 42, 164, 209
HeCol *see* heat from cold
HeCol cycle 345, 348, 352–356, 359, 360
Henry constant 18
HEx *see* heat exchanger
Hill–de Boer equation 18
host matrix 7, 72, 83, 107–109, 116, 124, 144, 145, 150, 152, 170, 172, 184, 185, 198, 205, 264, 382, 383
HTF *see* heat transfer fluid
humidity 197, 331, 368, 371, 382, 386, 389, 391
  absolute outdoor 363
  air outlet 370
  atmospheric 45
  comfortable 363

hydrate 69–71, 83–85, 92, 99, 100, 109, 111, 116, 117, 124, 126, 131, 134, 217, 249–255, 257, 260, 385
  appropriate 87, 240, 253
  bulk 73, 96, 97, 99, 111, 122, 126, 132, 213, 240, 245, 247–249, 250, 252–256
  confined 227, 247, 248, 255, 260
  dispersed 212, 213, 245, 249, 250, 252–256, 258, 259
  low-temperature 95
  melted 253
  salt 85, 92, 95, 111, 115, 118, 131–133, 227, 230, 242, 248, 269, 270
  stable 115
hydrated salt 73, 85, 100, 260, 261, 270
hydrated water 73, 246, 253, 256
hydration 89, 90, 93, 109, 111, 117, 129, 132, 133, 135, 380, 385, 392
hydration reaction 47, 92, 131, 133, 265, 325, 392
hydration route 114, 115
hydration temperature 73, 76, 88, 90, 92, 130, 133–135, 383
hysteresis 9, 73, 86, 91, 92, 115, 127, 131, 148, 184, 380, 382

IDS *see* isothermal differential step
IDS method 265, 266, 273
impregnating solution 57, 60, 62, 70, 74, 76, 88, 150
impregnation 57–59, 61, 64, 71, 74, 86, 99, 127, 205, 227, 408
inflection 98, 121, 123, 125, 156, 235
inhomogeneity 185, 201, 202, 205
isothermal differential step (IDS) 29, 264, 269

Knudsen diffusion 26, 269, 270
Knudsen diffusivity 27, 50, 93, 273
Knudsen mechanism 222, 272
KSK 77, 78, 86, 87, 89, 90, 94, 96, 97, 100, 119, 153, 229, 232, 235, 237, 238, 300–302
KSK pores 98–100, 133, 230, 232, 233, 235–237, 239
KSM 97, 228, 232, 246, 282
KSM pores 136, 231, 232, 237, 248

Langmuir approach 16
large temperature jump (LTJ) 291–293
LDF *see* linear driving force
LDF model 30, 295
linear driving force (LDF) 16, 30, 294
low-grade heat utilization 5
low-temperature heat 4, 9, 36, 117, 291, 293, 313, 323, 327, 329, 334
LTJ *see* large temperature jump
LTJ method 292, 293, 296, 302
Luikov–Bjurström model 216, 218–221

matrix 9, 58–62, 64–66, 86, 88, 89, 100, 101, 109, 112, 113, 115, 117–119, 121, 123–125, 131, 172–174, 185, 187, 198, 199, 201–203, 238, 239, 335, 383, 392
  carbonaceous 170
  dehydrated 150
  hydrophobic 59, 117
  mesoporous 93, 107, 118, 177, 184, 187
  microporous 242
  porous host 7, 221
matrix grain 58, 61, 62, 289
matrix pores 58, 59, 89, 93, 131, 133, 135, 148, 165, 170, 227, 263, 269, 271

mesopores 23, 24, 50, 51, 93–95, 98, 107, 111, 112, 116, 119, 123, 187, 216, 237, 240
methanol removal 144, 156, 165
methanol sorbents 143, 144, 146–148, 150, 152, 154, 156, 158, 160, 162, 164, 327, 353, 356
methanol sorption 143, 145, 147–157, 159, 161–164, 300–303, 305, 306, 354, 358, 396, 400
methanol vapor 7, 143–145, 147, 149, 152, 153, 159, 164, 302, 359, 398
moisture 43–45, 209, 335, 337, 364–370, 372, 374–376, 389, 391
Monte Carlo simulation 136

nanocomposite sobents "salt in porous matrix" (CSPMs) 57, 58, 60–62, 64, 67, 68, 71–73, 93–95, 101, 143, 147–153, 169–171, 209, 210, 216–218, 220–222, 273, 274, 291–293, 313, 314, 322, 333, 334, 396–399, 407
nanotailoring 5–7, 58, 66, 143, 148, 327, 382
neutron scattering 245, 256, 257, 259
NMR *see* nuclear magnetic resonance
NMR spectrum 246–248, 251
nuclear magnetic resonance (NMR) 231, 245, 246, 249–252, 257, 264, 281

Pake's model 247
Pake-type powder patterns 250
phase transition 210, 213, 229, 237, 238, 407, 410–412
Poiseuille diffusivity 50

Poiseuille flow 26, 50, 52
Poisson–Boltzmann equation 136
Polanyi approach 20
Polanyi invariance 397
Polanyi principle 36, 162, 348, 354, 397
pore size distribution (PSD) 86, 109, 112, 119, 121, 135, 152, 159, 239, 384, 392
pore size effect 72–75, 109, 118, 129
pore space 100, 119, 220, 234, 270, 302
pore structure 26, 27, 107, 118, 119, 129, 220, 273, 335, 396
pore tortuosity 259, 264, 301
porous materials 2, 3, 57, 64, 76, 274
porous structure 66, 78, 99, 109, 125, 159, 164, 205, 289, 328
pressure jump 52, 266, 281, 291, 307, 346, 350
pressure swing adsorption 366
PSD *see* pore size distribution
pseudo-boehmite 282, 287, 289

reagent 30, 50, 61, 64, 133, 134
regeneration temperature 36–38, 203, 204, 316, 331, 334–336
relative humidity (RH) 3, 4, 36, 43, 46, 111, 276, 368, 379, 380, 382–394, 407, 408
relative pressure 19, 23, 36–38, 43, 70, 76, 84, 87–89, 120, 127, 128, 147
RH *see* relative humidity
RH evolution 391, 393
RH variation 387

salt solution 58, 59, 61, 62, 64, 70–72, 98, 100, 114–117, 124, 126, 217–221, 235–237, 269, 270, 285, 287, 288, 301, 302
   aqueous 71, 120, 150, 222, 227, 242, 289, 399, 407

confined 131, 221
neutral 151
saturated 152
supercooled 231
salt swelling 86, 93, 116, 117, 169
Schreder–Le Chatelier equation 230
silica gel 2, 64, 65, 71, 99, 109, 124, 125, 145, 147, 148, 152, 153, 158–160, 187–189, 246, 268, 269, 273, 274, 324, 325, 331, 335, 336, 369, 370, 383, 384, 413
conditioned 379, 382
functionalized 144
industrial 86
low-density 86
mesoporous 9, 210, 248, 335
microporous 331, 367
regular-density 324
silica gel pores 151, 161, 214, 228, 229, 231, 233, 235, 237, 238, 245, 246, 253
silica pores 69–71, 73, 89, 90, 92, 125, 126, 129, 132, 133, 156, 158–161, 163, 236, 237, 246, 248, 260, 261, 300
solvate 7–9, 25, 143, 145, 148, 151, 153, 165, 192
sorbed water 122, 125, 215, 217, 222, 264, 271, 275, 281, 283, 285, 287–289
sorption 8, 9, 15, 16, 24, 46–48, 76, 87, 89–91, 98–100, 116, 147, 148, 151, 152, 155–157, 162, 163, 187–191, 201–203, 292–294, 296, 302, 303, 305, 307
sorption dynamics 264, 266, 268–270, 272, 274, 276, 281, 282, 284, 286, 288, 290, 292, 294–299, 301–303, 305

sorption enthalpy 178, 183, 184, 191

sorption equilibrium 25, 44, 48, 71–73, 131, 133, 135, 151, 162, 164, 185, 187, 191

sorption heat 157, 158, 201, 296, 297

sorption isobars 42, 95, 151, 153, 162, 296, 307, 319, 326, 328

sorption isotherms 24, 28, 72, 74, 77, 120, 124, 178, 185, 190, 267, 272, 273

sorption rate 28, 264, 268, 270, 273, 275, 286, 300, 303, 305

SWS (selective water sorbent) 9, 21, 91, 98, 222, 414

SWS-1A 121–124, 131, 220, 281, 282, 363, 367–370, 375

SWS-1C 122–124, 131

SWS-1L 9, 10, 37, 38, 94, 96–101, 118, 120, 125, 209, 210, 214, 216–220, 222, 223, 264–266, 268–269, 281, 282, 299, 300, 307, 308, 322–325, 333, 334

SWS-1S 125, 126, 131, 265, 282

SWS-1V 112–114, 265, 333, 334

SWS-2L 216–219, 222, 223

SWS-3L 216–219

SWS-8L 89, 90, 129, 215

SWS-9L 325, 326

SWS-9V 325–327, 333, 334, 336

Tammann temperature 64, 77

temperature dependence 17, 44, 135, 204, 249–251, 253–255, 269, 308

temperature invariance 19, 36, 162, 348, 354, 397

thermal conductivity 111, 116, 216–223

thermal coupling 319, 320, 349

thermal dispersion 57, 64, 65, 74, 77

thermal effects 30, 266, 267

thermal masses 41, 305, 318, 347, 355

thermal stability 395, 402, 403

thermodynamic cycle 21, 96, 329, 330

thermodynamic parameters 147, 183, 209, 211, 381

Tian–Calvet calorimeter 157, 201

tortuosity 26, 93, 258, 259, 267, 286

Trouton's rule 20–22, 43, 316, 354, 355

vacuum adiabatic calorimeter 210

van't Hoff equation 24, 85, 90, 170

vapor 39, 40, 263, 265, 269–271, 273, 275, 282, 283, 289, 290, 293, 304, 305, 307, 350, 351, 380, 382, 383

adsorbed 350

refrigerant 292

supersaturated 305

vapor adsorption isotherm 350, 351

vapor diffusion 93, 263, 267, 273, 283, 284, 290, 301, 302, 351

vapor diffusional flux 350

vapor diffusivity 272, 307

vapor pressure 67, 69, 98, 99, 101, 116, 117, 121, 123, 126, 127, 221, 223, 291, 292, 294, 346, 348, 350, 352

vapor transport 269, 276, 285, 303, 308

ventilation 364, 375, 387

mechanical 364

ventilation systems 209, 337, 363–365, 374

VENTIREG 363, 365, 374, 375

VENTIREG systems 365, 367

vermiculite 112, 115, 117, 170, 172–174, 176, 184–187, 189, 190, 264, 265, 325, 326, 329, 333, 335, 336, 392, 393

Volmer equation 18

water adsorption 97, 115, 127, 266, 297, 320, 326, 379, 382
water buffer 363, 365, 382, 383, 391
water desorption 46, 113, 117, 222, 298, 308, 333, 384
water diffusion 131, 258, 259, 273, 287, 288
water evaporation 96, 115, 239, 409, 411, 412
water extraction 46, 47
water molecule 4, 7, 83, 84, 95, 96, 124, 127, 131, 213, 214, 246, 251, 255, 256, 266–268, 270, 273
  hydrogen-bonded 251, 252
  motion of 254, 255
water sorption 9, 10, 20, 46, 47, 77, 78, 88–91, 93–99, 114–118, 120–126, 129, 130, 149, 150, 220, 264–267, 269–271, 273–276, 285, 287, 296
  isothermal 269
  quasi-equilibrium 264, 274
water sorption dynamics 261, 264, 284, 296
water sorption kinetics 265
water transport 51, 261, 270, 272, 273, 275, 276, 285, 287

water uptake 67, 70, 73, 76, 77, 87, 88, 90, 97, 123, 124, 128, 130, 214–217, 220–222, 272, 274, 307, 308
water vapor 37–39, 46, 64, 83, 84, 86–88, 92–101, 110–112, 120–122, 124, 128–132, 134–136, 198, 248, 249, 294, 295, 367, 368, 380, 408, 409

XRA *see* X-ray amorphous
X-ray amorphous (XRA) 69, 71, 72, 74, 77, 92, 150, 151, 159, 174, 177, 180
X-ray diffraction (XRD) 64, 68, 111, 119, 121, 123, 128, 172, 175, 201, 205, 403
XRD *see* X-ray diffraction
XRD analysis 117, 127, 155
XRD pattern 70, 158, 173, 188, 403
XRD technique 133, 260

zeolites 2, 4, 6, 21, 65, 88, 124, 128, 129, 197, 204, 333–335
  commercial 204, 205
  conventional 161
  microporous 128
  mineral 214
  natural 2